Defense against Bioterror

Detection Technologies, Implementation Strategies and Commercial Opportunities

NATO Security through Science Series

A Series presenting the results of scientific meetings supported under the NATO Programme for Security through Science (STS).

The Series is published by IOS Press, Amsterdam, and Springer Science and Business Media, Dordrecht, in conjunction with the NATO Public Diplomacy Division.

Sub-Series

A. Chemistry and Biology	(Springer Science and Business Media)
B. Physics and Biophysics	(Springer Science and Business Media)
C. Environmental Security	(Springer Science and Business Media)
D. Information and Communication Security	(IOS Press)
E. Human and Societal Dynamics	(IOS Press)

Meetings supported by the NATO STS Programme are in security-related priority areas of Defence Against Terrorism or Countering Other Threats to Security. The types of meeting supported are generally "Advanced Study Institutes" and "Advanced Research Workshops". The NATO STS Series collects together the results of these meetings. The meetings are co-organized by scientists from NATO countries and scientists from NATO's "Partner" or "Mediterranean Dialogue" countries. The observations and recommendations made at the meetings, as well as the contents of the volumes in the Series, reflect those of the participants in the workshop. They should not necessarily be regarded as reflecting NATO views or policy.

Advanced Study Institutes (ASI) are high-level tutorial courses to convey the latest developments in a subject to an advanced-level audience

Advanced Research Workshops (ARW) are expert meetings where an intense but informal exchange of views at the frontiers of a subject aims at identifying directions for future action

Following a transformation of the programme in 2004 the Series has been re-named and re-organised. Recent volumes on topics not related to security, which result from meetings supported under the programme earlier, may be found in the NATO Science Series

www.nato.int/science
www.springeronline.com
www.iospress.nl

Series B: Physics and Biophysics – Vol.1

Defense against Bioterror

Detection Technologies, Implementation Strategies and Commercial Opportunities

edited by

Dennis Morrison

Institute of Engineering Research & Applications,
New Mexico Tech, Albuquerque, NM, U.S.A.

Fred Milanovich

Lawrence Livermore National Laboratory,
Livermore, CA, U.S.A.

Dmitri Ivnitski

Institute of Engineering Research & Applications,
New Mexico Tech, Albuquerque, NM, U.S.A.

and

Thomas R. Austin

The Boeing Company, U.S.A.

Published in cooperation with NATO Public Diplomacy Division

Proceedings of the NATO Advanced Research Workshop on
Defense against Bioterror: Detection Technologies, Implementation Strategies
and Commercial Opportunities
Madrid, Spain
8-11 April 2004

A C.I.P. Catalogue record for this book is available from the Library of Congress.

ISBN-10 1-4020-3385-0 (PB)
ISBN-13 978-1-4020-3385-8 (PB)
ISBN-10 1-4020-3386-9 (HB)
ISBN-10 1-4020-3384-2 (e-book)
ISBN-13 978-1-4020-3386-5 (HB)
ISBN-13 978-1-4020-3384-1 (e-book)

Published by Springer,
P.O. Box 17, 3300 AA Dordrecht, The Netherlands.

www.springeronline.com

Printed on acid-free paper

Printed in the Netherlands.

Table of Contents

PREFACE

Instability in warfare arises when offense significantly outstrips defense. After a half century of vaccine and antibiotic successes in the war against infectious diseases, the advantage has shifted back to the pathogen. Infectious diseases are again the leading cause of human mortality worldwide. To compound matters, the possible intentional spread of disease is not only possible but also it is reality. For instance, in the past two decades, the United States alone has had three biological attacks or incidents against civilians: 1984 *salmonella*, 1999 West Nile-like Virus, and 2001 *anthracis*. In addition, several foreign natural epidemics (the recent United Kingdom foot and mouth virus pandemic and the mad cow disease outbreak) have shown the potential for both serious economic and political harm.

Indeed, the events of 2001 exposed civilization's vulnerability to the covert introduction of harmful biological agents. Bioterroism and biological warfare employs living agents or toxins that can be disseminated/delivered by infected individuals, insects, aerosols, and by the contamination of water and food supplies. Most biological agents can be thousands of times more lethal per unit than the most lethal chemical warfare agents. Unlike chemical agents, biological agents attack people stealthily with no observable reaction until after an incubation period (days to weeks). Current disease surveillance and response systems rely on post-symptomatic reporting. However, many infectious agents such as smallpox have a long latency to clinical symptoms, thereby eluding early detection potentially resulting in widespread, uncontrolled contagion.

Consequently, the threat of deliberate dissemination of biological agents is the most complicated and problematic of the weapons of mass destruction facing mankind today.

This volume contains papers presented at the NATO Advanced Research Workshop “Defence against Bioterror: Detection Technologies, Implementation, Strategies and Commercial Opportunities” held at the Hotel Wellington, Madrid Spain from 26 May to 29 May 2004. The objective of the workshop was to contribute to the critical assessment of state-of-the-art of emerging (“breakthrough”) biosensor technologies that will allow for the rapid identification of biological threat agents in the environment and human population; to identify directions for future research, and to promote close working relationships between scientists from different countries and with different professional experience. The volume is devoted to a comprehensive overview of the current state of biological weapons threat; challenges confronting biodetection technologies and systems; ongoing research and development; and, future requirements. Following the structure of the NATO Advanced Research Workshops, the special section starts from the threat overview and current art, then followed detection platforms, networked alarm-type biodetector systems, implementation strategies, electro-optical and electrochemical biosensors.

A strategy and commensurate technology to detect a bioagent release at the earliest moment is an essential element of a defense against bioterrorism. The strategy should include: (1) systems of networked biodetectors that provide wide area monitoring for the early warning of a bioagent release; (2) a medical surveillance system that provides early (detection of the presence of disease in the population at large; and, (3) a concept for integrating these technologies into the public sector. In (1) the development of networked alarm-type biodetection systems is extremely important for detecting, tracking and responding to threats. By fielding a network of alarm-type biodetectors, civilian and military defense officials can obtain early warning in the event of a biological attack. The networked alarm-type biodetectors will provide generic discrimination, i.e., pathogenic vs. non-pathogenic bacteria and may be used as a “trigger” for a more sophisticated detector/identifier system. A desired performance requirement of networked alarm-type biodetectors is real-time, pre-exposure detection, discrimination, and identification of biological threats. The sensing element should be able to detect the presence of biological agents at below threshold concentrations in 5-10 min and be sensitive to a broad range of bioagents (multiplex capable). Obviously the incorporation of all these features within one biodetector based on current technology is a very complicated task. Most commercially available biodetectors are inherently bulky, utilize complex instrumentation, multistep assays and other time consuming procedures. The solution may be based on application of new emerging sensor technologies such as array-based biochips, liquid arrays, artificial olfaction and microfluidic systems, ion-channel switches and magnetoresistance technology. In (2) a science base that will provide for biomarkers in human body fluids that indicate the presence of disease must be established. These markers must be able to distinguish pathogen type and be present at the very onset of disease (presymptomatic). Once established, detection technologies and application strategies need to be developed to bring presymptomatic detection to practical application. (3) strategies need to be developed to bring these advances to the public. Indeed, all levels of government are seeking to improve their capability for dealing with the effects and consequences of a biological incident or attack. In particular, cities recognize that their personnel will play a major part in a bioterrorist attack. Each small community is faced with the daunting problem of developing a bio-terrorism response plan with limited resources and limited local expertise.

Transportation bioterror security presents an extremely complex problem (alarms, indications, situational awareness, level of response, large probability of false positive detection, effectiveness/performance, cost, politics, limited protection). No accurate and rapid "silver bullet" technology or system in existence can meet this challenge., We recommend the continued development of sensing technologies and the approach of employing multiple, layered detection systems with orthogonal technologies. Biological Warfare Agent (BWA) sensor for defense purposes may also be designed to offer "dual use" capability in the civil sector, including public health environmental air and water monitoring as well as drug discovery. Approach and preparation for biological terrorism can be compared to existing civilian methods for earthquake protection – very low probability of occurrence but with very high consequence. Continued collaboration among NATO members is recommended due to risk of BWA attack against both existing member countries (USA, UK and others) as well as emerging member countries (Russia). We would like to acknowledge the NATO Science Committee for their contributions. Special acknowledgement goes out to Cynthia Hernandez for providing technical document production and preparing the camera-ready version of the text.

The editors: Dennis Morrison, Fred P Milanovich, Dmitri Ivnitski, and Tom R Austin

LIST OF CONTRIBUTORS

Dennis Morrison
Institute of Engineering Research & Applications 901 University Blvd. SE Albuquerque, NM 87106 USA Phone 505 272 7235
Fax 505 272 7203 morrison@iera.nmt.edu
Fred P. Milanovich
Lawerence Livermore National Labs 7000 East Avenue -174 Livermore, CA 94550 USA
Phone 925 422 6838 Fax 925 422 8020 milanovich1@llnl.gov
Tom R. Austin
The Boeing Company 2201 Seal Beach Blvd MC 110-SC45 Seal Beach, CA 90740 USA
Phone 562 797 3798 Fax 562 797 4778 tom.austin@boeing.com
Shawn H . Park
The Boeing Company 5301 Bolsa Ave MC-H013-B319 Huntington Beach, CA 92657 USA
Phone: 704 896 1606 Fax: 704 896 6417 shawn.h.park@boeing.com
John C Stammreich
The Boeing Company 2201 Seal Beach Blvd MC-110-SA32 Seal Beach, CA 90740 USA
Phone 562 797 3252 Fax 562 797 4778 john.c.stammreich@boeing.com
Oleg VIgnatov
Institute of Biochemistry & Physiology of Plants & Microorganisms Russian Academy of Sciences
Entuziastov av.,13 Sartov, 410049 Russia Phone 7 8452 970 383 Fax 7 8452 970 383
oignatov@ibppm.sgu.ru
Dmitri Ivniski
Institute of Engineering Research & Applications, New Mexico Tech
901 University Blvd. SE Albuquerque, NM 87106 USA Phone 505 272 7255
Fax 505 272 7203 ivnitski@nmt.edu
Laura M Lechuga
Biosensor Group Centro nacional de Microelectronica (IMM-CNM-CSIC)
Isaac Newton , 8 28760 Tres Cantos (PTM) Madrid, Spain Phone 34 91 806 0700
Fax 34 91 806 0701 laura@imm.cnm.csic.es
Anthony. Turner

Cranfield University Silsoe Bedfordshire, MK45 4DT UK Phone 44 0 1525 863005
Fax 44 0 1525 863360 a.p.turner@cranfield.ac.uk
Steven Kornguth
Institute of Advanced Technology 3925 West Braker Lane
Suite 400 Austin, TX 78759 USA Phone 512 232 4486 Fax 512 471 9103 steve_kornguth@iat.utexas.edu
Robert G. Barton
Midwest Research International 425 Volker Blvd. Kansas City, MI 64110 USA
Phone 816 360 5268 Fax 816 531 0315 bbarton@mriresearch.org
Larry D. Brandt
Sandia National Laboratories PO Box 969 MS 9201 Livermore, CA 94551 USA
Phone 925 294 2969 Fax 925 294 1559 lbrandt@sandia.gov
Andrey Bratov
Centro Nacional de Microelectronica Campus UAB. Bellaterra E -08193 Cerdanyola del Valles
Barcelona, Spain Phone 34 93 594 77 00 Fax 34 93 580 14 96 andrey.bratov@cnm.es
Victor D. Bunin
State Research Center of Applied Microbiology Obolensk Moscow region, 142253 Russia
Phone 7 0967 705716 Fax 7 0967 705716 vikbun@inbox.ru
Fatima T. Adilova
Inst. of Cybernetics Academy of Sciences 34, F. Khodjaev str. Tashkent, 700125 Uzbekistan
Phone 99871 162 71 62 Fax 99871 162 73 21 fatima_adilova@ic.uz
James M. Clark
GL Detection Dstl Porton Down Salisbury SP4 OJQ UK Phone 44 1908 613 405
Fax 44 1980 613 987 JMCLARK@mail.dstl.gov.uk
Bob V. Collins
Midwest Research International 425 Volker Blvd. Kansas City, MI 64110 USA
Phone 816 360 5322 Fax 816 531 0315 rvcollins@mriresearch.org
Serge Cosnier
Laboratory of Organic Electrochemistry and Redox Photochemistry UMR CNRS 5630
Institute of Molecular Chemistry FR CNRS 2607 Batiment Chimie Université Joseph Fourier Grenoble 1 301 rue de la Chimie,BP 53 38041 Grenoble Cedex 9 France serge.cosnier@ujf-grenoble.fr
Boris B. Dzantiev

Institute of Biochemistry RAS Leniskii Prospect 33 Moscow, 110971 Russia
Phone 7 095 954 2804 Fax 7 095 954 2804 dzantiev@inbi.ras.ru
Eugene V. Grishin
Shemyakin-Ovchinnikov Institute of Bioorganic Chemistry Russian Academy of Sciences Ul. Miklukho-Maklaya, 16/10, 117997 GSP Moscow, V-437 Russia
Phone 7 095 3305892 Fax 7 095 3350812 grev@ibch.ru
Ted L. Hadfield
Midwest Research International 1470 T reeland Blvd, SE Palm Bay, FL 32909 USA
Phone 321 723 4547 ext. 300 Fax 321 722 2514 thadfield@mriresearch.org
Jay Lewington
Smiths Detection 459 Park Ave Bushey Watford, Herts WD23 2BW UK
Phone 44 0 1923 658206 Fax 44 0 1923 658025
Jay.Lewington@smithsdetection.com
Marco Mascini
Universita di Firenze Sesto Fiorentino Dipartimento di Chimica, Polo Scientifico
Via della lastruccia 3 Firenze, 50019 Italy Phone 39 055 457 3283
Fax 39 055 457 3384 mascini@unifi.it
Sebastain Meyer-Plath
Smiths Detection 59 Park Ave Bushey Watford, Herts WD23 2BW UK
Phone 44 0 1923 658193 Fax 44 0 1923 221361
sebastian.meyer-plath@smithsdetection.com
Michael Moniz
Circadence Corporation 4888 Pearl East Circle, Ste. 101 Bolder, CO USA
Phone 303 413 8837 Fax 303 449 7099 mike@circadence.com
Louis Montulli
Midwest Research International 613 Rolling Hills Rd Vista, CA 92081 USA
Phone 760 977 9601 Fax 760 598 6488 LT@montulli.org
Petr Skladal
Masaryk University Kotlarska 2 CZ 61137 Brno Czech Republic Phone 420 5 41129402
Fax 420 5 41211214 skladal@chemi.muni.cz
Ronald Pethig
School of Informatics University of Wales, Bangor Dean Street Bangor, Gwynedd LL57 1UT UK
Phone 44 1248 382 682 Fax 44 1248 361 429 ron@informatics.bangor.ac.uk
Erling B. Myhre
Dept. of Infectious Diseases, University Hospital SE-221 85 Lund Sweden

Phone 46 46 17 18 67/17 11 30 Fax 46 56 13 74 14
Erling.Myhre@infek.lu.se
Arjan van Wuijckhuijse
Dr. Ir. A.L. van Wuijckhuijse TNO Prins Maurits Laboratory Detection Identification and Analytical Chemistry PO Box 45, 2280 AA Rijswijk, THE NETHERLANDS
Phone +31 15 284 3343 Fax: +31 15 284 3963 www.pml.tno.nl
Sergey D . Varfolomeyev
The M.V.Lomonosov Moscow State University Vorobiovy Gory 1, Bldg. 11 119992 Moscow Russia Phone 7 095 939 3589 Fax 7 (095) 939 5417
sdvarf@enzyme.chem.msu.ru
Itamar Willner
The Hebrew University of Jerusalem Jerusalem 91904 Israel Phone 972 2 658 5272
Fax 972 2 652 7715 willnea@vms.huji.ac.il
Yuri M. Yevdokimov
Head of Laboratory of condensed state of nucleic acids of Engelhardt Institute of Molecular Biology RAS Engelhardt Institute of Molecular Biology RAS Vavilov str.
32 Moscow, 119991 Russia Phone 7 095 135 97 20 Fax 7 095 135 14 05
Yevdokim@genome.eimb.relarn.ru

STRATEGIC ACTIONABLE NET-CENTRIC BIOLOGICAL DEFENSE SYSTEM

S. Kornguth

Director, Countermeasures to Biological and Chemical Threats Institute for Advanced Technology, University of Texas at Austin, USA

Abstract: Technologies required for strategic actionable net-centric biological defense systems consist of : 1) multiplexed multi-array sensors for threat agents and for signatures of the host response to infection; 2) novel vaccines and restricted access antivirals/bacterials to reduce emergence of drug resistant strains pre- and post-event; 3) telemedicine capabilities to deliver post-event care to 20,000 victims of a biological strike; and 4) communication systems with intelligent software for resource allocation and redundant pathways that survive catastrophic attack. The integrated system must detect all threat agents with minimal false positive/negative events, a seamless integrated broad-band communications capability that enables conversion of data to actionable information, and novel pre- and post-event treatments. The development of multiplexed multi-array sensors, appropriate vaccines and antibiotics, and integrated communication capabilities are critical to sustaining normal health, commerce, and international activities.

1. INTRODUCTION

The overarching objectives in developing effective countermeasures to biological threats are to protect the Defense community and citizenry from such threats, and to develop agile responses to unanticipated events considering that successful terrorists do the unexpected. The need for protection against and responses to biological threats has been strikingly demonstrated by the use of anthrax contaminated letters that were sent through the U.S. mail in October 2001. That attack resulted in human illness, the loss of life, and discontinuity of government operations because of contamination of federal office buildings in Washington, DC. A recent report prepared by the Center for Strategic and International Studies (CSIS) and supported by the Defense Threat Reduction Agency (DTRA) of the

D. Morrison et al. (eds.), Defense against Bioterror: Detection Technologies, Implementation Strategies and Commercial Opportunities, 17–27.

Department of Defense (DoD) came to the conclusion that the U.S. is at present not well prepared for a similar attack using anthrax1. The major problems include a lack of: 1) a clear chain of command, and 2) tools to provide the public with information that permits appropriate responses.

The incidence of Congo-Crimean hemorrhagic fever in Afghanistan, an area where coalition forces are being deployed, increases this need. The potential threat posed by emergent disease (e.g., Severe Acute Respiratory Syndrome [SARS] and West Nile Fever virus) or from a major release of a contagious biological agent such as smallpox, has been a growing concern at all levels of the international community. This article outlines and discusses a new strategy that is needed if we are to be fully capable of sensing, preventing, and managing biological threats.

2. NEW PARADIGM

The current paradigm addresses biological and chemical terrorist threats in a vertical (stove-piped) response. In the arena of developing sensors for the detection of biological agents, the paradigm has been to develop separate detectors for each agent or to develop a platform for detecting 12-24 threat agents using a single probe for each agent. There is a lack of an interactive networked communication system that is capable of managing a devastating emergent disease. To establish a highly networked system that is rapid enough to permit effective protection, it is necessary to evolve from the stove-piped, compartmentalized model currently in use to an integrated, fused, probabilistic, and frequently updated information model. Multiplexed multi-array sensor systems, capable of recognizing all bacterial or viral genomic materials related to pathogenicity or of recognizing antigenic domains that are specific indicators of pathogens are one component of a network needed for rapid detection and identification of biological threats. With respect to therapeutics, modern technologies for vaccine and antibiotic production provide decided advantages over older methods. The traditional vaccines require extensive development times before they become available for human use and undesired side effects commonly result from vaccines produced by these protocols. The cost associated with developing and testing vaccines, using traditional technology, approximates 50-100 million dollars per vaccine. The dissemination of antibiotics and antivirals through the world markets has resulted in the appearance of pathogenic bacteria and viruses that are resistant to these drugs. One approach to reduce the development of antibiotic resistance is to restrict the distribution of newly developed antibiotics. Such an approach presents ethical and social dilemmas. The consideration of options available for reduction of drug resistance, prior to a threat event, may permit development of a rational policy. A major problem facing our nations in the event of a biological attack or emergent disease is the large numbers of patients that can be anticipated to require medical treatment. Although improvements in emergency medical care and hospital equipment have been achieved during the past two decades, the ability of any community to manage an outbreak of infectious disease affecting >10,000 people is lacking. Rapid progress has been achieved whereby medical care can be provided to patients at sites that are distant from primary caregivers using telecommunication systems (e.g., the Armed Services in theater, large scale HMO providers such as Kaiser Permanente, or the correctional institutions in the U.S.). The funds needed to acquire telecommunication equipment for such distributed medical care delivery are estimated to be less than 100 million dollars for the entire U.S. At the present time such a distributed care system is not readily available.

The new paradigm couples a network centric integrated sensor alert system that can detect all threat agents simultaneously, with a seamlessly integrated communication software capability that converts large scale data to actionable information. For this to be effective, the sensor system must yield minimal false positive and false negative results. The new paradigm incorporates large-scale databases on normative values of threat agents in many regions of the world so that significant changes in real time can be detected. The paradigm also includes the development and implementation of novel pre- and post-event treatment capabilities.

Attention must be paid to the ability of high level decision makers and operators to recognize that a new state of threat has emerged, based upon output of the sensors, data fusion system, and iconographic display. Ambiguity of data, lack of an autonomous processing system, and high stress on the operator (e.g., sleep deprivation, lack of training) may all compromise the utility of a highly networked system of systems. What is needed for this new paradigm to succeed? The needs include multiplexed multi-array sensors for biological agents that infect people, livestock, and edible crops. The agents of concern include many on the Militarily Critical Technologies List prepared for the Office of the Secretary of Defense. We need multiplexed multi-array sensor systems with high specificity and selectivity for the rapid detection of host responses to infection. We need a new generation of effective therapeutics, including vaccines, antibiotics, and antivirals. A network centric intelligent communication system that can provide accurate comprehensible information to decision makers (from command officers to unit operators) is required. To minimize morbidity and mortality and optimize containment of disease, a biosurveillance system based on archival health databases, statistical models, and data mining strategies that can provide an early alert to a disease outbreak is required.

In many cases the operator may be required to understand the meaning of acquired data in very short time periods (seconds to minutes) if the response is anticipated to change the outcome of an engagement. The Tactical Decision Making Under Stress (TADMUS) program is one example of such a situation. In the biological threat arena, the detection and identification of toxins require rapid analysis and operator comprehension. The large increase in numbers of sensors (for high explosives [HX], biological and chemical agents, meteorological conditions) together with the rapid changes in op tempo required to manage emergence of clinical disease would suggest a need for the development of systems capable of autonomous generation of an alert when threat conditions arise.

3. CURRENT STATE OF TECHNOLOGY NEEDED FOR THE NEW PARADIGM

In the sensors area, the genomes of most biological threat agents have been sequenced and the signatures of toxins described. Novel multiplexed multi-array sensor-platform systems utilize the genomic datasets to detect the appearance of threat levels of these agents. In the therapeutics area, researchers are working towards identifying critical antigenic epitopes of these agents. New therapeutics can emerge that have an antigen binding capacity significantly greater than antigen-cell receptor binding, resulting in the potential for agent neutralization. Technologies have been developed over the past decade for the development of new drugs and DNA based vaccines. Restricted access antivirals/antibacterials will need to be developed to reduce the emergence of drug resistant strains pre- and post-event.

A significant development in our program at The University of Texas at Austin (UT-Austin)2 has been the novel design and production of an antibody that binds the anthrax PA antigen 1000 times stronger (Kd<10-11) than any antibody to date. The antibodies were produced using phage display technology for selection of the antibodies. In tests with experimental rodents in a controlled facility, administering the Bacillus anthracis PA antigen to the animals resulted in 100 percent fatalities, whereas the co-administration of the newly developed antibody against the PA antigen resulted in 100 percent survival2. Research is also being conducted to determine unique nucleic acid sequences in the genome of pathogenic bacteria and viruses that contribute to the pathogenic properties of the organisms. This information is being used to develop multiplexed assay systems that can detect selected agents simultaneously. By quickly screening for multiple pathogenicity island sequences or pathogenic factors, end-users will have the capability to detect the first signs of emergent disease without requiring screening for each particular organism. In the communications area, researchers are developing 'belief maintenance' software to provide decision makers some estimate of the validity of incoming data. An estimate of information credibility is critical to effective decision-making in crisis situations when one must rely on an 80 percent solution (i.e., 80 percent of needed information is available). Waiting for a 100 percent solution could have a catastrophic impact on response effectiveness.

In the area of telecommunications, researchers are developing the means to provide effective medical triage to victims in a contaminated hot zone. The hot zone in this case refers to a region experiencing an outbreak of a highly contagious disease that causes death in a significant percentage of infected individuals. Advanced telecommunications technology can permit physicians and other medical experts at remote locations to provide medical information and support care delivery to personnel in the hot zone. Prior training of these personnel (from physicians to local citizens) is required. An extensive use of local persons will be necessary if it is deemed inadvisable to introduce health care workers and communications experts into the hot zone; the external support teams are required to provide 'back-fill' support to the overburdened local community.

Biosurveillance systems are being developed to serve as an early warning of emergent disease. A variety of databases are being developed that are health related. Examples of these databases include school absenteeism, over-the counter drug sales, hospital emergency clinic visits, and archival records on the incidence of diseases in different geographic regions, CONUS and OCONUS. Each database must be statistically characterized regarding parameters such as variance, confidence intervals, seasonality, etcetera, and be integrated into validated predictive models. Once the reference databases are in place and suitably modeled, statistically significant departures from baseline values can be detected and transmitted in real time to decision makers through intelligent communication systems.

3.1 Technology Gaps

A number of critical technology gaps exist that must be addressed if we are to recognize, prevent, and minimize the effect of biological agents. These gaps include: deficiencies in the availability of multiplexed multi-array agent sensors and platforms; critical reagents; capability for large-scale production of effective vaccines, antibiotics and antivirals; ability to treat a large number (10,000) of infected people 24/7 for several weeks in a biological hot zone; archival biosurveillance databases and intelligent and secure communications networks. With the new capabilities and devices anticipated during the next decade, approaches that address these gaps include the use of autonomous (e.g., cell phone-based) microelectronic detectors for the transmission of data on agent exposure, development of novel antibodies, antibiotics and antivirals to manage disease outbreaks, and establishment of global surveillance systems for emergent diseases (e.g., SARS, West Nile Fever, Congo-Crimean Hemorrhagic Fever).

3.2 Research Areas

Because of the broad scope of needs for technology to prevent and minimize biological threats, a number of research areas have been identified as critical. These include: the scientific validation that a biological incident has occurred (requisite tools/capabilities include situation awareness systems, sensors and signatures); the availability of medical countermeasures (vaccines, pharmaceuticals, and medical transport); and a highly effective communications network for the secure transmission of data and the conversion of such data to comprehensible information so that decision makers can take appropriate actions.

3.3 Sensors Research

For effective sensors, a variety of materials are being developed that include effective high-affinity binders of biological threat agents. The high affinity binders include antibodies, cDNA gene probes, polynucleotide aptamers, and combinatorial chemicals. Using phage display methods, antibody fragments can be selected that have a high affinity for agents such as anthrax toxin and brucella. Another binding system that has been examined uses polynucleotide aptamers about 31 nucleotides long that have good binding affinity to ricin toxin. These sensor materials require opto/electronic transduction platforms. Sensor platform research currently is being focused on micro-electro-mechanical systems (MEMS) devices, microelectronics technology, microfluidics (laboratory-on-a-chip), DNA/protein microarrays, and transduction devices. Efforts are also being directed toward the development of multiplexed multi-array systems that detect approximately 100 biological threat agents of concern. For military application, it is essential that sensor systems can detect and identify agents present in samples rapidly, using platforms that are small and have low power requirements.

3.4 Therapeutics Research

With current approaches, the development-to-market of new vaccines, antibiotics, and antivirals is in the order of 5-10 years. A paradigm shift to newer culture and DNA-based technology is needed if we are to have an effective response to a major biological or chemical event. Current estimates regarding the time required for developing and fielding new vaccines and antibiotics/antivirals to specific threat agents, using new technology and expedited approval, is in the order of three years.

3.5 Communications Research

While current computer/informatics research includes the development of telecommunications assets, a critical need in the communications area is the development of seamless integrated communication networks. These network centric systems enable the conversion of data to actionable information. Research is being conducted to provide intelligent agent software designs for such communication systems. This will enable an enhanced accuracy in critical decision-making and resource allocations. The integrated system must have redundant pathways that can survive a catastrophic attack. The communication system must be capable of integrating data on an emerging threat in a timely manner, and provide useful information for public safety coordination and perimeter management.

3.6 Telemedicine Needs

Telemedicine capabilities can aid in the delivery of post-event care to 10,000-20,000 victims of a biological strike in a densely populated area for 24 hours a day, seven days a week, for several months. In the event of smallpox attack in which 10,000 people develop clinical symptoms of infection within 7-10 days following exposure, local hospitals and medical response capabilities would be overwhelmed if current treatment protocols were used. Telemedicine allows physicians at a remote location from the hot zone to provide medical support via telemedicine capabilities (visual, audio) to aid local physicians in treating patients. A treatment level of 50 patients per day per physician would require 200 physicians to provide telemedicine care for 10,000 patients. Each physician would require telemedicine devices; hence 200 telemedicine devices would be required at the remote location, and a similar number in the hot zone. A national telemedicine system could include the establishment of approximately eight telemedicine response centers nationally, interconnected via satellite to telepresence and telemedicine/robotic systems. The remote care capability reduces the likelihood of the dissemination of disease to physicians and communities in which the physicians reside. A telemedicine system would also retain health care delivery in communities providing health care back-fill.

3.7 Networked Operations at the Coordinating Level and Lower Echelons of Command

A decision making component is required for coordinating the delivery of actionable information. Figure 1 illustrates such a flow chart. In threat situations, data developed from sensor arrays, surveillance systems, and therapeutics inventories can be electronically encrypted and transmitted via intelligent communication networks to decision makers for appropriate actions. The decision makers include individuals, government authority, medical care experts (doctors, hospitals, the Centers for Disease Control and Prevention [CDC], etc.), and the military.

3.8 Network Centric Response to Threat Data Fusion and Human

1. Perception/Comprehension
2. Primary issue of concern
3. Full situation awareness is contingent on at least four elements including:
4. Large scale acquisition of data (i.e., SIGINT, MASINT, HUMINT, sensors)
5. High fidelity communication of data sets to autonomous processing centers
6. Data fusion involving weighting of data and marked reduction in data volume to yield information that provides users a common operational picture
7. Rapid comprehension of time dependent information by operators facilitated by new iconographic displays, training, measures of vigilance

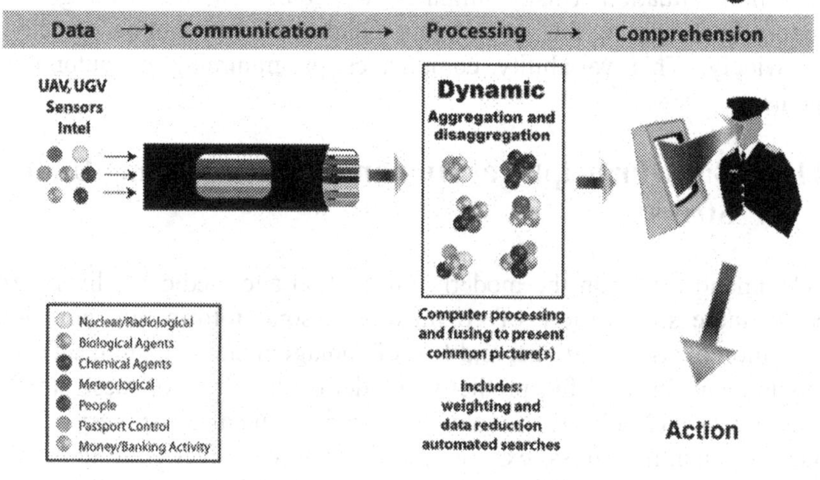

Figure 1.

3.9 Proposed Goal of the Autonomous Data Processing

The end goal is achievement of a rapid appropriate action in response to detection of threat (within minutes of threat identification). In order to meet this time constraint in a real world scenario, it is probable that the man-in-the-loop may have a denial capability rather than an approval function.

3.10 A Strategy to achieve this Goal

The tools required for autonomous weighting of data and subsequent reduction of data elements for particular missions remain to be developed and agreed upon. The technologies required for data acquisition, communication, and autonomous processing fundamentally differ from that required for comprehension by an operator. The three technologies are systems that must be developed prior to deployment and will have known probabilities of accuracy and reliability. The fourth element involves the training and state of vigilance/alertness of the operator as well as the development of software (e.g., icons, data mining, and compression) used to display threat conditions. The point of the fourth element is to permit an operator to have rapid comprehension of the state of threat in a rapidly changing environment. Because there is a time factor involved in the comprehension of threat conditions by an operator, and in the translation of the information into action, the fourth element must include temporal qualities. Since the time dependence (seconds to minutes) of an intense threat situation will differ from that involved in a peacekeeping or supply distribution situation (many minutes to hours on the average), the autonomous weighting factors and data reduction factors can be expected to vary widely. This variability complicates programming of autonomous systems.

3.11 Basic Principles of Human Decision Making Under Stress

Command teams on the modern information-age medical delivery front face an increasing variety of cognitive stressors: information ambiguity, information overload, fatigue, and fear of contagion and quarantine. There is a requirement for a useful, predictive model of the effect of these stressors on individual and collective cognition within the medical delivery team. A model to quantify stress experienced by the caregiver and to identify countermeasures and mitigators of stress, develop organizational strategies for optimum performance under stress is needed. Psychological assessments that can predict individual and team cognitive functioning and physiological markers that can determine quantitatively and objectively the effect of stress experienced on an operators vigilance have been identified (Table 1). The identification of specific physiological markers that are predictive of stress-induced changes in complex cognitive functioning will aid in the construction of autonomous weighting systems.

Table 1. Stressors and Measures

	Stressors				
Measurement Technologies	Information Ambiguity	Information Overload	Fatigue	Social Isolation	Danger
Cognition	+++	+++	++	++	++
Behavioral Markers	+	+	+	+	++
Emotional Assessment †	+++	+++	+++	++	+++
Observer/Controller Ratings	+++	+++	+++	++	++
Auditory/Visual Evoked Potential (P300, P600)	+++	+++	+++	+	++
Respiration / Ventilation Rate	++	++	+	+++	+++
Cardiac q-t Interbeat Interval	+++	+++	++	+++	+++

+ Degree of reactivity to measurement

† (e.g., Voice Stress Analysis)

4. ACKNOWLEDGEMENTS

The research reported in this document was performed in connection with Contract number DAAD13-02-C-0079 with the U.S. Edgewood Chemical Biological Command.

REFERENCES

1. Lessons from the Anthrax Attacks: Implications for U.S. Bioterrorism Preparedness. CSIS Funded by the Defense Threat Reduction Agency of the Department of Defense. David Heyman investigator. March 2004.
2. Maynard, J.A., Maassen, C.B.M., Leppla, S.H., Brasky, K., Patterson, J.L., Iverson, B.L., and Georgiou, G., "Protection against Anthrax Toxin by Recombinant Antibody Fragments Correlates with Antigen Affinity," Nature Biotechnology, vol. 20, pp. 597-601, June 2002.

NATURAL TOXINS: THE PAST AND THE PRESENT

E. Grishin
Deputy Director of Shemyakin-Ovchinnikov Institute of Bioorganic Chemistry, Russian Academy of Sciences, ul. Miklukho-Maklaya, 16/10, Moscow 117997, Russia

Abstract: Different bacteria, viruses and toxins constitute a potential menace for people. The number of toxins that could be applied for a bioterrorist attack with real public health risk, though, is relatively limited. However, these natural toxins could cause difficult troubles. The objective of this paper is focused on the toxins of various origins that might be used as a biological weapon. To be used in such a way the toxin should be highly lethal and easily produced in large quantities. Our current knowledge on natural toxins is conducive to select the toxin list threatening public health. At present this list includes a few bacterial and plant toxins, as well as a set of toxins produced by algae and molds. Novel methods of toxin detection should be able to monitor the presence of many toxins at the same time.

1. INTRODUCTION

Natural toxins vary widely in animal, plant and microbial organisms. Over the past decade we can witness the remarkable progress in our knowledge about chemical nature of toxin molecules and their biological action. In fact the natural toxins represent an immense number of various compounds that causes injury, illness or death of a living organism. Unfortunately today we are a witness to bioterrorism. Bioterrorism is the intentional use of biological agents to cause illness of people. Some of the natural toxins can be used as potential biological weapons. The basic knowledge and biological weapon expertise are very needed to select a set of natural toxins that could be utilized for bioterrorism. Civilian population should be informed about the potential biological weapons of bioterrorism. The number of toxins that could be applied for a bioterrorist attack with real public health risk is relatively limited. However, these natural toxins could

D. Morrison et al. (eds.), Defense against Bioterror: Detection Technologies, Implementation Strategies and Commercial Opportunities, 29–45.

cause difficult troubles. Regular national or international meetings are devoted to toxin research. The balance between openness and biosecurity is still under discussion. The use of biological weapony has been known for centuries, culminating in special research programs run by several countries. The objective of this paper is focused on the toxins of various origins that might be used as biological weapons. Our current knowledge on natural toxins is conducive to select the toxin list threatening public health. At present this list includes a few bacterial and plant toxins, as well as a set of toxins produced by algae and molds, and some animal toxins. To be included in this list the toxin should fit some criterions. There are at least four such criterions: availability and simplicity of manufacture, simplicity of application, efficacy of action (toxicity) and reasonable stability.

2. BIOLOGICAL ACTIVITY OF NATURAL TOXINS

Poisonous organisms have been known for centuries. Many scientists have been involved in the investigation of toxic molecules from these organisms. Their attention has been directed on identification, isolation, structure determination, biological and toxicological studies of toxic molecules which constitute a great many of natural toxins. Natural toxins can generally be divided into several broad groups according to their origin or their chemical nature or their biological activity. When the potential weapons for bioterrorism are reviewed for needed characteristics the biological activity/toxicity attracts special attention. A number of selected toxins and their toxicity are presented in Table 1. Natural toxins listed in the Table 1 possess a wide range of toxic activity with LD_{50} value for 10^{-5} mkg/kg up to 200 mkg/kg.

Potential agent for bioterror should be highly lethal and easily produced in large quantities. Only botulinum toxin and ricin meet these requirements. Some other natural toxins as saxitoxins, microcystins and mycotoxins could be included in the list of potential biological agents for bioterror.

Table 1. Toxicity of some natural toxins.

Name	LD_{50} (mkg/kg)	Mr	LD_{50} (mol/kg)
Botulinus	$1.1x10^{-5}$	150 000	$0,6x10^{-16}$
Tetanus	$2.8x10^{-5}$	140 000	$2.0x10^{-16}$
Abrin (ricin)	2.7	65 000	$4.3x10^{-11}$
Palytoxin	0.15	3 300	$4.5x10^{-11}$
Taipoxin	2.0	42 000	$4.8x10^{-11}$
Cobrotoxin	75	7 819	$9.6x10^{-9}$
Scorpion toxin	9.0	7 249	$1.2x10-^{9}$
Batrachotoxin	2.0	538	$3.7x10^{-9}$
Tetrodotoxin	8.0	319	$25x10^{-9}$
Curare	200	696	$29x10^{-8}$
Potassium cyanide	1000	65	$1.5x10^{-4}$

3. BOTULINUS

Botulinum toxin (BTX) is the most potent biological toxin yet known [1]. BTX is a protein complex consisting of 150 kDa di-chainal toxins referred as botulinum neurotoxins associated with non-toxic companion proteins. It is produced by *Clostridium botulinium, C. baratii*, and *C. butyricum*, anaerobic, spore forming gram-positive bacteria, which are the causative agents of botulism. Respiratory failure secondary to paralysis of the respiratory muscles during botulism development can lead to death. BTX acts preferentially on peripheral cholinergic nerve endings to block acetylcholine release. Due to the severity and potency of this neurotoxin, its importance as a biological weapon is of major concern to public health officials [2].

The ability of BTX to produce its effects is largely dependent on its ability to penetrate cellular and intracellular membranes. Thus, toxin that is ingested or inhaled can bind to epithelial cells and be transported to the general circulation. BTX is structurally organized into three domains endowed with distinct functions (Fig.1): high affinity binding to neurons, membrane translocation and specific cleavage of proteins controlling exocytosis of neurotransmitter [3, 4]. Toxin that reaches peripheral nerve endings binds to specific receptors on the cell surface, which may comprise gangliosides. The binding mediated by the toxins heavy chain is followed by endocytotic internalization of neurotoxin/receptor complex. Thus, it penetrates the plasma membrane and the endosome membrane upon acidification of vesicles, when the light chain of the neurotoxin is translocated into the cytosol. Here, the light chain (Zn^{2+}-endopeptidase) cleaves one or two among three synaptic proteins (VAMP-synaptobrevin, SNAP 25, and syntaxin) (Fig.2). All three protein targets of BTX play a major role in the fusion of synaptic vesicles at the release sites. Subsequently, their cleavage is followed by blockage of neurotransmitter exocytosis.

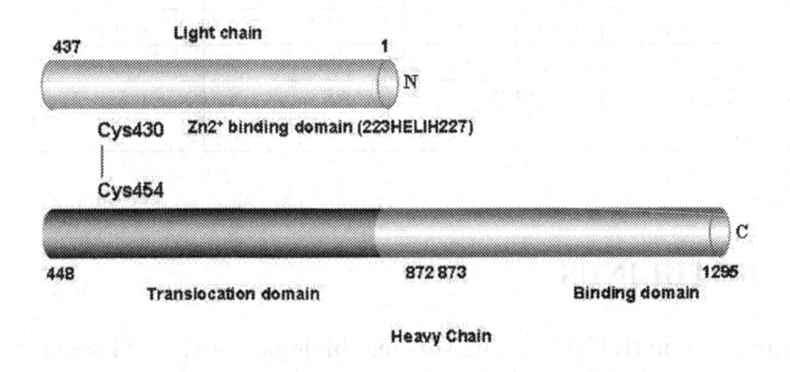

Figure 1. Schematic structure of Botulinum toxin.

Type A BTX is the most widely used in human drug trials [5]. It has become the treatment of choice for blepharospasm, hemifacial spasm, cervical and laryngeal dystonia. It may also be used in the treatment of patients with oromandibular dystonia and limb dystonia, and has been used successfully in the treatment of spasticity and cerebral paralysis. The toxin also alleviates pain and may be used in therapeutic trials for prediction of the response to surgical elongation.

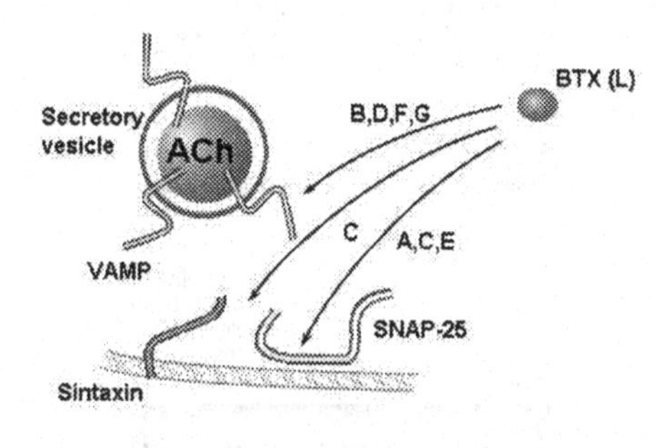

Figure 2. Cleavage of synaptic proteins (VAMP, SNAP 25 and syntaxin) by the light chain of Botulinum toxin.

4. RICIN

Ricin is a naturally occurring toxin derived from the beans of ubiquitous worldwide castor oil plant *Ricinus communis* [6, 7]. Taking into account the volume of annual castor beans processing in the world and fairly easily production of the toxin from the waste mash with an output reaching 5% ricin by weight, the toxin should be considered as being potentially widely available. The ready availability of ricin, coupled to its extreme potency, has identified this protein toxin as a potential biological warfare agent. Therapeutically, its cytotoxicity has encouraged the use of ricin in "magic bullets" to specifically target and destroy cancer cells, and the unusual intracellular trafficking properties of ricin potentially permit its development as a vaccine vector.

The most hazardous routes of ricin exposure are with inhalation and injection. It is quite stable and extremely toxic. The lethal dose by inhalation (breathing in solid or liquid particles) and injection (into muscle or vein) has been estimated, approximately 5-10 mkg/kg, that is 350-700 mkg for a 70 kg adult. Ricin intoxication can manifest as gastrointestinal hemorrhage after ingestion, severe muscle necrosis after intramuscular injection, and acute pulmonary disease after inhalation. Death had ensued within hours of deliberate subcutaneous injection.

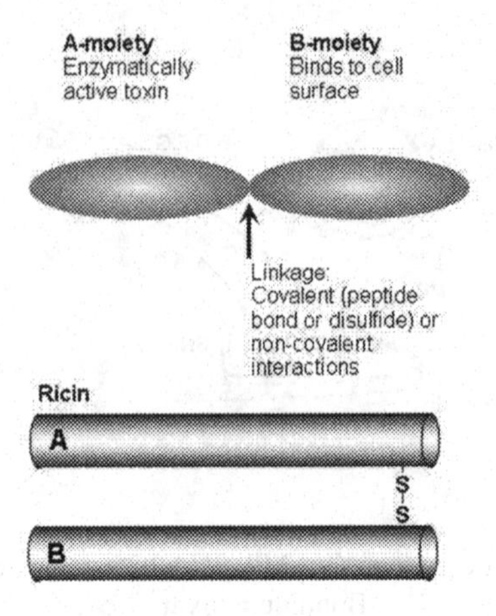

Figure 3. Schematic structure of ricin

Ricin is a heterodimeric protein made up of two hemagglutinins and two toxins being able to fatally disrupt protein synthesis by attacking the ribosome [8-10]. The X-ray structure of ricin has been elucidated. The toxins consist of two polypeptide chains (A and B chains), which are joined by a disulfide bridge (Fig. 3). The A chain is a specific N-glycosidase with a prominent active site cleft. The B chain is a two domain lectin.

To enter the cytosol and reach its target, the toxin must cross an internal membrane and avoid complete degradation without compromising its activity in any way [11, 12]. Cell entry by ricin involves a series of steps: 1) binding, via the ricin B chain, to cell surface glycolipids or glycoproteins having beta-1,4-linked galactose residues; 2) uptake into the cell by endocytosis; 3) entry of the toxin into early endosomes; 4) transfer, by vesicular transport,of ricin from early endosomes to the trans-Golgi network; 5) retrograde vesicular transport through the Golgi complex to reach the endoplasmic reticulum; 6) entry of the A chain in the cytosol, 7) interaction of A chain with the ribosome to catalyse the depurination reaction (Fig. 4). It has been reported that even a single molecule of ricin reaching the cytosol can cause the cell death. Combining our understanding of the ricin structure with ways to cripple its unwanted properties will also be crucial in the development of a long awaited protective vaccine against this toxin.

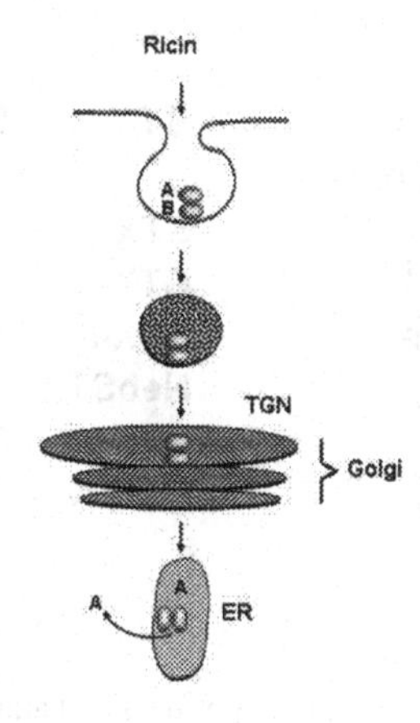

Figure 4. Some stages of ricin action: binding to cell membrane, endocytosis, transport to Goldgi apparatus and the ER, translocation of A subunit to the cytosol.

5. SAXITOXINS

Paralytic shellfish poisoning is caused by a group of toxins produced by both marine and freshwater algae. Marine dinoflagellates *Alexandrium catenella, A. minutum, A. ostenfeldii, A. tamarense, Gymnodinium catenatum and Pyrodinium bahamense* produce saxitoxins. In freshwater, blue-green algae, namely *Anabaena circinalis*, also manufacture the toxins. At the time of algal blooms (red tides), shellfish accumulate saxitoxins leading to paralytic shellfish poisoning (PSP) which, in extreme cases, causes death due to respiratory paralysis. People walking near the surfline during red tide events often complain of sore throats and difficulty breathing. This is as a result of inhaling a toxin-laden aerosol generated by the pounding surf. A debate continues about whether bacteria produce saxitoxin, or influence dinoflagellate production of the toxins [13]. There are about 20 derivatives of saxitoxin responsible for paralytic shellfish poisonings. Saxitoxin was named after the butter clam *Saxidomonus giganteus,* from which it was first isolated in 1957 [14]. The chemical structure of saxitoxin was determined in 1975 [15]. The toxin is a tricyclic molecule containing two guanidino groups (Fig 5). The polar nature of saxitoxin allows it to readily dissolve in water and lower alcohols being insoluble in organic solvents. Chemically, saxitoxin is very stable although it can be inactivated by treatment with strong alkali. It can be kept in dilute acidic solutions for years without loss of its activity. Saxitoxin itself is highly toxic. It is toxic by ingestion and by inhalation, with inhalation leading to rapid respiratory collapse and death.

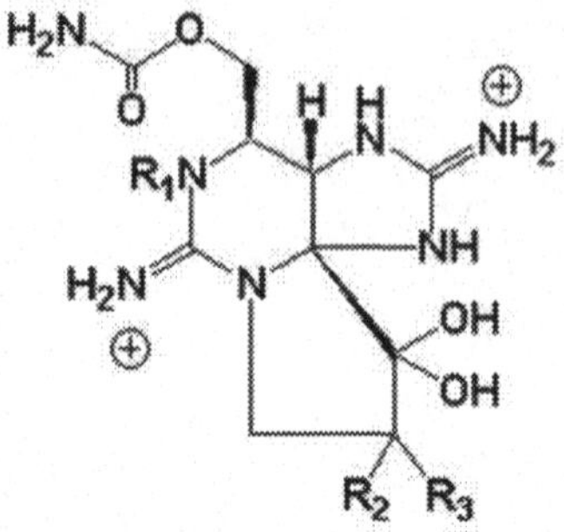

STX	R_1	R_2	R_3
STX	H	H	H
GTX-II	H	H	OSO_3^-
GTX-III	H	OSO_3^-	H
NeoSTX	OH	H	H
GTX-I	OH	H	OSO_3^-
GTX-IV	OH	OSO_3^-	H

Figure 5; Chemical structure of saxitotin (STX) and five saxitoxin derivatives.

Saxitoxin exerts it's toxic effect by interfering with the transmission of signals through the nervous system. The toxin binds specifically to the voltage-gated sodium ion channels and prevents their normal function. This blocks the creation of a proper action potential and the nerve cell no longer has the means to transmit a signal. Saxitoxin blocks the sodium channel selectively without any effect on other types of ion channels. The toxin binds to a site on the external surface of the sodium channel. Owing to the potent and specific sodium channel blocking action, saxitoxin has become a useful tool in neuroscience research in various studies of ion channels [16].

6. BLUE GREEN ALGAE TOXINS (CYANOBACTERIA TOXINS)

Cyanobacterial toxins or cyanotoxins are produced by fresh- and sea-water photosynthetic prokaryotes (blue-green algae) of cosmopolitan occurrence [17-19]. When environmental conditions are suitable for their growth, cyanobacteria may proliferate and form toxic blooms in the upper, sunlit layers. The occurrence of cyanobacteria and their toxins in water bodies used for the production of drinking water poses a technical challenge for water utility managers. The cyanotoxins differ from the intermediates and cofactor compounds that are essential for cell structural synthesis and energy transduction. They present acute and chronic hazards to human and animal health and are responsible for or implicated in animal poisoning, human gastroenteritis, dermal contact irritations and primary liver cancer in humans. These low molecular weight toxins (microcystins, nodularins, saxitoxins, anatoxin-a, anatoxin-a(S), cylindrospermopsin) are structurally diverse and their effects range from liver damage, including liver cancer to skin irritation and neurotoxicity. There are indications of the accumulation of cyanobacterial toxins in fish, and crop plants and they are also associated with the deaths of human dialysis patients.

6.1 Microcystins

Microcystins are a family of more than 50 structurally similar monocyclic heptapeptide hepatotoxins produced by species of freshwater cyanobacteria, primarily *Microcystis aeruginosa*. They are characterized by some invariant amino acids, including one of unusual structure, which is essential for expression of toxicity (Fig. 6) [20]. Microcystins are chemically stable, but suffer biodegradation in reservoir waters.

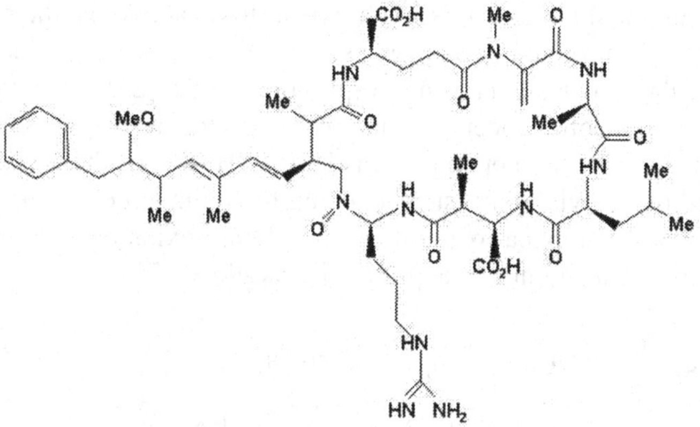

Figure 6: Chemical structure of Microcystins

The most common member of the family is microcystin. It has an LD_{50} in mice and rats of 36-122 mkg/kg by various routes, including aerosol inhalation [21]. Acute liver hemorrhage and death occur with high doses of microcystin, which is also a potent tumor promoter. Although human illnesses attributed to microcystins include gastroenteritis and allergic reactions, the primary target of the toxin is the liver, where disruption of the cytoskeleton, consequent on inhibition of protein phosphatases, causes massive hepatic hemorrhage. Uptake of microcystins into the liver occurs via a carrier-mediated transport system, and several inhibitors of uptake can antagonise the toxic effects of microcystins. Microcystins are very strong protein phosphatase inhibitors, which are needed for regulation of structural proteins of the cell [22, 23]. They covalently bind the serine/threonine protein phosphatases 1 and 2A (PP1 and PP2A), thereby influencing regulation of cellular protein phosphorylation.

6.2 Anatoxins

Anatoxins are a group of low molecular weight neurotoxic alkaloids produced by different species of fresh-water cyanobacteria (*Anabaena flos-aquae, Anabaena planktonia, Oscillatoria, Aphanizomenon)*, in isolation or in combination with hepatotoxins [24]. Not all strains of the causative species are toxic and there are no systemic indications to the presence or absence of anatoxin in a particular strain. Three common anatoxins have been described: anatoxin-a and homoanatoxin-a are secondary amines and anatoxin-a(S) is a phosphate ester of a cyclic N-hydroxyguanine structure (Fig. 7)

Anatoxin-a and homoanatoxin-a are neurotoxic alkaloids, the main effect of which is production of a sustained postsynaptic depolarizing neuromuscular blockade causing respiratory arrest [25]. The toxin binds strongly to the nicotinic acetylcholine receptor [26]. It mimics acetylcholine action and is thus able to open the ion channel. However, the toxin cannot be deactivated by acetylcholineesterase which force the ion channel to be open. Anatoxin-a and homoanatoxin-a are potent neurotoxins causing rapid death in mammals (LD_{50} in mice of approx. 250 mkg/kg) [27]

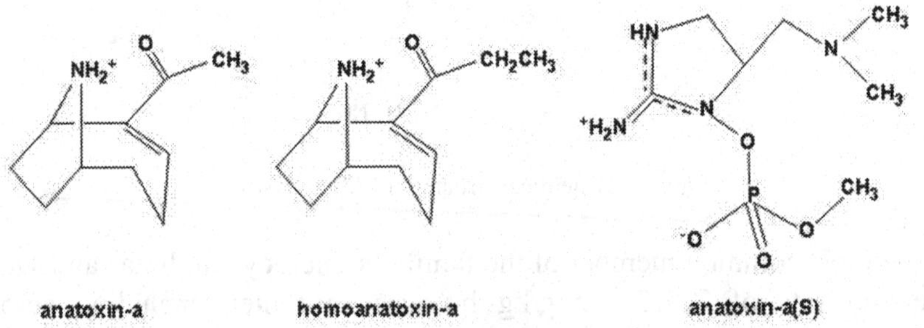

Figure7: Chemical structures of anatoxins.

Anatoxin-a(S), the second type of anatoxins, is the only natural organophosphate known. It is more water-soluble than synthetic organophosphates. Anatoxin-a(S) is more potent towards mice (LD_{50} in mice is approximately of 50 mkg/kg) [28]. It binds to acetylcholineesterase and acts as its inhibitor [29], thereby causing toxicity through the exhaustion of muscle cells. However the toxin is unable to cross the blood-brain barrier to influence brain cholinesterase activity. In addition to respiratory difficulty, anatoxin-a(S) unlike anatoxin-a, induces hypersalivation in mammals as well as other symptoms more typical of neurotoxicity such as diarrhea, tremors and nasal mucus discharge.

7. MYCOTOXINS

Mycotoxins are low molecular weight secondary metabolites of mould or fungi [30-32]. These compounds are non-volatile and may be sequestered in spores or secreted into the growth substrate. Numerous but not all fungi species produce mycotoxins with variation of toxin production depending on the substrate. There are over 200 recognized mycotoxins, however, the study of mycotoxins and their health effects on humans are still in the progress.

Many mycotoxins are harmful to humans and animals when inhaled, ingested or brought into contact with human skin [33]. Diseases in animals and human beings resulting from the consumption of mycotoxins are called "mycotoxicoses". The mechanism of micotoxicity involves interference with various aspects of cell metabolism, producing neurotoxic, carcinogenic or teratogenic effects. The toxins vary in specificity and potency for their target cells, cell structures or cell processes by species and strain that produces them. They are nearly all cytotoxic, disrupting various cellular structures such as membranes, and interfering with vital cellular processes such as protein, RNA and DNA synthesis [34]. Most mycotoxins possess immunosuppressant properties that vary according to the compound. The toxicity of certain fungal metabolites such as aflatoxin, ranks them among the most potently toxic, immunosuppressive and carcinogenic substances known [35].

7.1 Aflatoxin

Aflatoxins are a group of closely related mycotoxins that are widely distributed in nature. Aflatoxin is one of the most potent carcinogens known to man and has been linked to a wide variety of human health problems [36]. Aflatoxins B_1, B_2, G_1, and G_2 are produced in grains in both field and storage by *Aspergillus flavus* and *A. parasiticus* which are common in most soils and are usually involved in decay of plant materials. Aflatoxin persists under extreme environmental conditions and is even relatively heat stable at temperatures above the boiling point of water. Roasting and some microbial treatments may sharply reduce but not eliminate the aflatoxin content.

The toxins pose a significant public health concern as diseases resulting from ingestion of aflatoxins, include acute liver disease to cancer development. Mutagenesis by aflatoxins is caused by the reaction of these electrophilic chemicals with one of the purine or pyrimidine bases in DNA resulting in irreversible changes in normal cells [37, 38].

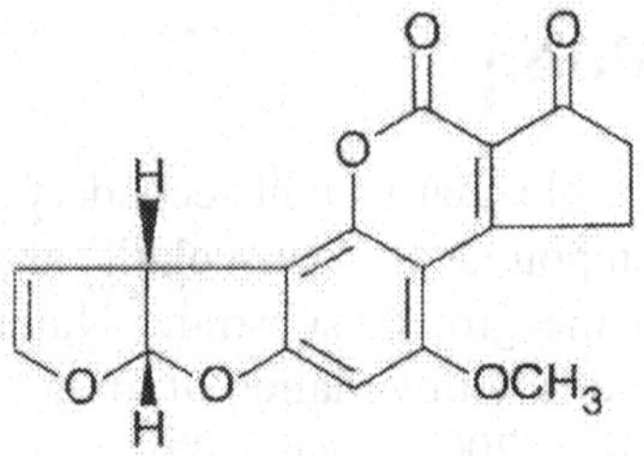

Figure 8. Chemical structure of Aflatoxin B_1

Aflatoxin B_1 (Fig. 8) is the most potent carcinogen known for the liver. It is a specifically metabolized into epoxide form by the action of the mixed function mono-oxygenase enzyme systems (cytochrome P450-dependent) in the tissues (in particular, the liver) of the affected animal [37]. This epoxide is highly reactive and can form derivatives with several cellular macromolecules, including DNA, RNA and protein.

7.2 Ochratoxin

Ochratoxin is primarily produced by *Aspergillus ochraceus, Penicillium viridictum*, which have been found in some samples of food and feed grains [30, 39].

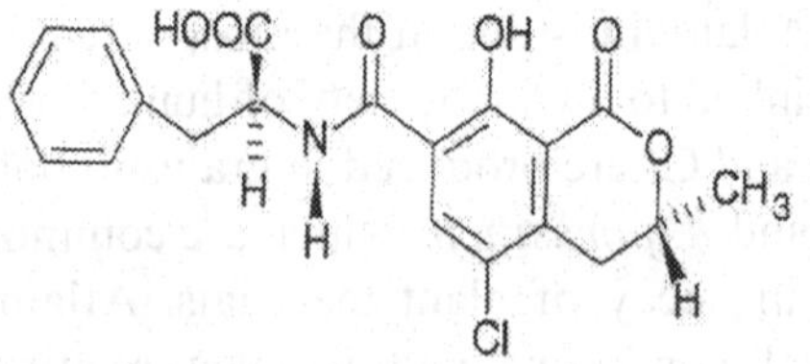

Figure 9: Chemical structure of Ochratoxin A.

Ochratoxin is damaging to the kidneys and liver and is also a suspected carcinogen. Ochratoxin is absorbed from the small intestine and, in plasma, binds to serum albumin. The prolonged half-life of the toxin results from its absorption by proximal tubules and enterohepatic circulation. [40]. The chemical structure is composed of an isocumarin part linked to l-phenylalanine (Fig. 9). Ochratoxin inhibition of phenylalanine hydroxylase and other enzymes that use phenylalanine as a substrate is based on this structural homology. The inhibition is competitive to phenylalanine resulting in protein synthesis inhibition, and is reversed by an excess of this amino acid. The effect of ochratoxin A on protein synthesis is followed by an inhibition of RNA synthesis, which might affect proteins with a high turnover. Recently, ochratoxin was also found to enhance lipid peroxidation both *in vitro* and *in vivo*. This inhibition might have an important effect on cell or mitochondrial membranes and be responsible for the effects on mitochondria that have been shown by several authors [24].

7.3 Trichothecenes, T-2 Toxin

Fusarium tricinctum and some strains of *F. graminearurm, F. equiseti, F. sporotrichioides, F. poae*, and *F. lateritium* commonly found in grains, produce T-2 toxin and other toxic trichothecenes [30, 41]. Feed contaminated with these toxins must be handled carefully because these toxins can cause severe skin irritation and bleeding from lungs. Upon ingestion, trichothecenes cause a modulation effect on cell-mediated immunity and alterations in gastrointestinal propulsion, and can produce tremors, incoordination, depression, and headaches [42]. Very little information is available relating to their toxicokinetics and toxicodynamics in humans. As with most other mycotoxins, the only control is to avoid contaminated feeds.

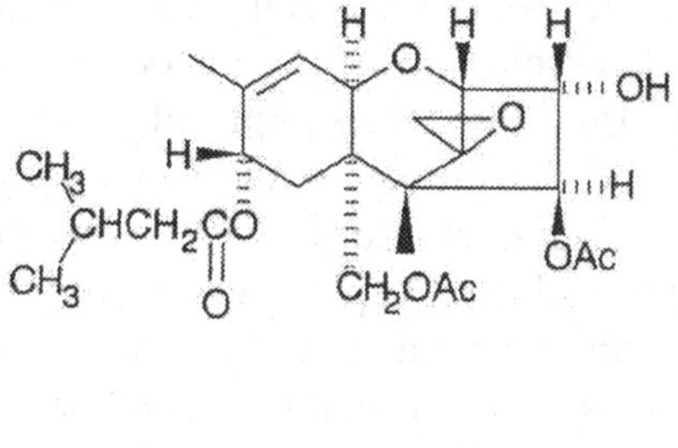

Figure 10. Chemical structure of T-2 toxin

T-2 toxin (12,13-epoxytrichothecene) is one of the more deadly toxins (Fig. 10). If ingested in sufficient quantity, T-2 toxin can severely damage the entire digestive tract and cause rapid death due to internal haemorrhage. T-2 toxin has been implicated in the human diseases alimentary toxic aleukia and pulmonary hemosiderosis. The toxin mainly affects mitotic cells of the gastrointestinal tract and the lymphoid system and promotes a transient and reversible aberration in a single enzymatic reaction to cell death. Regardless of the end point measured, the toxic response involves the interactions of virtually all-subcellular processes: membrane transport and permeability, chemical metabolism, DNA function, and energy production/expenditure. T-2 toxin rapidly crosses the cell membrane of cells and binds to the intracellular targets. It affects the polyribosome structure and thus inhibits protein synthesis being irreversible inhibitor.

8. BIOTOXIN DETECTOR

Development of fast and sensitive methods for the detection of natural toxins is of great importance because of the threat of bioterrorist attacks. In addition, biotoxin assays are used in food and medical industry and environmental monitoring. At the present time, laboratory methods for the detection of biotoxins involve animal tests, microbiological methods, PCR-based DNA assay, and immunological methods (direct and competitive immunofluorescence methods or ELISA). Immunoassays are the most sensitive, convenient, and least time-consuming and may be used as a method of choice for rapid detection of biotoxins in field conditions. However, these methods do not allow one to carry out simultaneous test for the presence of several toxins. In contrast, microchip based methods allow parallel analysis of different biotoxins.

An array biosensor for the simultaneous detection of several toxins has been developed by Ligler et al. [43-44]. In this sensor, biotinylated capture antibodies were immobilized in columns on avidin-coated waveguides on the surface of glass slide. Electronic microchips with capture antibodies were proposed to detect fluorescein-labeled staphylococcal and cholera toxins (direct immunoassay) [45]. Another example of biochips for the assay of pathogenic organisms and their toxins is the technology that uses surface-enhanced Raman scattering microscopy (SERS) for the detection of the antibody-antigen complexes [46]. Despite good sensitivity and low detection time that are achieved with the mentioned above systems, these biochips are produced by rather sophisticated technologies and/or complex instruments are needed for the detection of signals.

Microchips on the basis of three-dimensional hydrogels have been developed by Mirzabekov et. al. [47]. Gel microchips with immobilized antibodies were already used for quantitative assay of tumor-associated markers with the sensitivity as low as 0.2 ng/ml.

9. CONCLUSION

Some natural toxins can be used as agents for bioterrorism, and their potential for future use is a major concern. Although the list of potential agents is rather short some natural toxins should be under special attention. Sensitive methods of toxin detection have to be developed.

REFERENCES

1. L.L. Simpson, Identification of the major steps in botulinum toxin action, Annu Rev Pharmacol Toxicol. 44 (2004) 167-193.
2. D. Josko, Botulin toxin: a weapon in terrorism, Clin Lab Sci. 17 (2004) 30-34.
3. G. Lalli et al., The journey of tetanus and botulinum neurotoxins in neurons, Trends Microbiol. 11 (2003) 431-437.
4. B. Poulain and Y. Humeau, Mode of action of botulinum neurotoxin: pathological, cellular and molecular aspect, Ann Readapt Med Phys. 46 (2003) 265-275.
5. C. Singer, Indications and management of botulinum toxin, Rev Neurol. 29 (1999) 157-162.
6. S.M. Bradberry et al., Ricin poisoning, Toxicol Rev. 22 (2003) 65-70.
7. M.J. Lord et al., Ricin. Mechanisms of cytotoxicity, Toxicol Rev. 22 (2003) 53-64.
8. J. Robertus, The structure and action of ricin, a cytotoxic N-glycosidase, Semin Cell Biol. 2 (1991) 23-30.
9. J.M. Lord, L.M. Roberts and J.D. Robertus, Ricin: structure, mode of action, and some current applications, FASEB J. 8 (1994) 201-208.
10. S. Olsnes and J.V. Kozlov, Ricin, Toxicon 39 (2001) 1723-1728.
11. K. Sandvig et al., Ricin transport into cells: studies of endocytosis and intracellular transport, Int J Med Microbiol. 290 (2000) 415-420.
12. J. Wesche, Retrograde transport of ricin, Int J Med Microbiol. 291 (2002) 517-521.
13. M.R. Watters, Organic neurotoxins in seafoods, Clin Neurol Neurosurg. 97 (1995) 119-124.
14. E.J. Schantz et al., Paralytic shellfish poison. VI. A procedure for the isolation and purification of the poison from toxic clams and mussel tissues, J.Am.Chem.Soc. 79 (1957) 5230-5235.
15. E.J.Schantz et al., The structure of saxitoxin, J.Am.Chem.Soc. 97 (1975) 1238-1239.
16. T. Narahashi, M.L. Roy and K.S. Ginsburg, Recent advances in the study of mechanism of action of marine neurotoxins, Neurotoxicology 15 (1994) 545-554.
17. J.F. Briand et al., Health hazards for terrestrial vertebrates from toxic cyanobacteria in surface water ecosystems, Vet Res. 34 (2003) 361-377.
18. P.V. Rao et al., Toxins and bioactive compounds from cyanobacteria and their implications on human health, J Environ Biol. 23 (2002) 215-224.

19. G.A. Codd, C.J. Ward and S.G. Bell, Cyanobacterial toxins: occurrence, modes of action, health effects and exposure routes, Arch Toxicol Suppl. 19 (1997) 399-410.
20. R.M. Dawson, The toxicology of microcystins, Toxicon 36 (1998) 953-962.
21. K. Bischoff, The toxicology of microcystin-LR: occurrence, toxicokinetics, toxicodynamics, diagnosis and treatment, Vet Hum Toxicol. 43 (2001) 294-297.
22. B.M. Gulledgea et al., The microcystins and nodularins: cyclic polypeptide inhibitors of PP1 and PP2A, Curr Med Chem. 9 (2002) 1991-2003.
23. M.M. Gehringer, Microcystin-LR and okadaic acid-induced cellular effects: a dualistic response, FEBS Lett. 557 (2004) 1-8.
24. Toxic Cyanobacteria in Water: a guide to their public health consequences, monitoring and management. Chorus, E. & Bartram, J. (Eds.) World Health Organisation 1999, E&FN Spoon London & New York.
25. W.W. Carmichael, D.F. Biggs and P.R. Gorham, Toxicology and pharmacological action of anabaena flos-aquae toxin, Science. 187 (1975) 542-544.
26. C.E. Spivak, B. Witkop and E.X. Albuquerque, Anatoxin-a: a novel, potent agonist at the nicotinic receptor, Mol Pharmacol. 18 (1980) 384-394.
27. N.B. Astrachan, B.G. Archer and D.R. Hilbelink, Evaluation of the subacute toxicity and teratogenicity of anatoxin-a, Toxicon 18 (1980) 684-688.
28. N.A. Mahmood and W.W. Carmichael, The pharmacology of anatoxin-a(s), a neurotoxin produced by the freshwater cyanobacterium Anabaena flos-aquae NRC 525-17, Toxicon 24 (1986) 425-434.
29. N.A. Mahmood and W.W. Carmichael, Anatoxin-a(s), an anticholinesterase from the cyanobacterium Anabaena flos-aquae NRC-525-17, Toxicon 25 (1987) 1221-1227.
30. P.S. Steyn, Mycotoxins, general view, chemistry and structure, Toxicol Lett. 82-83 (1995) 843-851.
31. D. Bhatnagar and K.C. Ehrlich, Toxins of filamentous fungi, Chem Immunol. 81 (2002) 167- 206.
32. J.W. Bennett and M. Klich, Mycotoxins, Clin Microbiol Rev. 16 (2003) 497-516.
33. R. Kappe and D. Rimek, Fungal diseases, Prog Drug Res. Spec No (2003) 13-38.
34. S.E. Browne and M.F. Beal, Toxin-induced mitochondrial dysfunction, Int Rev Neurobiol. 53 (2002) 243-279.
35. R. Goldman and P.G. Shields, Food mutagens, J Nutr. 133 Suppl 3 (2003) 965S-973S.
36. H.N. Mishra and C. Das, A review on biological control and metabolism of aflatoxin, Crit Rev Food Sci Nutr. 43 (2003) 245-264.
37. M. McLean and M.F. Dutton, Cellular interactions and metabolism of aflatoxin: an update, Pharmacol Ther. 65 (1995) 163-192.
38. C.P. Wild and P.C. Turner, The toxicology of aflatoxins as a basis for public health decisions, Mutagenesis. 17 (2002) 471-481.
39. G. Dirheimer and E.E. Creppy, Mechanism of action of ochratoxin A, IARC Sci Publ. 115 (1991) 171-186.
40. V. Berger et al., Interaction of ochratoxin A with human intestinal Caco-2 cells: possible implication of a multidrug resistance-associated protein (MRP2), Toxicol Lett. 140-141 (2003) 465-476.
41. D.L. Sudakin, Trichothecenes in the environment: relevance to human health, Toxicol Lett. 143 (2003) 97-107.
42. P.P. Williams, Effects of T-2 mycotoxin on gastrointestinal tissues: a review of in vivo and in vitro models, Arch Environ Contam Toxicol. 18 (1989) 374-387.
43. J.B. Delehanty and F.S. Ligler, A microarray immunoassay for simultaneous detection of proteins and bacteria, Anal. Chem. 74 (2002) 5681-5687.

44. F.S. Ligler et al, Array biosensor for detection of toxins, Anal. Bioanal. Chem. 377 (2003) 469-477.
45. K.L. Ewalt et al, Detection of biological toxins on an active electronic microchip, Anal. Biochem. 289 (2001) 162-172.
46. A.E. Grow et al., New biochip technology for label-free detection of pathogens and their toxins, J. Microbiol. Methods 53 (2003) 221-233.
47. A.Yu. Rubina et al., Hydrogel-based protein microchips: manufacturing, properties, and applications, BioTechniques 34 (2003) 1008-1022.

BIOLOGICAL WEAPONS INSPECTIONS- THE IRAQ EXPERIENCE

E. B. Myhre
Professor, Lund University, Department of Infectious Diseases University Hospital Medical Centre, Lund SE-22185, Sweden

Abstract: Prior to the 1990 Iraq-Kuwait conflict it was well known that Iraq had developed weapons of mass destruction but the extent of its programs were unclear. After the Iraqi defeat in the ensuing Gulf war 1991 the UN Security Council authorized the creation of UN Special Commission for Iraq (UNSCOM) with the purpose of ridding Iraq permanently of weapons of mass destruction. Several conclusions can be drawn from more than ten years of biological weapons inspections in Iraq. Firstly, UNSCOM managed to get a rather clear picture of the past weapons programs. Secondly, it was not possible for Iraq to restart a substantial program with UNSCOM being present in the country. Thirdly, a full and final and complete account of the weapons program could not be established despite the use of the best intellectual and technical capabilities available at the time.

1. DEVELOPMENT OF BIOLOGICAL WEAPONS

Development and production of biological weapons of mass destruction is an illegitimate activity. In the era of globalization such weapons could end up in the hands of terrorist groups. The dissemination of weapon technology through migration of scientists is another concern. Insight into and control of state-sponsored weapons programs is highly relevant with regards to prevention of international terrorism.

Prior to the 1990 Iraq-Kuwait conflict it was well known that Iraq had developed weapons of mass destruction but the extent of its programs were

D. Morrison et al. (eds.), Defense against Bioterror: Detection Technologies, Implementation Strategies and Commercial Opportunities, 47–50.

unclear. After the Iraqi defeat in the ensuing Gulf war 1991 the UN Security Council authorized the creation of UN Special Commission for Iraq (UNSCOM) with the purpose of ridding Iraq permanently of weapons of mass destruction. It was expected to be a quick and easy task. Iraq was supposed to produce a full final declaration of their weapons programs, to hand over remaining weapons and to destroy them under supervision. Instead of full open cooperation Iraq engaged in a policy of concealment and deception. Friendly walk-through inspections were soon transformed into unannounced intrusive on-site visits. Instead of being provided with requested information international UNSCOM experts had to collect the information themselves. This resulted in an atmosphere of mutual distrust. Soon these inspections became high profile activities, closely watched by the mass media and wrongly portrayed as the hunt for weapons of mass destruction. The objectives were to fully understand the Iraqi weapons programs, to map all its components and to ensure that the weapons program was not restarted. In 1999 UNSCOM was transformed in into a new organisation, the UN Monitoring, Verification and Inspections Commission for Iraq (UNMOVIC). Facing the same restraints and an unchanged Iraqi attitude UNMOVIC would not be more successful than UNSCOM in providing a full final and complete account of the Iraq's programs of weapons of mass destruction. Shortly before the Second Gulf War in 2003 UNMOVIC withdrew its staff from Iraq and the inspection work came to an end.

Initially Iraq vehemently denied the existence of any offensive biological weapons program. But 1995 became a turning point with the defection to Jordan of Kamal Hussein, Saddam Hussein's son-in-law. Kamal Hussein had first class insight into the weapons programs and UNSCOM was immediately notified about the existence of an inconspicuous farm house outside Baghdad. When UNSCOM searched the farm house cases of documents were found revealing among other things an offensive biological research program and a secret production facility. This facility had been visited by UNSCOM expert teams several times but samples collected had failed to detect any biological agents. The Iraqis had successfully sanitized the facility. Yet the location, the construction of the buildings as well as the equipment found there had always roused suspicion.

UNSCOM on-site inspections were highly sophisticated information gathering operations. The work was carefully planed and organized by the permanent staff at UNSCOM headquarter in New York City but provision was made for local initiatives for quick responses. Operational plans were often kept secret even from some UNSCOM staff members in order to prevent unwanted dissemination of information. Sites were visited without prior notice yet UNSCOM teams were always accompanied by Iraqi minders. With access to their own helicopters UNSCOM staff members could quickly reach any part of Iraq. Vast amounts of data were accumulated in a highly systematic way. Detailed information had been collected on institutions of interest, their technical equipment and their human resources. The capability of various sites to engage in research and production of biological weapons was critically evaluated.

Iraqi minders were always present during site visits including interview sessions with the Iraqi scientists, administrators and laboratory workers. At times the minders could be disruptive and the situation could become ugly. Iraqi government officials normally videotaped entire inspections, a procedure which was very stressful both to the inspectors and the Iraqi who were working at the site. The minders were apparently instructed to assess the area of competence of each new UNSCOM inspector. Occasionally the minders tried to hinder inspections by simply outnumber the UNSCOM team. UNSCOM own linguists were extremely useful on site as their presence made studies of various Arabic documents possible.Dual-use equipment provided a special analytical problem. Such equipment were tagged and only allowed to be moved after prior notice.

There were many constraints, some of which were of critical importance. Access to all locations, for example, was one. Presidential palaces, mosques, private homes and facilities used by various security organizations were all off limits. Another constraint was the issue of timing. Inspections were never carried out during the night. The Iraqis could move and hide equipment during the night if they wanted. There was evidence of such activities.

UNSCOM withdrew twice from Iraq as the US retaliated with cruise missile strikes for Iraq's failure to cooperate with UNSCOM. A revisit to a site after an attack was suggested that the Iraqi had known exactly which sites were to be attacked. The Iraqis allowed their staff to take home items of special importance. Items too heavy to be transported to private homes were stored in a caravan some hundred meters away from the buildings as the Iraqis knew from past experience that this was the safest place.

Unparalleled experience in conducting weapons inspections was gained from 10 years of work in Iraq. Even with access to the best intellectual capabilities and technical equipment UNSCOM/UNMOVIC could not confirm nor deny that Iraq had retained weapons of mass destruction and had restarted its offensive weapons programmes. Both the intelligence and the scientific community were equally divided on these issues. Different governments made different conclusions based on how they interpreted available data. There were worries that Iraq would restart their program once the sanctions were lifted and UNSCOM gone. We knew that the Iraqi Government kept the weapons groups intact for years. The Iraqi scientists who had been involved in the past program showed no regret. An ongoing compartmentalized activity possibly under the auspices of security agencies was also suspected as were transmigration of scientists and movements of equipment to other countries.

As soon as the hostilities had ended the Iraqi Survey Group stared to search for proscribed weapons and illegitimate weapons activities. In October 2003, the Iraqi Survey Group presented to the Congress in an interim report that no weapons of mass destruction had been found so far. What was the actual situation in Iraq at the onset of the last Gulf War 2003? Many questions remain unanswered. Did Iraq still possess weapons of mass destruction? Had Iraq restarted its biological weapons program? The answer today is that we still do not know and we might never know.

2. CONCLUSION

Several conclusions can been draw from more ten years of biological weapons inspections in Iraq. Firstly, UNSCOM managed to get a rather clear picture of the past weapons programs. Secondly, it was not possible for Iraq to restart a substantial program with UNSCOM being present in the country. Thirdly, a full and final and complete account of the weapons program could not be established despite the use of the best intellectual and technical capabilities available at the time.

INTEGRATED, SECURE AND SUSTAINABLE DISEASE SURVEILLANCE SYSTEM IN UZBEKISTAN: ASPECTS OF LABORATORY RESEARCH NETWORKS

F. T. Adilova
Professor, Institute of Cybernetics of the National Academy of Sciences, Tashkent, Uzbekistan

Abstract: Epidemiological analysis by the WHO showed that the spread of infectious diseases in all Central Asia states, including Uzbekistan have worsen. Uzbekistan is one of the most populated Central Asian countries with 25 million people, approximately 70% of whom live in rural areas. Many infectious diseases are spread from animal sources. In Uzbekistan, rotavirus, coroavirus, parovavirus, herpes virus, enterovirus infections, and brucellosis, anthrax, foot and mouth diseases, salmonellosis, kolibacteriosis, plaque diseases and others are wide spread. Epidemiological monitoring and the measures systematically undertaken by the local health service ensure a stable level of morbidity for the first and second groups of infectious diseases. Natural breeding grounds for infectious diseases such as plague, anthrax, and brucellosis and others in the arbovirus-group infections like the Crimean–Congo hemorrhagic fever (CCHF), tick-borne encephalitis (TBE), and fever syndromes of unknown origin dictate the need for more in-depth and comprehensive study in a cross section of every region in Uzbekistan. In October 2003 within the Cooperative Threat Reduction Program (CTR) and the Biological Weapons Proliferation Prevention Program (BWPPP) sponsored by the U.S. Department of Defense, a program began to establish an Integrated, Secure and Sustainable Disease Surveillance System in Uzbekistan. Main goals of this project are: detect deliberate or accidental release of biological materials relevant to the bioterror threat; create, strengthen and integrate existing surveillance systems; facilitate integration of host nation scientists and institutes into the international scientific community; and to create a potential to integrate national surveillance systems into an international system. Key elements of the project are modern, standardized, reliable diagnostics methods, e.g. PCR-based diagnostics; improved communications, transport, and integration, e.g. computerization; data analysis and sharing. We plan to select optimal prototypes of worldwide known

D. Morrison et al. (eds.), Defense against Bioterror: Detection Technologies, Implementation Strategies and Commercial Opportunities, 51–65.

Laboratory Research Networks and contribute new sophisticated methods of available data analysis.

1. INTRODUCTION

In October 2003 within Cooperative Threat Reduction Program (CTR) and Biological Weapons Proliferation Prevention Program (BWPPP) sponsoring by Department of Defense, of the United States began to establish an Integrated, Secure and Sustainable Disease Surveillance System in Kazakhstan and Uzbekistan.

The main goal of this project is to detect deliberate or accidental release of biological materials relevant to the bioterror threat by: strengthening and integrating existing surveillance systems; facilitating integration of host nation scientists and institutes into the international scientific community; and, potentially integrating national surveillance systems into an international system.

Key elements of the project are modern, standardized, reliable diagnostics methods, e.g. PCR-based diagnostics; improved communications, transport, and integration, e.g. computerization; data analysis and sharing.

Key requirements

Laboratory: secure and safeguarded central reference laboratory; safe, secure, and efficient pathogen transportation systems and capabilities; verifiable training in biosafety/biosecurity, diagnostics; modern diagnostics methods; reduced need for pathogen collections and rapid diagnoses.

Epidemiology and Surveillance: standardized and repeatable human and animal disease monitoring systems; mobile epidemiological response teams and secure transportation of disease elements; and verifiable epidemiology training.

Communications and Information Technology: design, develop, deploy and sustain a robust and secure electronic communicable disease reporting system; create and sustain secure communications and data storage systems.

Distinguish feature of the project is designing 2 parallel systems of sentinel stations: for human disease , and for animal disease (vector and agricultural).

Clinical sentinel stations will deal with:

- Diseases: (plague, anthrax, brucellosis, tularemia, CCHF, and tick-borne encephalitis);
- Syndromes: (fevers of unknown origin, flu-like symptoms (requiring hospitalization), and hemorrhagic fevers);

- Antibiotic resistant bacteria;
- Animal sentinel stations
- Vectors: (plague, tularemia, and CCHF);
- Livestock: (foot and mouth, rinderpest, avian influenza, Newcastle disease, brucellosis, glanders, and poxviruses);
- Antibiotic resistant bacteria

2. NATURAL SOURCES OF DANGEROUS ANIMAL PATHOGENS AND EXISTING PATHOGEN COLLECTION OF UZBEKISTAN

Epidemiological analysis by the World Health Organization showed that the spread of infectious diseases in all Central Asia states, including Uzbekistan is going from bad to worse. Uzbekistan is one of the more highly populated Central Asian countries with 25 million people, approximately 70% of whom live in rural areas. Many infectious diseases are spreading from animal sources. Also because of infectious diseases the agricultural animal population of Uzbekistan has decreased approximately twice during the last decade. In Uzbekistan, rotavirus, coroavirus, parovavirus, herpes virus, enterovirus infections, and brucellosis, anthrax, foot'n mouth diseases, salmonellosis, kolibacteriosis, plaque diseases and others are wide spread.

Epidemiological monitoring and the measures systematically undertaken by the Republic of Uzbekistan Health Service ensure a stable level of morbidity for the first and second groups of infectious diseases. Natural breeding grounds for such infectious diseases as plague, anthrax, and brucellosis and such arbovirus-group infections as Crimean–Congo hemorrhagic fever (CCHF), tick-borne encephalitis(TBE) , and fever syndromes of unknown origin dictate the need for more in-depth and comprehensive study in a cross section of every region in Uzbekistan. According to official statistics, each year up to 2 million people in Uzbekistan suffer from fever illnesses where the nature of the disease remains etiologically undetermined. For example, scientific research of the last 2–3 years, involving the use of serologic, virological, entomological, and clinical laboratory studies, showed that out of 1227 patients suffering from fever of an unknown nature in Surkhandar'ya region alone more than 15% who were ill, 4.93% were infected with arbovirus infections, including CCHF, while the figure is 5.05% for Tamdy, 3.1% for Karshy, and 2.2% for the Syrdar'ya Valley.

Since 1936 tests of bacteriological weapons have been carried out on "Vozrojdenie" island, in Uzbekistan, , and in 1954 the Island became one of the former Soviet Union's largest military bioranges. as In opinion of the experts, tests of the bacteriological weapons, despite of all safety measures undertaken, have left changes in fauna, and carriers of dangerous diseases on island of "Vozrozhdenie" can be mice and jerboas. With the collapse of Soviet Union, military personnel of the biorange were disbanded, and the range - demounted, however the territory of range was not disinfected, and burial places of the bacteriological weapon were not destroyed. Because of a drought resulting from constant shallowing of the Aral Sea and narrowing of its water space, the island has became a peninsula, and its territory has increased greatly. Thus in 1950 the territory island averaged 200 square kilometers, and now it is 2 thousand square kilometers. Thus, buried traces of anthrax have become a serious problem. According to various data it is clear, that in several places of burial (6 of 11) spores have kept viability. In view of the international situation - growth and development of bioterrorism, adverse ecological conditions in Karakalpakistan and nearby areas of Central Asia, burial place of anthrax and the possibility of their distribution from territory of "Vozrozhdenie island" becomes of special danger. It is necessary to mention, that spores of anthrax resistant against factors of an environment, could be transferred among animals (rodents etc.) and therefore periodically cause flashes of this dangerous disease. While gathering biomaterial for continuous studies, all compound components of biogeocenosis should be covered: water, air, plants, animal, and also materials and the equipment, left on the place of former bioarange. All researched samples should be checked for presence of the strain of anthrax by microbiological verification. Therefore, monitoring of the biogeocenosis of the "Vozrozhdenie" island for presence of anthrax with the purpose of revealing, localization and subsequent destruction of anthrax is an extremely pressing and vital problem.

At present, The Research Institute of Veterinary has created and stores (in museums of 6 laboratories) the Uzbek National Collection of over 300 of the most interesting vaccine and virulent cultures, bacteria and viruses – agents of infection, the most hazardous diseases of animals.. This collection is deemed to be the property of the Institute and the Republic.

The Brucellosis Laboratory conducts continuous work with over 220,000 virulent and vaccine cultures of brucella of abortus, melitensis and suis types. Collection of Tubercullosis laboratory has 19 cultures of mycobacteria used for the preparation of diagnosticums and preparations for specific prophylaxis of tubercullosis among animals.

Collections of laboratories of radiobiology, diseases of young animals and poultry have 9 cultures of pasterella, 8 salmonella, 21 esherichia coli, 2 diplococcus. The Leptospirosis laboratory has 4 cultures of leptospira.

3. STATUS OF HUMAN AND ANIMAL INFECTIONS LABORATORY RESEARCH

Considering the current status of infection dissemination described above, the organization and support of laboratory diagnostics should make it possible to identify etiologic structures with extreme accuracy, especially for such arbovirus diseases as CCHF, TBE, West Nile Fever (WNF), and other infectious diseases of viral nature that are recorded in the Republic of Uzbekistan. The importance of laboratory research has been confirmed quite convincingly by the notion that the arbovirus-group infections mentioned above are recorded in many regions of Uzbekistan. At the present time, there are 17 virology laboratories in operation at republican, regional, and municipal centers of Gossanepidnadzor [the State Sanitary and Epidemiological Inspectorate] of Uzbekistan Ministry of Health, as well as the Laboratory of Transmissible Viral Infections of the Scientific Research Institute of Virology. Well-trained virologists work at all virology laboratories in the Republic and by existing intellectual potential in terms of training, they, at any moment, could be involved in an emergency situation with group I and II infectious diseases. Against this background, the Republic needs to solve such urgent problems as setting up laboratories that meet all requirements for working with group I and II infections, the necessary physical facilities must be provided. Of these laboratories, just three have permission to work with groups I, II pathogens: the Laboratory of Transmissible Viral Infections of the Scientific Research Institute of Virology , the laboratory of the Surkhandar'ya Viloyat (Region), and the laboratory of the Bukhara Viloyat. Other laboratories work with pathogens of groups 3–4 because they do not have the proper facilities and equipment.

The Laboratory of Transmissible Viral Infections (TVI) of the Scientific Research Institute of Virology performs a full range of virological and serologic research, they have the requisite set of physical facilities, and is equipped with qualified personnel.

The Bukhara Viloyat's laboratory has a full set of the necessary physical premises, is outfitted with minimal equipment of old models, and is staffed with virologists. It performs virological research only on cell cultures and chick embryos. Among serological methods, the complement-fixation reaction (CFR) is used when there is a diagnosis of TBE virus.

The Surkhandar'ya Viloyat's laboratory has insufficient space to perform the full range of virological and serological research for the first and second groups. It is outfitted with minimal equipment in old models yet they have qualified personnel.

To perform the full range of virological and serological research on arboviruses, both regional laboratories must be supplied with biological protection boxes and PCR apparatus, diagnostic kits, and reagents for research using the complement-fixation reaction, indirect hemagglutination, diffuse precipitation reaction in agar (DPRA). Within the DTRA-Uzbekistan project the full range of research should be concentrated on arbovirus infections at the Central Reference Laboratory, placed in Virology Institute (Tashkent), with selection of the following existing regional labs: in Surkhandar'ya, Bukhara, Khorezm, Samarkand and Karakalpakistan regions, and for Fergana Valley – Andijan.

Therefore, to control and prevent the sources of animal infectious diseases and their spreading among populations, it is important to develop molecular genetic tools to analyze pathogens and study the molecular basis of tolerance, and to create a database system for rapid analysis of newly formed mutant strains.

4. ICT STATUS IN UZBEKISTAN

The methodology applied in monitoring the ICT development by UNDP experts is based on the research outputs of the International Development Centre at Harvard University 'Readiness for Information World: Guide for Developing countries'. According to the methodology the monitoring was focused on the indicators of 24 components such as availability, speed and quality of network access, use of information and communication technologies (ICT) in educational establishments, various institutions and organizations, economy, governance and everyday life, and implementation of ICT state policies. The above indicator categories are rated according to a 4-point scale: from phase 1 – least developed to phase 4- most developed . The monitoring of ICT development in Uzbekistan was based on the above methodology assessing specific indicators in various areas such as telecommunications infrastructure, efficiency of state policies in this area, availability of an enabling environment for the electronic business, availability of human resources, staff training, etc.[http://www.ddi.uz]

For purposes of monitoring and assessment of the development level of information society in Uzbekistan, a special set of indicators of following components was traced.

Table 1. Indicators and their components

COMPONENTS	INDICATORS	Rating 05.2003
	Share of population using the Internet regularly	**1**
	People's awareness about the Internet	**3**

People and organizations in the network	People's access to cable and mobile telephone communication	2
	Share of officially registered organizations working in the ICT area in the total number of organizations	1
ICT at work places	Share of organizations with LANs	2
	Share of state agencies connected to the Internet	2

Uzbekistan considerably lags behind the other countries in the number of users and the amount of information in the country on the Internet. People's awareness of the possibilities and advantages of information technologies and the Internet is insufficient. As in many other countries, there is a considerable disparity in the availability of information technologies between the capital and the provinces, especially rural areas. The indicator of people's access to cable and mobile telephone communication shows that 7.5% of the population have some communication means , and there are 6.7 primary telephone lines per 100 residents. Only 8 of 1,000 residents are mobile telephone subscribers, and most of them live in Tashkent. Of 195,000 companies operating in the country, only 892 work in the ICT area, which constitutes 0.45% of their total number.

Provision of work places with computers is not sufficient. As a rule, enterprises use unsophisticated and inexpensive software designed by either organizations themselves (at major state agencies and ministries) or by local software companies. Application of electronic systems for management, relationship with clients and planning (ERP, ERM and CRM) is limited. Electronic data exchange (of the EDI type) systems are almost absent from relationship among various organizations. Only an insignificant number of companies apply production management systems corresponding to world standards such as SAPR/3, ORACLE APPLICATIONS, etc.

The number of organizations having their own websites has increased in the past few years. However, few of them have a genuine corporative portal. A small number of organizations have LANs used for internal data exchange. Some of them have certain client-server supplements with a central database. According to expert estimates, of the 195,000 enterprises only around 10% (i.e. 19,500) use IT-based management systems. The most computerized are the banking system and the taxation and customs services. Overall, the number of employed population using ICT means in their professional activity constitutes Only 2.8% of the total number of those are employed. Most of them live in Tashkent where approximately half of the country's computers are concentrated. LANs exist at 3.9% of economic entities having ICT means. Most of them (36.1%) are located in Tashkent.

The efficiency of ICT application is also characterized by connection of state agencies to the Internet. The share of such agencies in the country is 1.1% of their total number.

The existing local networks in urban and rural areas sharply differ by types of stations (digital, analogous and quasi-electronic), the quality of their services and possibilities for providing additional services, and they do not meet the growing requirements of the population.

Mobile communication networks include the networks of cellular communication providers, radial zoning (including trunking) networks and personal radio call networks. Cellular mobile communication networks are rapidly developing. There are 15 operators in the cellular mobile service market and only third of them using AMPS (DAMPS) and GSM standards have a possibility to cover the country's entire territory.

Despite the competitive market, Internet access services remain rather expensive. The cost of leasing a dedicated line with speed of 64 Kb/sec is high yet. Naturally, only major banks and foreign representative offices can afford such lines. Most clients use the dial-up connection (the average ratio of these to those connected by dedicated lines is approximately 100:1).The number of regular Internet users in Uzbekistan is about 511,000, i.e. approximately two users per 100 residents.

The cost of local telephone calls is also important for Internet access since the most widespread way of accessing the Internet remains the dial-up connection, although some providers have started to use new technologies such as ADSL, ZDSL, ISDN and Radio Ethernet.

Along with the general carrying capacity of external access channels, one of the major indicators of the quality of Internet access is the total capacity of modem pools. The total capacity of modem pools in the country is estimated at 3,500 units, i.e. 13.8 of the modem pool per 100,000 residents. The total carrying capacity of external Internet access channels as of May 2003 was 24 Mb/sec against 6.7 Mb/sec in May 2001, which testifies to a practically four-fold increase in the carrying capacity of external Internet access channels.

Therefore we can mark the following advantages of ICT promotion in this country: extension of digital (optical fiber and radio-relay) channels to all oblast administrative centers and to some rayon centers; establishment of a competitive environment in the Internet service market; dynamic growth of the number of public Internet access centers and the global web users; and considerable increase in the carrying capacity of external channels of Internet access. Unfortunately, high prices on Internet access, international communication and cellular communication services, insufficient utilization of the existing digital transport networks capacity limits the growth of the public Internet access.

5. LABORATORY RESPONSE NETWORK FOR BIOTERRORISM

To facilitate rapid detection of a future bioterrorist attack, an increasing number of public health departments are investing in new surveillance systems that target the early manifestations of bioterrorism-related disease. Whether this approach is likely to detect an epidemic sooner than reporting by alert clinicians remains unknown. The detection of a bioterrorism-related epidemic will depend on population characteristics, availability and use of health services, the nature of an attack, epidemiologic features of individual diseases, surveillance methods, and the capacity of health departments to respond to alerts. Predicting how these factors will combine in a bioterrorism attack may be impossible. Nevertheless, understanding their likely effect on epidemic detection should help define the usefulness of syndromic surveillance and identify approaches to increasing the likelihood that clinicians recognize and report an epidemic[1].

Broadly labeled "syndromic surveillance," these efforts encompass a spectrum of activities that include monitoring illness syndromes or events, such as medication purchases, that reflect the prodromes of bioterrorism-related diseases [2-6]. Establishing a diagnosis is critical to the public health response to a bioterrorism-related epidemic, since the diagnosis will guide the use of vaccinations, medications, and other interventions

Increasing computational power in the last 10 years has resulted in the development of mathematical algorithms to routinely and rapidly detect significant clusters within large amounts of surveillance data [7,8].

Automated electronic laboratory reporting is frequently promoted to improve data quality and timeliness of collection[9]. More recently, the general availability of the Internet permits feedback to many users, who can have continuous, simultaneous, and even interactive access to information. The Internet allows for immediate communication of signals of possible outbreaks to relevant professionals for interpretation and action. Let us briefly overview the methodology and main types of algorithms.

One of two mathematical foundations of early detection is signal detection theory, the concepts of which are familiar to many researchers in public health, the objective of detection is to recognize from input data (signal) the occurrence of an event such as an epidemic . A detection method processes the signal and produces as output a determination of whether an event is present or not [10] .

In case of signal presented as a set of historical data one can calculate an expected total value for the current epidemiologic week, a regression line is plotted through the totals in the nine epidemiologic weeks centered on the same epidemiologic week in the previous 5 years. For example, to calculate an expected value for week 20, a regression line is plotted through the values at weeks 16–24 of the previous 5 years. The detection method may be as simple as comparing the amplitude of the signal with a threshold in case of on-line analysis. The accuracy of the detection methods is reported using various parameters such as sensitivity, specificity, and positive predictive value.

Detection methods in both cases (historical data or on-line signal) can often be adjusted to increase or decrease the sensitivity of the detection. For example, the threshold can be lowered. However, improvement on sensitivity usually occurs at the cost of loss of specificity. The optimal level of sensitivity relative to specificity depends on the consequences of false alarms and the benefits of true alarms. These factors are not fundamental properties of the detection method itself, but rather on the use to which the detection method is being applied. Therefore researchers often report the specificity of a detection method over a range of sensitivities (by manipulating the threshold of detection) and report the results using a receiver-operating characteristic (ROC) curve, which plots sensitivity as a function of specificity.

Timeliness can be treated as a property of a detection method, similar to the properties of sensitivity and specificity. Timeliness can be measured by subtracting the time of detection from the time of the event itself. It can be improved by adjustment of threshold and at the expense of the other parameters. Although it is rarely done, timeliness, sensitivity, and specificity can be plotted on a generalized ROC curve.

The second mathematical foundation of early detection is decision theory. Decision theory is a mathematical formalism that can be used to identify an optimal sensitivity, specificity, and timeliness of detection for a specific application (such as detection of anthrax).

Decision theory and the related field of decision analysis provide methods for estimating the benefit of true alarms and the costs of false alarms and for using these quantities analytically to identify optimal sensitivity, specificity, and timeliness (and threshold) for that application.

The third strategy suggested by detection theory is to improve the detection method itself. The available signals may contain information that detection methods are not capable of utilizing fully. For example, algorithms have yet to be developed that can detect an outbreak of inhalation anthrax by searching for temporal and spatial associations in the affected population that are consistent with recent weather patterns and the incubation period of the disease.

The last strategy involves tuning the detection system for improved timeliness at the expense of specificity or sensitivity. Detection systems typically produce a numerical output (e.g., number of cases of X observed per day) and this output is compared against a threshold to determine whether to alarm. This strategy involves the relatively simple-minded idea of lowering the threshold so that alarms are sounded earlier. A lower threshold, however, will result in more false alarms, creating a tradeoff between the cost of the additional false alarms and the potential benefit of earlier detection.

Below we presented two new methods which can be used for detection algorithm development.

Method1 We assume that initial features describe the investigated phenomena (objects) rather indirectly, and conditions of gathering of information can unregulated(I DON'T UNDERSTAND THE USE OF THIS TERM) change. In these cases results of classification are strongly deformed: some classes are represented by single objects. If to exclude such objects which we name special results of classification sharply improve. Special objects indicate presence of sharp changes in an investigated source of the data and consequently are considered as sources of outbreaks (in this case, bioattacks). We offer a method of selection of special objects in a separate subset, simultaneously with classification of typical objects [11, 12].

Variation statement of a problem, with corresponding criterion of quality of extremized function and the parameters determining capacity of the separated subset of special objects is formed. For acceleration of work of algorithm the preliminary evaluation of importance of objects and features is used.

Let M is initial set of objects for a classification. A' priory defines d, indicating the number of special objects and ℓ, designating the number of ($|M|$-d) typical objects classes, and set ϖ consider as desired set of special objects.

Then a criterion of classification with isolation of special objects is:

$$J(\kappa,\varpi)=\sum_{q=1}^{l}\sum_{S\in K_q}\mu(S,K_q) \qquad (1)$$

$\mu(S,K_q)$ - measure of objects S similarity to class Kq.. Formula (1) implies that J(κ, ϖ) can be calculated on set M\ϖ and therefore the classification procedure expects to be two step procedure:

$$M=(\kappa=\{K1, K2\ldots Kl\};\varpi)$$

At the first step current pair (κt-1, ϖt-1) transforms to next more specific pair (κt, ϖ t-1), consisting of new classification κt and fixed set

ϖt-1. Then, on the second step pair (κt, ϖt-1) turns to pair (κt, ϖt) under condition:

$$J(\kappa t, \varpi t\text{-}1) \le J(\kappa t, \varpi t)$$

This heuristic algorithm of classification belongs to class of algorithms of aggregation [13] ; it has many modifications and effective applications on large scale data.

In 60 computer simulation experiments the method has been tested on specially developed data when number and kind of special objects were various. It is shown, that the program separated special objects precisely the same as it the people does. Sensitivity and specificity of a method during test evaluations was of 100 %. Using the data of infectious disease in regions of the country (10 years of monitoring) the method has shown (on the average) sensitivity of 84, 5 %, specificity of 61, 3 %.

Method2 is method of automated development of artificial neural networks (ANN) for rapid detection of outbreak.The weight of qualitative feature is calculated according to the following equation:

$$\omega_{jc} = \left(\frac{\lambda_c}{\lambda_{\max}}\right)\left(\frac{\beta_c}{\beta_{\max}}\right)\omega_{\max}$$

Where

$$\omega_{\max} = \max(-2\omega_{j0} / r)$$

$$\lambda_{\max} = \sum_{t=1}^{l} |K_t| (|K_t| - 1)$$

$$\beta_{\max} = \sum_{t=1}^{l} |K_t| (m - |K_t|)$$

$\lambda_c, \beta_c, \lambda_{\max}, \beta_{\max}$ is calculating proceed from the assumption of their contribution to possible differentiation ability of this feature. On order to define the optimal sample for ANN training the special genetic algorithm is developed. This method is very simple and straight-forward; it optimizes all the weights (i.e. whole neural network) once and then freezes them. Moreover, it optimizes all the objects involved to training sample, selecting those which provide correct solution of classification task [14] method works on mixed large scale qualitative and quantitative data records. Using this algorithm we provide fast and correct detection of analyzing data.

We selected the well-known benchmark to test our approach, described above. This benchmark is the acute myocardial infarction (AMI) outcome predicting. The data set in this case were real-world data obtained from Russia, Krasnoyarsk city [http: //neurocom.chat.ru/cardbase.htm], and contained 1700 records. Each example consisted of 126 – element real valued vectors, 11 elements of which were quantitative. The purpose was to predict an auricle fibrillation (AF), as either will be such complication or not. We compare the results of the proposed algorithm with the results of other studies such as the Neyman-Pearson method, Back Propagation (BP) type of neural network. The table shows the results of comparing methods mentioned above for AF predicting.

Table 2. Comparing methods for AF predicting.

Authors	Best results (% of errors)	
	Patients with AF	Patients without AF
Back Propagation NN	15	40
Neyman-Pearson method	33.3	26.5
Present method	0	0

As mentioned above the input weights in the neural network are freezing. It is known that freeze weights reduce the computational expense and training time. As far as the input connection weights are fixed, this implies that one could freeze those nodes for reducing the computational expense, which is directly proportional to the number of weights updated by network. The weight freezing not only reduces the computational cost, but also improves the convergence rate. It was found that while weight freezing improves the convergence rate; its effect on classification accuracy is not known. Our simulation results covered this gap: weights selecting and freezing in conjunction with designing patterns for training have been given an absolute accuracy of classification.

Therefore, we proposed an efficient algorithm for developing compact networks. The novelty of this is that it can determine the number of input nodes automatically. By analyzing a networks output, a connections weighting technique has been introduced. Also the special set of patterns for training was designed by using the pruning process of objects selection.

The experimental results for the acute myocardial infarction complication problem are shown that the training time was reduced, and also the classification error was minimized.

6. CONCLUSION

Laboratory Research Network in Uzbekistan will be designed to complement any conventional methods of out break detection (e.g., clinician – based surveillance of infection diseases). Laboratory – based surveillance will be less timely and sensitive than conventional methods in detecting many local out breaks of disease particularly those clearly associated with a certain setting, and in detecting many wide spread outbreaks of disease with unusual signs and symptoms.

As project plans to develop 6 local labs in country come to fruition, local outbreaks may also be rapidly detected by these laboratories. Otherwise the data will be transferred to a Central Reference Laboratory where they will be processed by more sophisticated algorithms, including those described above. Much is still needed to be done in Uzbekistan, and efforts are now concentrated on increasing the data available to LRN, system evaluation, and improvement, with the aim of having a flexible, automated outbreak detection system for laboratory reported pathogens.

7. ACKNOWLEDGMENTS

I thank the manager of DTRA – Uzbekistan project, Dr. Michael Balady for his support; I thank prof. Sh. Khodjaev, Director of Virology Institute, Ministry of Health, prof. M. Butaev, Director of Veterinary Institute, Ministry of Agriculture, prof. A. Abdukarimov, Director Institute of Genetics, Academy of Sciences of Uzbekistan for useful scientific information used in this paper.

REFERENCES

1. James W. Buehler, Ruth L. Berkelman, David M. Hartley, and Clarence J. Peters Syndromic Surveillance and Bioterrorism-related Epidemics Emerg Infect Dis [serial online] 2003 October [http://www.cdc.gov/ncidod/EID/vol9no10/03-0231.htm]
2. Lazarus R, Kleinman KP, Dashevsky I, DeMaria A, Platt R. Using automated medical records for rapid identification of illness syndromes (syndromic surveillance): the example of lower respiratory infection. BMC Public Health 2001;1:9.

3. Mostashari F. BT surveillance in New York City. Presentation at the CDC International Conference on Emerging Infectious Diseases, 2002, Atlanta. [Cited January 2003] Available from: URL: ftp://ftp.cdc.gov/pub/infectious_diseases/iceid/2002/pdf/mostashari.pdf
4. Centers for Disease Control and Prevention. Draft framework for evaluating syndromic surveillance systems for bioterrorism preparedness. [Cited December 2002] Available from: URL: http://www.cdc.gov/epo/dphsi/phs/syndromic.htm
5. Henning KJ. Syndromic surveillance. In: Smolinski MS, Hamburg MA, Lederberg J, editors. Microbial threats to health, emergence, detection, and response. Washington: Institute of Medicine National Academies Press; 2003. Appendix B.
6. Broad WJ, Miller J. Health data monitored for bioterror warning. New York Times, Jan 27, 2003, page A1. [Cited Feb 2003]. Available from: URL: http://www.nytimes.com/2003/01/27/national/27DISE.html
7. Hashimoto S, Murakami Y, Taniguchi K, Nagai M. Detection of epidemics in their early stage through infectious disease surveillance. Int J Epidemiol 2000;29:905–10.
8. Stern L, Lightfoot D. Automated outbreak detection: a quantitative retrospective analysis. Epidemiol Infect 1999;122:103–10
9. Centers for Disease Control and Prevention. Electronic reporting of laboratory data for public health: meeting report and recommendations; 1997. Available from: URL: http://www.phppo.cdc.gov/
10. Wagner MM, Tsui F, Espino JU, Dato VM, Sittig DF, Caruana RA, et al. The emerging science of very early detection of disease outbreaks. Journal of Public Health Management Practice 2001;7:51–9.
11. Shortliffe EH, Buchanan BG, Feigenbaum EA Knowledge engineering for medical decision-making. A review of computer-based clinical decision aids Proc IEEE, 1979:1207-24
12. Mottle V.V. The method of partial approximation in experimental curves analysis. Avtomatika I Telemechanika, 1977, 1, p 96-104
13. Braverman E.M., Muchnik I.B. Structure methods of data processing, Moscow,"Nauka",1983, 464p.
14. Adilova FT, Ignat'ev NA, Matlatipov G.R.,Chernysh P.P. Knowledge discovering from clinical data based on classification tasks solving Proceedings of MEDIFO 2001V.Patel et al.(Eds) Amsterdam:IOS Press,2001 p.1354-58

APDS, A NETWORK-READY, BROAD SPECTRUM, ENVIRONMENTAL PATHOGEN DETECTION SYSTEM

F. P. MILANOVICH*, J. DZENITIS, B. J. HINDSON,
A. J. MAKAREWICZ, M. T. McBRIDE, and B. W. COLSTON
Lawrence Livermore National Laboratory, 7000 East Avenue, P.O. Box 808 ,Livermore, CA 94550

ABSTRACT: The Autonomous Pathogen Detection System (APDS) 1 is a stand-alone pathogen detection system capable of rapid, continuous, low cost environmental monitoring of multiple airborne biological threat agents. Its basic design comprises aerosol sampling, in-line sample preparation, multiplex detection and identification immunoassays, and orthogonal, multiplexed PCR (nucleic acid) amplification and detection. Its primary application is to warn civilians and emergency preparedness personnel of a terrorist attack, the same system could also have a role in protecting military personnel from biological warfare attacks. APDS instruments can be used at high profile events such as the Olympics for short-term, intensive monitoring or more permanent installation in major public buildings or transportation nodes. All of these units can be networked to a single command center so that a small group of technical experts could maintain and respond to alarms at any of the sensors. The APDS has several key advantages over competing technologies: (1) the ability to measure up to 100 different agents and controls in a single sample, (2) the flexibility and ease with which new bead-based assays can be developed and integrated into the system, (3) the presence of an orthogonal, real-time detection module for highly sensitive and selective nucleic acid amplification and detection, (4) the ability to use the same basic system components for multiple deployment architectures, and (5) the relatively low cost per assay (<$2 per 10-plex or $0.20 per assay) and minimal consumables.

D. Morrison et al. (eds.), Defense against Bioterror: Detection Technologies, Implementation Strategies and Commercial Opportunities, 67–75.

1. BACKGROUND

By the mid-1990s, the U.S. Congress began to assess the vulnerability of the U.S. civilian population to biological terrorism and found us considerably lacking in our ability to cope with even a small-scale biological event. Initial thinking was that the DOD technology could be readily transferred to the civilian arena. However, upon further reflection, it was concluded that although there was overlap between military and civilian defense needs, in the case of a biological threat, there are marked differences: (1) the soldier is trained and equipped with protective gear so he may respond to a threat quickly enough to prevent a lethal dose; (2) military intelligence usually reduces the potential threat to a relatively small number of biological agents; and, (3) military battlefield tactics are designed to minimize the density of soldiers. The civilian population, however, is neither trained nor equipped, is vulnerable to any conceivable pathogen and often gathers in large crowds (special events, sporting venues, etc.) where a small release could potentially infect thousands. (2) In response to these differences, federal agencies, including DOE, have recently begun funding directed research efforts to reduce civilian biological terrorist vulnerabilities. A near term goal of these efforts is an integrated network of sensors and analytical software that will help us protect critical assets such as subway systems or major events.

The Lawrence Livermore National Laboratory (LLNL) is a major participant in the above efforts and, as a result, has intimate knowledge as to the performance parameters of all requisite technologies and techniques that can be applied to civilian, biological, counter-terrorism. At present there are more than 30 pathogens and toxins on various agency threat lists. Public health personnel rarely see most of the pathogens so they have difficulty identifying them quickly. (3) In addition, many pathogenic infections aren't immediately symptomatic, with delays as long as several days, limiting options to control the disease and treat the patients. (3) The lack of a practical monitoring network capable of rapidly detecting and identifying multiple pathogens or toxins on current threat lists translates into a major deficiency in our ability to counter biological terrorism. (4) APDS addresses that deficiency. (4)

APDS instruments can be used to implement a number of different monitoring applications. The Postal Service, for example, is implementing a mail-screening strategy in their Central Processing Facilities. Available systems are not fully automated, have difficulty operating in the relatively dirty environment, and are expensive to maintain. The Department of Transportation is actively seeking biomonitoring systems for protection of transportation hubs, with airports residing at the top of this list. The military continues to evaluate options to their current biowarfare detection systems. A variation of the APDS platform can also be used to monitor a variety of other environmental or clinical pathogens. Mobile units could be transported to suspected "sick buildings" to test for mold or fungal spores that might be causing tenant illnesses. Units with reagents for animal diseases could be placed in livestock transport centers or feedlots to rapidly detect airborne pathogens and protect against disease outbreaks. Finally, monitors in hospitals could be used to test for airborne spread of contagious materials among patients.

2. SYSTEM OVERVIEW

(a)

(b)

Figure 1. (a) A prototype APDS with aerosol collector above a Luminex flow cytometer followed by the sample preparation module. (b) The Research International/ LLNL bioaerosol collector

There are four basic elements of the APDS, (1) a two-stage aerosol collector; (2) a sample preparation/fluidics module; (3) a detection module (bead based multiplexed flow cytometry immunoassay and real-time PCR) and, (4) computer and software for control and networking. The aerosol collector continuously samples the air and traps particles in a swirling buffer solution. Particles of a given size distribution can be selected by varying the flow rate across a virtual impactor unit. At given time intervals, the collected sample is added to a reagent containing optically encoded microbeads. Each color of microbead contains a capture assay that is specific for a given bioagent. Fluorescent labels are then added to identify the presence of each agent on the bound bead (secondary labeling). Each optically encoded and fluorescently labeled microbead is individually read in a flow cytometer, and fluorescent intensities are then correlated with bioagent concentrations. We have also developed a second detection system for confirmation based on nucleic acid amplification and detection. An archived sample is mixed with the PCR master mix and then introduced by the fluidic system into the real-time PCR module. Specific nucleic acid signatures associated with the targeted bioagent are amplified and detected using fluorescence generated from nucleic acid replication from Taqman probes. All modules of APDS have been integrated into a self-contained "ATM" style chassis (fig. 1a). All fluids and reagents are contained in the instrument, so the only utilities required are AC power and a network connection for remote communication. The central computer uses a simple serial-based Labview control system to control all instrument functions. A custom software system has been developed for data acquisition, real-time data analysis, and result reporting via a graphical user interface.

2.1 Bioaerosol Collection

One of the easiest methods of rapidly exposing a large population to a biowarfare agent is through an aerosol (witness the effect of the recent, relatively small-scale anthrax mailroom releases). Current aerosol collectors contain dry, matrix type filters that are difficult to couple to autonomous systems, are relatively unselective in the types and sizes of particles collected, and simply do not collect enough particles over a given period to produce a sensitive enough detection capability. At LLNL, we have designed a two-stage aerosol collector (Fig. 1b) that utilizes an LLNL-designed virtual impactor pre-concentration stage in front of a commercial wetted wall cyclone collector (Research International SASS 2000). The virtual impactor captures particles 1-10 μm in diameter, which is the size of particles most likely to be captured in the human lung. Particles are collected in a fluid, making downstream processing much easier. The fans and inputs to the SASS 2000 have all been replaced to obtain much higher collection rates, up to 3000 liters of air per minute flow through the detection system, allowing many more particles to be collected over a shorter. The enhancements also improve the system sensitivity and reduce the collection times. An on-board computer controls air flow rates and the size range of particles collected, while a commercial particle counter provides real-time feedback on the size and quantity of particles collected.

2.2 Autonomous Sample Preparation/Fluidics

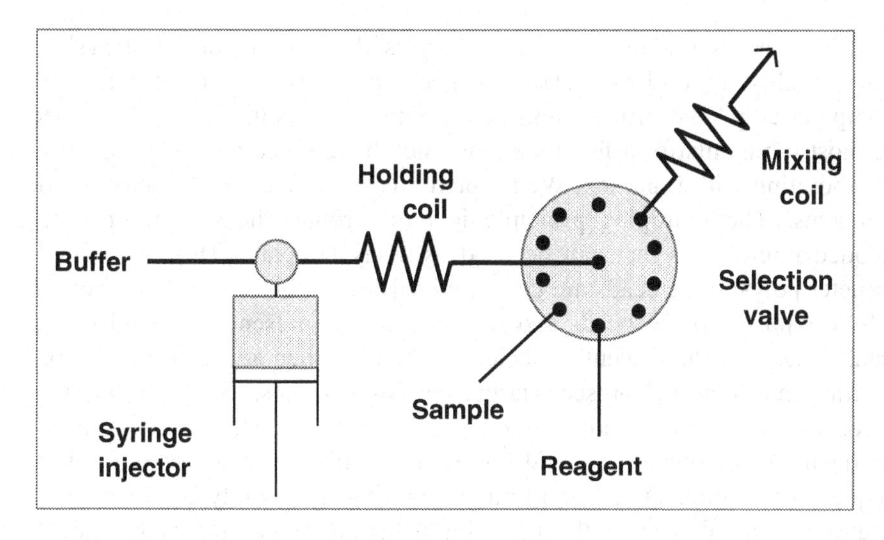

Figure 2. Schematic diagram of the fluidic handling system

The APDS sample preparation module reproduces the function performed by biologists who routinely perform "wet" chemistry on the bench: moving the sample from the aerosol collector, preparing the sample (mixing, filtering, incubation, etc.), and delivering the sample reaction volume to the immunoassay and nucleic acid detectors. Conventional sample preparation instrumentation used, for example, in high-throughput drug discovery analyses, use robotic manipulation of micropipettes coupled to disposable filter wells. Since robotics are inherently complex (and difficult to scale), we chose a powerful, highly flexible technique called sequential injection analysis (SIA) as the basis for our sample preparation module.(5) Global FIA (Gig Harbor, WA) have extended SIA to a more versatile and powerful fluid handling technology called "zone fluidics", and they have been consulting with LLNL on this technology and testing components for potential use (Fig. 2). Automation is achieved through the manipulation of small solution zones under conditions of controlled dispersion in narrow bore tubing. Zone fluidics makes use of a multi-position selection valve and a high precision, bi-directional pump to construct a stack of well-defined sample and reagent zones in a holding coil of narrow bore tubing. By appropriate manipulation of this zone stack, a wide range of sample handling unit operations can be accommodated. The pump is used to move the sample from one device to the next achieving the required sample manipulation in the process. Once a detectable species has been formed, the zone stack is transported to the immunoassay and nucleic acid detectors.

2.3 Multiplex Immunoassay Detection

The heart of our detection capability is the use of "liquid arrays", a novel, highly multiplexed assay that competes (in bead format) with "computer chip" platforms. Luminex Corporation (Austin, Texas) developed the most robust platform for these new microbead-based assays, the Lx100. Collaborating with Luminex, We adapted this technology to the detection of pathogens. The multiplex principle is built around the use of optically encoded microbeads that can be used as assay templates (Fig. 3). Small diameter polystyrene beads are coated with 1000s of antibodies. The sample is first exposed to the beads and the bioagent, if present, is bound to the bead. A second, fluorescently labeled antibody is then added to the sample resulting in a highly fluorescent target for flow analysis. Since the assay is performed on a microbead matrix, it is possible to measure all types of pathogens, including viruses and toxins. Each microbead is colored with a unique combination of red and orange emitting dyes. Only the number of uniquered colored beads in the array, limits the number of agents that can be detected from a single sample. We have currently demonstrated the use of this technology for measuring a wide

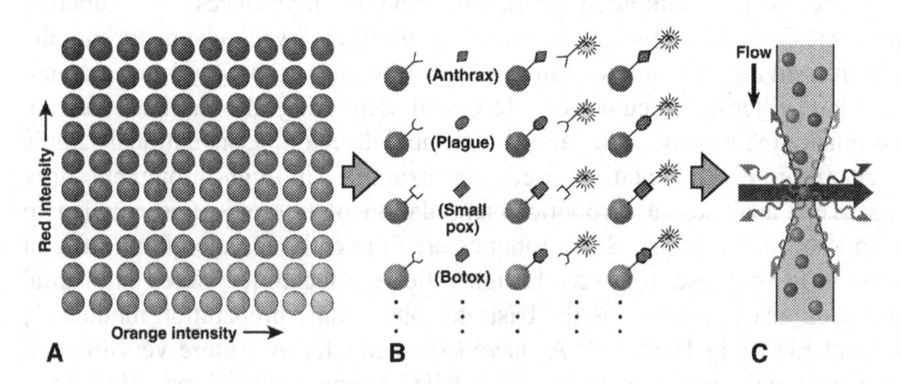

Figure 3. Multiplex detection is achieved by equipping the 100-plex array (A) with specific capture antibodies (B) and measuring them rapidly with a laser diagnostic (C)

range of bioagents (6) at sensitivities and selectivities comparable to non-automated conventional immunoassay techniques (such as enzyme-linked immunosorbent assays) that take 4-6 times as long. A novel component of this system is the use of additional beads in the array for internal positive and negative controls to monitor each step in the sample preparation process. This imparts a measure of quality control noticeably lacking in most other approaches, and imperative in an autonomous system. Sample preparation followed by LX-100 analysis can be completed within 25 minutes with the APDS fluidic system.

The versatility of the bead array is illustrated in Fig. 4 below. Here a series of 12 experiments were conducted with the assay being a seven-plex for MS2, a virus; Bg, a bacterial spore; Eh, a live bacteria; and, OV a protein. In addition the assay contained three reagent controls, NC, AC, and FC. The experiments increased in complexity (from left to right in the figure) with experiment 11 indicating the simultaneous detection of a spore, bacteris, virus and protein.

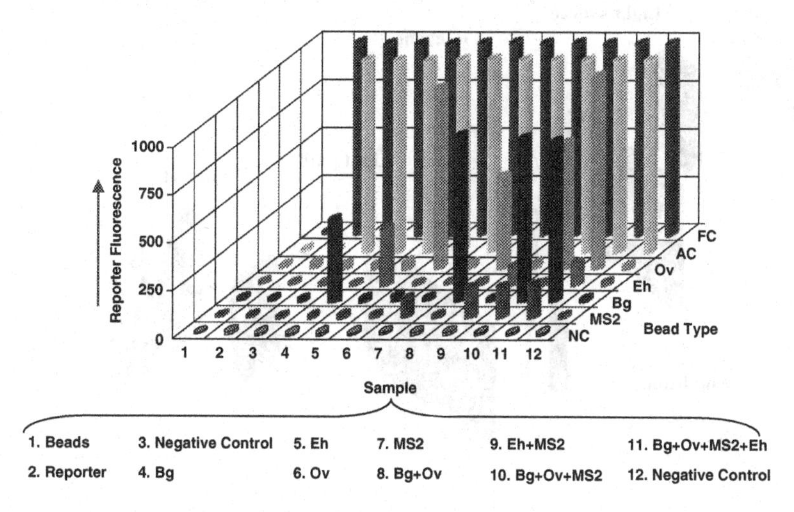

Figure 4. Seven-plex assay run on 12 samples of increasing complexity (see text above)

2.4 DNA Confirmation with Flow-through PCR Technology

The APDS contains a second detection system that is based on nucleic acid amplification and detection.(7) An archived sample is mixed with the TaqMan reagent and introduced by the SIA technique into the flow-through polymerase chain reaction (PCR) system. Specific nucleic acid signatures associated with the targeted bioagent are amplified up to a billion-fold and detected as a fluorescence change from a TaqMan probe. The addition of this flow-through PCR component provides a complementary detection technology to the multiplex bead assay significantly increasing system reliability and minimizing the possibility of false positives. This is particularly important for Homeland Defense applications, where the impact of evacuating a major event or office building is significant.

This prototype flow-through PCR module consists of an LLNL-designed, silicon-machined, thermocycler that is mounted in-line with our sample preparation unit (Fig. 5). The use of silicon components allows thermocycling to proceed very rapidly (less than 1 minute per heat/cool cycle). The thermocycler contains appropriate light sources and detectors to perform real-time TaqMan assays. The APDS system automatically identifies a positive immunoassay result, and then initiates the PCR analysis for confirmation.

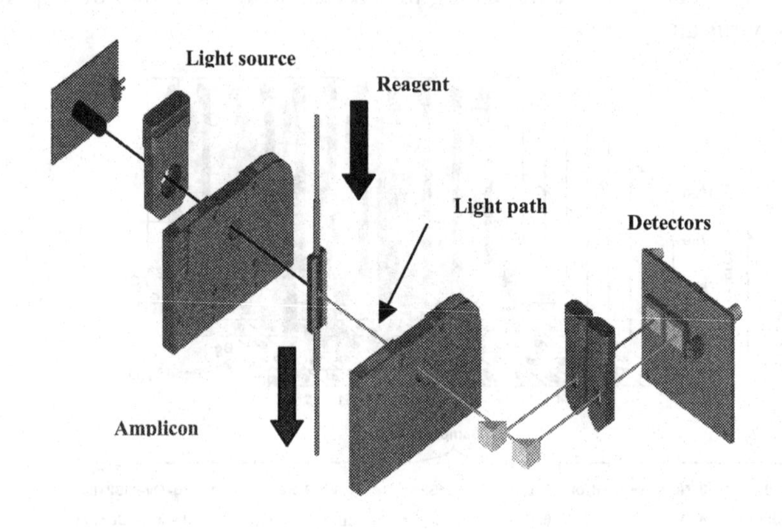

Figure 5. The flow-through PCR module contains a silicon thermocycler with a disposable, molded tubing insert.

The flow through module has been demonstrated to be very consistent run to run. Fig. 6 above shows the results of fourteen PCR measurements run consecutively with the fluidics module automatically mixing reagent with sample and then delivering the resulting mixture to the PCR chamber

3. SUMMARY

Currently domestic biosurveillence strategies rely on aerosol sampling followed by laboratory analysis. Modest coverage of the US population

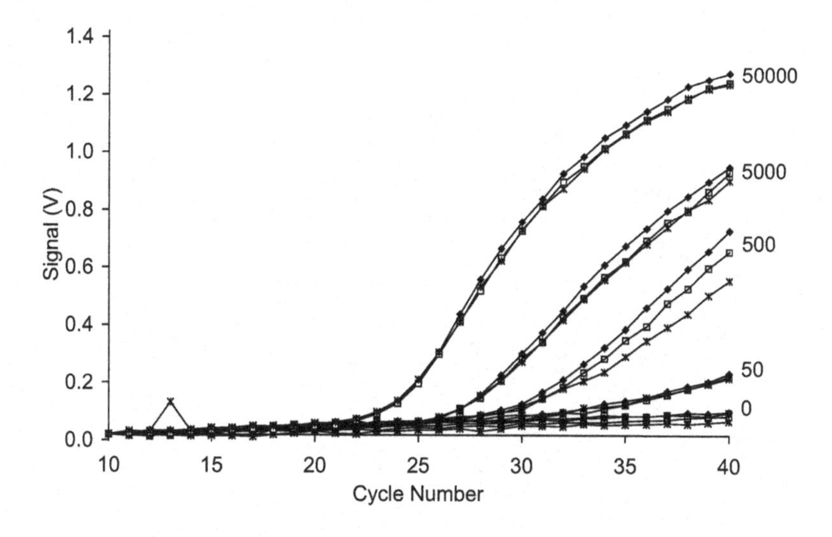

Figure 6. Flow-through PCR amplification curves generated with B. anthracis genomic DNA. The initial DNA concentration ranged from 0 to 50000 copies per reaction

under such a strategy requires a significant labor force and will be costly to maintain. The APDS offers an alternative with the potential of reduced cost and higher performance. The key performance parameters are the availability of redundancy in analysis to reduce the potential of false positives, the frequency of maintenance, sensitivity to threat agents, and cost per analysis. Extensive field-testing of APDS is in progress and commercialization activities have commenced.

REFERENCES

1. McBride, M. T. et. al., Anal. Chem. 2003, 75, 5293-5299.
2. Koplan, J.P.,. MMWR, 2000. 49(No. RR-4): p. 1-14.
3. Bradley R.N. (2000) Prehospital Emerg. Care 4(3) 261-269
4. Franz D.R. (1997) JAMA 278(5) 399-411
5. Lenehan,C.E.; Barnett, N.W.; Lewis, S.W. Analyst, 2002, 127, 997-1002.
6. McBride, M.T.; Gammon, S.; Pitesky, M.; O'Brien, T.W.; Smith, T.; Aldrich, J.; Langlois, R.G.; Colston, B.; Venkateswaran, K.S Anal. Chem. 75, 1924 (2003)
7. Belgrader, P.; Benett, W.; Hadley, D.; Richards, J.; Stratton, P.; Mareilla, R. Jr.; Milanovich, F. Science, 1999, 284, 449-450.

CONCEPT DESIGN OF ANAUTONOMOUS BIOLOGICAL AGENT DETECTOR SYSTEM (ABADS)

R. BARTON, R. COLLINS, AND R. STARNES
Midwest Research Institute, Kansas City Missouri, USA

ABSTRACT: Midwest Research Institute has begun the development of an autonomous biological agent detector system based on the enhancement and integration of proven technologies. The design concept incorporates a high volume air to liquid sampler, fluid conditioning and concentration systems, and organism and toxin detection platforms. More than 20 threat agents identified by the US Centers for Disease Control are addressed using an integrated microfluidic system. Organisms are detected using a combination of advanced multiplexed PCR techniques. Protein toxins are detected using amperometric transduction of an immunoreaction. Core technologies from established organizations are used. The device is designed to be used in an urban area. The 0.3 m^3 device has a targeted purchase price of US $ 25,000 in quantities of 3,000 units with annual total operating costs of US $ 10,000.

1. INTRODUCTION

Automatic detection systems will eventually play a key role in the detection and prevention of biological attacks. However, the capabilities of current technology still lag far behind the needs defined by government agencies charged with protecting the public and military forces. These agencies have defined several conceptual missions for biological detection systems and the associated technology requirements.

One such mission is the rapid detection of an open-air attack on a population center. This mission requires a biodetection system that can detect an attack within a few hours of its occurrence and locate the subset of the population with the greatest exposure to initiate a response prior to the

D. Morrison et al. (eds.), Defense against Bioterror: Detection Technologies, Implementation Strategies and Commercial Opportunities, 77–90.

onset of clinical symptoms in the affected population. Tens to hundreds of systems networked together would be required for any large population center. Thus, each system must be affordable and require minimal routine maintenance. Each biodetector must be capable of detecting low concentrations of a wide range of target materials in air with a high level of reliability and a very low false positive rate. There are no commercially available or validated prototype devices that meet all of the requirements for this mission.

Midwest Research Institute (MRI) has assessed numerous technologies with potential applicability to this mission. Based on our assessment, we have developed a conceptual design that will address all of the requirements outlined by the US government for this mission as verbalized by the Department of Homeland Security[1]. We have assembled a team of private sector organizations that possess a wide range of expertise to supplement the public sector work being performed in national laboratories to develop a system for this critical mission area. This paper describes our technology evaluation and summarizes the final technology approach our team developed.

2. TECHNOLOGY EVALUATION

Biodetection technologies perform four primary functions — collect material dispersed in the air, prepare the collected material for analysis, sense the presence of the targeted biological material, and transducer the presence of the biological material into an electronic signal. Our technology evaluation considered each of these component systems.

3. SAMPLE COLLECTION

Sample collection technologies are relatively simple mechanical devices that are responsible for transferring target materials from an environmental matrix into the biodetection device. The complexity of these systems varies from a simple tube to a complex electromechanical system such as an electrostatic precipitator. Sampling technologies are typically derived from devices long used in air and water cleaning and environmental protection. The theory of operation and design of these devices are supported by substantial data[2].

Sampling requires an understanding of the physical nature of the target material and the matrix from which it is to be removed. For the mission of interest, the target materials will consist of solid particles dispersed in urban air. The concentration of these particles can be very high near the point of release (in excess of 10^5 particles/m^3) to low values on the edges of the cloud or far downwind (1 particle/m^3). The particles range in size from 0.5 to 10 μm and have densities of 1,000 to 2,000 kg/m^3 depending on the quantity and nature of the non-biological material used to produce the biological weapon[3]. Urban air typically contains significant levels of entrained particles in this size range including pollen, latex, crystal material, droplets and photochemical smog.

The most commonly used collection mechanism is inertial separation. Inertial devices include filters, cyclones, and impactors. These devices use the substantial difference between the inertia of the entrained particles and that of the surrounding air to transfer particles to either a stationary solid matrix or a liquid matrix. Filters are the most common of these devices. Filters have high particle capture efficiencies, are simple, and require little maintenance. However, it is difficult to remove materials from a filter once they are captured. In addition, filters require relatively large amounts of energy to draw air through them and their use is difficult to automate.

Impactors can transfer particles to either a fluid or a solid surface. They do not require as much energy as filters but are typically less efficient. A great variety of these devices are available. Variations include the methods used to handle the substrate onto which the particles are impacted and the number and shape of the air jets used.

Cyclones are the largest of the inertial collectors for any given collection efficiency. However, some research groups have addressed this issue through the use of micromachining. Cyclones can be either wet or dry.

Table 1. Classes of Air Sampling Equipment

Technology Class	Examples	Capabilities	Limitations
Dry Inertial	Filter, impactor, cyclone	High capture efficiency possible, simple mechanical design	High pressure drop and energy use, difficult to remove target from substrate, stresses on organisms captured
Wet Inertial	Impinger	Fluid sample obtained, sample transport automation simplified, viable samples possible	Increased expendables due to fluid needed, high pressure drop and energy use
Dry Non-Inertial	ESP	High capture efficiency at a low pressure drop, simple design	Difficult to remove target from substrate, high potential stresses on organisms captured
Wet Non-Inertial	Wet ESP	High capture efficiency at a low pressure drop, fluid sample obtained, sample transport automation simplified, viable samples possible	Mechanical design complexities, fluid needed, increased expendables due to fluid needed,

MRI's team selected the SpinCon® technology for use in its developmental biodetection system. Figure 1 illustrates the mechanism of operation of the SpinCon. Air is accelerated through a narrow slit in the wall of a cylinder and impacted into a film of liquid. The particles remain in the liquid while the air is exhausted. The film is maintained on the wall of the cylinder by the centrifugal force imparted by the high velocity air stream and surface tension. The fluid continually cycles over the slit.

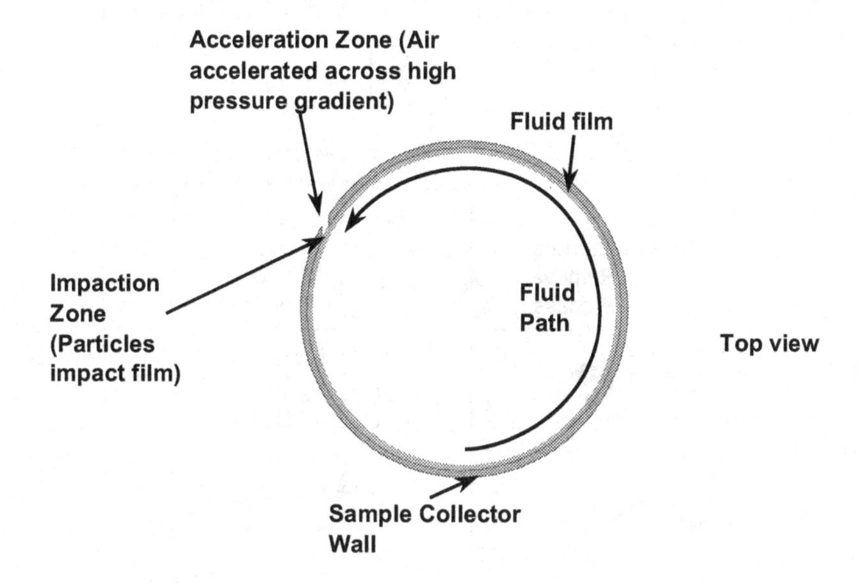

Figure 1. Particle collection mechanism in the SpinCon air sampler

4. SAMPLE PREPARATION

There are no well-defined and tested devices available for sample processing in autonomous biological detection systems. This system must remove environmental contaminants and concentrate target materials. In addition, it should free the targets from any sort of binding or encapsulation. This includes lysing cells to free target materials and separating cells from flowing agents. Typically each biodetection system integrator addresses this area *a priori* and often designs unique processes. The sample processing steps used are strongly linked to the biosensor technique envisioned and the sample collection method selected.

Two sample preparation approaches have been used. Figure 2 illustrates the basics operations performed in each of these approaches. Different processing implementations may perform these steps in a different order.

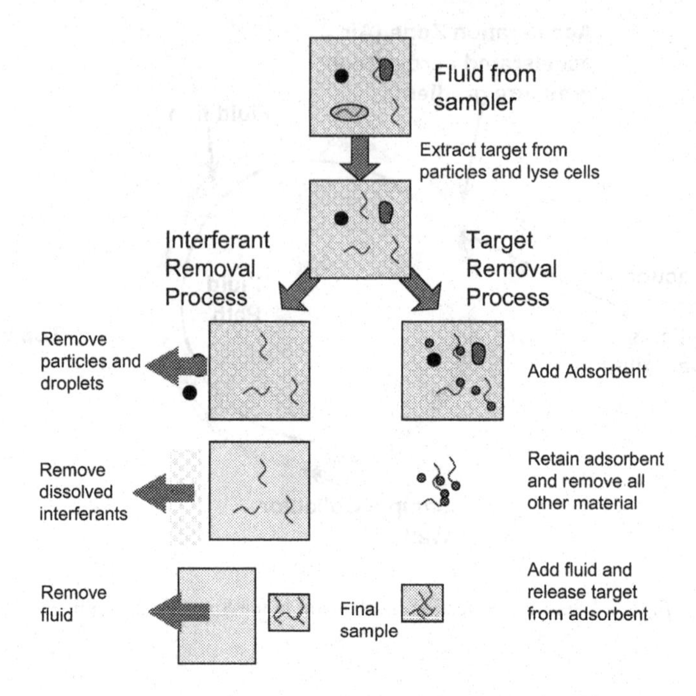

Figure 2. Sample processing steps.

Target materials can be present in free form, bound to other solids, and contained within cells. The fluid will contain a variety of extraneous solids, dissolved compounds and immiscible fluids. In the first step illustrated, all of the target material is liberated into solution. This involves extracting the target from the binding solids and lysing the cells. Two paths are then possible. In a process focused on removing interferants, suspended materials are next removed. Third, the dissolved contaminants are removed. In the final step, the target materials are concentrated into a much smaller volume of fluid.

As an alternative, once the target material is liberated, an adsorbent selective for the target could be added. The adsorbent and attached targets are retained while the fluid and contaminants are removed. A small quantity of fluid is then added back to the adsorbent and the target is released into the fluid.

The methods used to liberate target material depend on the ruggedness of the target and the strength of any binding. Sonication is one of the most widely used methods for lysis and extraction. Pressure cycling is also common. More complex chemical processes can be used but they require additional reagents and mechanical complexity.

Separation of particles, debris and immiscible fluids is primarily a mechanical process. Settling, filtration, and centrifugation are possible approaches. Removal of dissolved compounds is typically a chemical process using selective absorption. Reverse osmosis and active membranes can be used to remove water.

Our targets include both protein toxins and DNA. However, the DNA analysis tends to be less sensitive and more strongly affected by interferants than protein analysis.(in all field trials that I am aware of , DNA techniques have been shown to be more sensitive that immundiagnostics) Thus, our design directed an aliquot of the unprocessed fluid to the protein analysis system and selectively extracted DNA from the remaining fluid. Our design uses the coated magnetic bead-based technology developed by Promega and is discussed in greater detail below.

When DNA is targeted for analysis, the polymerase chain reaction (PCR) is often used to enhance detection. This reaction makes used of the natural replication capabilities of DNA and is supporting enzyme systems to selectively increase the concentration of specific target sequences. Development of advanced PCR platforms and techniques is an area of significant research efforts over the past decade. Laboratory PCR systems and reagent mixtures (primers, buffer solutions, and enzymes) are widely used. Recent advances in these systems have focused on reducing cycle times and equipment size[4] and on multiplexed reactions[5].

However, of equal or greater importance to the design of biodetection systems such as those described in this paper, significant effort has been invested in the production of microfluid PCR platforms. MRI's developmental system makes use of PCR to amplify target sequences. The PCR reactions are carried out on a microfluidic platform developed by Caliper technologies. The operations of the system will be described in greater detail below.

5. SENSING

The sensing technologies interact with a given target in a sensitive and selective manner. This is an area of extensive current research. A detailed discussion of the broad variety of techniques is beyond the scope of this paper. However several major classes can be identified. These are summarized in Table 2.

Table 2. Summary of Biological Sensing Technologies

Technology Class	Examples	Capabilities	Limitations
Structural Identification	Immunoassay	Stable, development known, flexible	Sensitivity, selectivity, interference
Sequence identification	PCR, labeled probes	Highly selective, established technology	Inhibitors, speed
Molecular weight/ion fragment pattern	Mass spectrometry	Sensitive	Large equipment, difficult for large molecules
Selective adsorption	Electrophoresis	Flexible, established technology	Sensitivity, selectivity
Metabolic processes	Dye uptake, ion channel	Viable determination, some inherent amplification	Selectivity, flexibility

Structural identification occurs through the use of chemical interactions that depend on the shape and charge distribution pattern of large complex organic molecules. Immunoreactions are a naturally occurring example of this approach. A large number of immunoassays have been developed. These target proteins of interest such as the toxins produced by some biological agents or used as biological agents themselves. Denaturing of the target proteins due to environmental stresses such as heat, cold, or radiation and changes in charge distributions in response to fluid composition can degrade the sensitivity of the technique. In addition, the reactions tend to be slow and irreversible. Both of these characteristics make use of the approach in a continuous monitor problematic. Synthetic versions of the recognition chemicals are known and molecularly imprinted polymers are under investigation by several groups.

Sequence identification is specific to DNA targets. The sequence of bases in a segment of DNA can be used as the basis for a sensing reaction. DNA chips in which a single strand of anti-sense DNA is bound to a specific location on a silicon chip are an example of an application of this sensing approach. A variety of closely related approaches have been developed. Another example of the application of this sensing principle are the real-time PCR detection systems used in many laboratories. A single strand of anti-sense DNA is attached to a fluorescent dye. The DNA segment will then bind to its complimentary sequence only.

The PCR reaction itself is a form of sequence-based identification. Because it increases the concentration of the selected target sequence only due to the use of specific initiator sequences, the reaction products can be combined with a nonspecific dye or binding agent. (the use of an additional sequence as a probe, as in TaqMan, is the current state of art for detection of PCR products). The concentration of the target sequence will then far exceed the concentration of any other sequence present.

Highly sensitive mass spectrometers can be employed by fragmenting an organic molecule in a well-controlled manner and then examining the charge to mass ratio of the resulting fragments. Chemical and physical fragmentation methods can be used. This approach is more effective at protein identification than DNA identification.

Selective adsorption is an adaptation of chromatographic principles. A molecule is placed in an environment that interacts differently with different types of molecules. Some are bound tightly, others loosely. A flow thought the matrix is initiated and the molecules are separated based on the strength of their interactions. Gel electrophoresis is an example of this approach.

Finally, metabolic processes may be used. In these, a dye is attached to a compound used in a metabolic process that is indicative of an organism. Due to the similarities between organisms within a species, this mechanism has limited discriminating power. However, it can be combined with other markers to provide more definitive identification. Metabolic processes are one of the marker types used in many flow cytometry systems.

In our system, we combined the use of PCR with selective dyes to provide high sensitivity and low false positive rates for organisms and an immunoreaction for protein toxins.

6. TRANSDUCERS

Transducers convert the interaction between the target and the sensor into an electronic signal. Table 5 summarizes some of the key transducer technologies that are available. Selection of a transducer technology is closely linked to the selection of the sensor. While research is active in all of these technologies, optical approaches are becoming some of the most common transducers. In our system, fluorescent dyes are integrated using an excitation laser and the emitted light striking a photosensor creates the electronic signal.

Table 3. Types of Transducer Technologies Under Investigation

Technology Class	Examples	Capabilities	Limitations
Electrical properties	Amperometry, potential	Robust, mechanically simple	Sensitivity
Fluorescence	Dyes	Sensitive, little interference	Selectivity, spectral overlap
Optical Absorption	IR spectroscopy	Flexible	Sensitivity, mechanical complexity
Evanescent Waves	Fiber optic cables, metal coated prisms	Robust, sensitive	Interference

7. SYSTEM DESCRIPTION

In this section, we describe in detail the system design that we have developed based on the extensive review of available technologies described above. We have used an open architecture in our design to allow for the integration of more capable technologies as they become available.

Air sampling will be performed using Sceptor Industry's patented SpinCon system. This system will operate at 450 lpm and concentrate particles into a 1 ml liquid sample. Over a three-hour integration period, the sampler will achieve a concentration factor (CF) of 10^5 (CF is defined as the ratio of the concentration of the target in the fluid sample to the concentration of the target in the air). Organisms are then lysed and particles are removed from the capture fluid. The captured target materials are concentrated further using paramagnetic beads developed by Promega. Toxin detection is performed using an immunoassay/amperometric technique. Organisms will be detected using a PCR assay technique implemented on Caliper Technologies' microfluidic PCR platform. Figure 3 illustrates our concept for the ABADS.

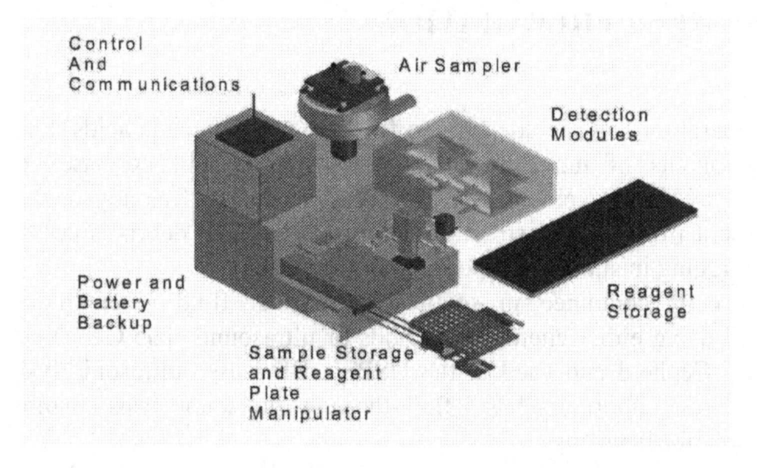

Figure 3. Conceptual rendering of the ABADS.

8. SAMPLE COLLECTION

The air sampler uses the SpinCon technology developed at MRI and sold by Sceptor Industries. Particles are transferred from the air to the collection fluid by impaction as described above. The fluid remains in the contactor while the air exits. In this way, particles from a large volume of air are concentrated into a small volume of fluid.

This technology was selected based on its ability to concentrate airborne particles in a fluid and its proven environmental performance. The SpinCon technology is the best commercially available biological air sampler that produces a liquid sample. It has been used in the Department of Defense's Joint Biological Remote Early Warning System (JBREWS) and Portable Biological Air Sampler (PBAS). In December, 2003, Sceptor Industries was awarded a multi-million dollar contract to provide over 800 units of SpinCons for the US Postal Service Biological Detection System (BDS).

The current SpinCon sampler has demonstrated concentration factors of 10^4 (for a 3 hr sampling period), where concentration factor is defined as the ratio of the concentration of the target in the fluid sample to the concentration of the target in the air. Tests indicate that over 95 percent of the biological material captured by the SpinCon is retained for more than 3 hours. Versions of the SpinCon have been operated continuously in pre-production tests of the USPS's BDS for more than 30 days. The system has been challenged with a variety of surrogate materials including polystyrene latex beads ranging in size from 1 to 10 μm, *Bg* spores, *Eh* cells, and *MS2* phage (a surrogate for pathogenic viruses).

9. SAMPLE PREPARATION

Fluid preparation will include lysis to free any target nucleic acids. Each of the components of the fluid preparation module will be based on commercially available technologies. Lysis is needed to free any nucleic material present from cellular structures. Removal of solid debris is needed to prevent plugging in subsequent portions of the system.

Lysis of cells contained in small quantities of fluid is commonly accomplished using either chemical methods or ultrasound. The GeneXpert developed by Cepheid and used in the USPS's BDS uses ultrasonic lysis. Pacific Northwest Laboratory has a flow-through ultrasound lysis chamber that is available for licensing.

The concentration of the nucleic acids in the fluid will be increased using paramagnetic beads. Promega's MagneSil® technology is used in this module. This technology is based on silica clad paramagnetic micro-particles. In an optimized chaotropic solution, nucleic acids in a complex mixture selectively adsorb to the surface of the particles. An external magnetic field is applied to the particles to capture and concentrate MagneSil particles and adherent nucleic acids, and wash solutions are used to eliminate contaminating material. The highly purified target material is then released in a small volume of aqueous solution.

This technology was selected because of the proven ability of paramagnetic beads to concentrate nucleic acids and the ease with which systems to manipulate magnetic beads can be developed. While several manufacturers have commercial paramagnetic bead products, Promega is a leader in the field and has commercial paramagnetic particle systems for the following applications:

- Genomic DNA isolation
- RNA Isolation
- Plasmid purification
- Sequencing Reaction Cleanup
- PCR Reaction Cleanup
- Pre-quantitated DNA Isolation for forensics

Many of these applications involve removal of DNA and RNA from complex matrices similar to those needed in the ABAD. With this detailed experience, we have high confidence that an effective concentration system can be developed.

10. SENSING AND TRANSDUCTION

An available system that appears to meet the requirements for the toxin detection module is marketed by AnzenBio. The detector component of the system is an amperometric-based technology that will respond to biological toxins when appropriate ligands are attached to the electrochemical surface. The sensor routinely detects immunologic binding of an antigen to antibodies bound on the surface of the electrochemical sensor. Experiments detecting prostatic secretory antigen (PSA) in raw urine samples showed a sensitivity of 10 picograms of PSA/mL of urine. These data suggest the detector array has the inherent capability to meet the sensitivity requirements of the detector. A variety of target-specific capture molecules will be identified and tested with this system. Such molecules may consist of antibodies, ligands, aptamers, or other unique materials. These capture molecules will be tested for target specificity (to maintain low false positive rate) and avidity.

A multiple-step process will be used to detect the presence of target nucleic acids. First, reverse transcription will be used to convert any RNA to DNA. This will be followed by a non-specific whole genome amplification (WGA) step. MRI has extensive experience using Multiple Displacement Amplification (MDA), in addition to other techniques including Degenerate Oligo Primer (DOP) PCR and Primer-Extension-Preamplification (PEP) PCR, and will adapt the best one suited for this application. The implementation of a WGA step will result in non-specific amplification of all genetic material present in each reaction by up to two to three orders of magnitude. Following WGA, the fluid will be transferred to the microfluidic chip for agent-specific multiplex analysis using specific amplification of identified target nucleic acid signatures.

The microfluidic chip will use multiple flow channels simultaneously. The parallel channels speed the flow through the heat cycling zone and provide for parallel analyses that facilitate attainment of the false positive targets. The total time for all reactions needed is about 40 minutes.

Each reagent mixture will contain primers and probes for two loci. Sufficient mixtures will be prepared to provide enough loci for each target to attain the required false positive level. As the samples proceed down the parallel channels, they will undergo a series of PCR cycles. The channels will be long enough to provide 30-50 cycles. At the end of the temperature cycling, the channels may be combined for dual color fluorescence detection. Thus, multiplexing in this system is attained by a combination of limited chemical multiplexing (two targets per organism per reaction) and extensive mechanical multiplexing. Reaction volumes for different targets follow one another down the microchannels. This produces a robust system that is resistant to interference, inhibition, and false positives.

The microfluidic chip developed by Caliper was selected for use in the system because of its extensive development and validation to date. The system also has a very rapid cycle time facilitating the extensive mechanical multiplexing discussed above. Caliper uses this platform on a daily basis to evaluate various PCR assays and has documented a detection limit of approximately 3 to 5 copies per reaction volume under laboratory conditions. The chip is small, reliable, and amenable to use in an automated system.

Inhibition is potentially a significant concern in systems that routinely handle environmental samples. The use of sample preparation and magnetic bead concentration (including multiple washes) are measures used to counter environmental inhibitors employed in the sample preparation and concentration steps. We are responding to this concern in the assay development section in three specific ways: WGA amplification, PCR assay conditions and the use of an internal positive control. The use of WGA and PCR additives described below are employed to counteract the adverse effects of poorly characterized environmental inhibitors. The use of an internal positive control reaction will be used in the system to monitor the possible adverse effects of inhibition in addition to acting as a QA step for the entire system.

It is our experience that the use of a WGA step can actually increase copy number of targets hence diminishing the impact of inhibition. MDA is a very robust application and can result in greater concentrations of target thus permitting a 3-10 fold dilution of material to be used in subsequent multiplex assays. In this way dilution of inhibitors can be achieved.

REFERENCES

1. *Detection Systems for Biological and Chemical Countermeasures (DSBCC)*, Research Announcement 03-01, US Department of Homeland Security, Advanced Research Projects Agency, September 23, 2003.
2. Wark, K. and C.F. Warner, *Air Pollution: Its Origin and Control*, Harper and Row, New York, 1981.
3. Cox, C.S. and C.M. Wathes, *Bioaerosols Handbook*, Lewis Publishers, New York, 1995.
4. *smallTalk 2003: The Microfluidics, Microarrays and BioMEMS Conference*, San Jose, CA, USA, 2003
5. Fan, Z, Ricco, A., Tan, W., Zhao, M., and Koh, C. "Integrating Multiplexed PCR with CE for Detecting Microorganisms", *7th International Conference on Miniaturized Chemical and Biochemical Analysis Systems*, Squaw Valley, CA, USA, 2003

ROLE OF PROTOTYPE SYSTEM DEMONSTRATIONS IN THE DEVELOPMENT OF DETECTION-BASED WMD DEFENSES

L. BRANDT
Sandia National Laboratories, Livermore, CA, USA

Abstract: Demonstration of system prototypes in realistic user environments is a critical element in the maturation of detection-based systems employed for defense against Weapons of Mass Destruction (WMD) attacks. The United States Department of Homeland Security utilizes structured efforts called Domestic Demonstration and Application Programs (DDAPs) to overcome the diverse barriers associated with moving laboratory technologies into useable, end-to-end systems. In this talk, specific examples, drawn principally from the major DDAPs concerned with the chemical and biological defense of transportation facilities, will illustrate the key issues and payoffs. System demonstrations and prototype deployments play important roles for the diverse participants in the development process, including technologists, commercial suppliers, end users, and government funding agencies. The deployments identify environmental and operational problems that impact system utility. The involvement of users is key to determination of realistic operational concepts and requirements. The prototype deployments also provide incentives for private sector investments in detector development and in other enabling technologies. In some cases, prototype deployments have evolved directly into full-scale, operational systems. These initial operational systems have then provided the basis for subsequent technology changes to improve operability and reduce costs. Two DDAPs that have already grown into operational, deployed systems are the Biological Aerosol Sentry and Information System (BASIS) for wide-area, detect-to-treat, biological defense and the Program for Response Options and Technology Ehancements for Chemical Terrorism (PROTECT) system for subway, detect-to-warn, chemical defense. The PROTECT program has been succeeded by the PROACT program which is focused on the biological defense of airports and BASIS is the foundation for the nation's BioWatch defense system. The importance of

D. Morrison et al. (eds.), Defense against Bioterror: Detection Technologies, Implementation Strategies and Commercial Opportunities, 91–104.

DDAPs and other demonstration programs is underscored by their relative growth as a fraction of the US Department of Homeland Security Science and Technology effort.

1. INTRODUCTION

Defensive systems that utilize advanced detection capabilities can enable effective responses to terrorist release of chemical or biological agents. While development of detectors that meet the stringent requirements of such scenarios is a major R&D challenge, the integration of these detectors within an operating system can be just as demanding. This paper addresses the steps and issues that surround the introduction of detectors and supporting systems into operating defenses against chemical and biological terrorism. The focus is not on the specific performance required of the detectors to support successful operation of the overall system. Instead, issues that affect the deployment of new detection capabilities into realistic use environments and the role of demonstration programs in addressing these issues are reviewed. The very different, and sometimes conflicting, needs of the various participants are key considerations in the design and execution of demonstration programs. Two successful demonstration programs sponsored by the U.S. Department of Homeland Security have evolved to become a focus for many current activities that address the chemical and biological threat. These system demonstrations have served several vital roles, including solution of difficult technical problems, maturation of key commercial suppliers, establishment of confidence in the user community, and provision of operational security assets.

2. ROLE OF DEMONSTRATION PROGRAMS

Technologists within the research and development community have often faced difficulties in the transition of technology from the laboratory into deployed systems. This gap between development and application has sometimes been called "the Valley of Death" by technology developers who find their products disregarded by the very users who might reap large benefits from their technical advances. Over the last ten years, the U.S. Department of Defense has increasingly employed an approach called the Advanced Concept Technology Demonstration (ACTD) to evaluate the performance of maturing technologies in realistic use environments. These demonstrations allow users to understand and develop operational concepts for the employment of new technologies and can accelerate the acquisition process for systems that meet user needs [1].

When initial work on domestic defenses against the emerging threat of chemical and biological terrorism began within the Chemical and Biological National Security Program in the U.S. Department of Energy, the concept of technology demonstrations to move laboratory and commercial components into defensive systems was adopted. The resulting programs were termed Domestic Demonstration and Application Programs (DDAPs) [2]. The DDAPs shared some of the same goals with ACTDs. However, the applicable user community consisted of the many federal, state, and local agencies and responder groups charged with the security of potential domestic targets. Specifically, the goals of the demonstration programs have included:

1. To prototype near-term solutions to critical problems
2. To engage users throughout the program
3. To provide guidance to R&D efforts leading to more capable future systems
4. To improve defensive capabilities, including enduring or "leave-behind" elements
5. To transfer technology to a broad set of suppliers and users

The objective of engaging users throughout the cycle is particularly important in the domestic defense setting. Domestic governmental and private organizations having some role in security operations comprise an extremely diverse group with differing needs and constraints. As compared to the centralized processes involved in military requirements definition and acquisition, domestic security system acquisition has few provisions for coordination of performance standards or operational protocols among the diverse user community. This has motivated the development of tools and templates within the demonstrations to allow users and developers to jointly define realistic concepts of operation and enhance communication regarding the needs and capabilities of the defensive systems. This is particularly important for WMD attacks due to their potentially large scope and impacts.

Two initial DDAPs provided chemical and biological defense capabilities through the use of emerging detection technologies. The first was the Program for Response Options and Technology Enhancements for Chem/bio Terrorism (PROTECT), which developed a system that provided detect-to-warn capabilities against chemical warfare agent attacks on subways and other enclosed facilities. The need for such a system was motivated by the 1995 Tokyo subway attack. A detect-to-warn defense is one that attempts to reduce exposure of personnel by rapid sensing of the release followed by responsive measures such as evacuation or airflow control that reduce the contaminated area or move people quickly out of danger. The second was the Biological Aerosol Sentry and Information System (BASIS), which provided a deployable system that sampled for airborne pathogens to provide a detect-to-treat response capability for special events. The importance of such a capability was underscored by the anthrax incidents in the United States in the fall of 2001. A detect-to-treat defense is one that identifies the presence of an airborne pathogen early enough to permit effective medical prophylaxis measures to be taken. More specific discussions of these DDAPs, focusing on the ways in which each addressed the needs of key participants, are included below.

3. NEEDS OF KEY PARTICIPANTS IN DEMONSTRATION PROGRAMS

A demonstration program, such as the DDAPs introduced above, is the focusing activity in which the diverse, and sometimes conflicting, needs of users, sponsors, technology developers, and commercial suppliers are linked with technical and operational solutions to create a workable defense system. In this section, the nature of the needs and constraints associated with each key category of participants is discussed.

Users

Addressing the needs and constraints of users is the primary consideration in structuring a demonstration program. In the DDAP programs, an initial user partnership provided the framework for joint system development. The initial use environment was expected to be representative of a larger class of potential applications. This initial venue also represented a site or class of sites where the potential threat is seen as very significant. Successful contribution to a user in a real operational environment must meet a number of requirements. These include:

1. End-to-end Operational Concept: Users who participate in demonstrations require that actionable information be generated by the prototype system. This requires an understanding of the response options available and their utility. Operational responses that have significant impact on the outcome of potential attacks must be developed.
2. Clear Protocols for Response to Alarms: The problem of false positives in a detection-based defense system and the appropriate response to alarms is particularly important to users. Protocols and templates that guide validation of detection alarms are essential.
3. Enhancement of Immediate Preparedness: Demonstration partnerships can involve joint efforts to support a broad array of WMD preparedness measures. Detection-based systems must often be supplemented with

other ancillary security or response measures to achieve a balanced overall approach to WMD preparedness.

4. Information Control: Demonstration teams utilize sensitive operational, vulnerability, and capabilities information in the execution of detection-based defenses at specific sites. In some cases, the need to control sensitive information can conflict with the federal sponsor's desire to provide wide dissemination of demonstration results to encourage technology transfer and investment by other users.
5. Leave-behind Capabilities: For demonstrations that address protection of critical infrastructures, the desire of the infrastructure owners is for the largest possible leave-behind capability, and one with a proven operational record. This can conflict with the needs of the sponsor to demonstrate key elements of the architecture only to the level needed for transition to the commercial sector. Early plans that specify final system disposition and level of operational validation are essential to align expectations of the user and the sponsor.

Federal Sponsor

Sponsorship of domestic chemical and biological defense demonstrations in the US is largely in the hands of the federal science and technology programs that fund development of new defensive systems. The government role is to identify and provide initial developmental funding for key technologies and to promote the adoption of those technologies by commercial suppliers and by users of defensive systems employing these technologies. It is important to note that the government can also be the primary, long-term user of the demonstrated system. In the case of deployable, special event defensive systems (one of which is discussed below), the federal sponsor may choose to own and maintain the capability to provide security at events or locations deemed of national importance.

Technology Developers

Contributions to technological goals occur in two areas as a result of the demonstration programs. First, component developers (e.g., detector developers) gain the ability to test their technologies in realistic environments, an understanding of performance requirements based on an end-to-end architecture, and user confidence that can enhance subsequent commercialization steps. Second, the demonstration environment provides systems engineers with a venue to understand overall user and application needs and to develop tools and information systems that permit effective end-to-end systems integration. Information networks that collect and interpret detector status, detection alarm information, and response recommendations are essential features of each demonstration program.

Tools that permit effective training and exercises are also generally needed to develop and implement joint concepts of operation between the demonstration team and users.

Commercial Suppliers

Both direct and indirect benefits accrue to the commercial supplier community. In some cases, suppliers whose components are used in the demonstration directly participate in the modification, testing, and validation of those components. The indirect benefits to the broader supplier community include the more specific prescription of the architecture and requirements associated with detection-based defenses and a reduction in cost and performance uncertainties that might otherwise deter both funding agencies and customers from investments in the development and deployment of the technologies.

4. EXAMPLE 1 - TECHNICAL AND OPERATIONAL FEATURES OF PROTECT

4.1 Summary

The PROTECT program has served to combine detection and supporting technologies into an integrated system for defense of domestic facilities. The initial focus was on the protection of subway systems against chemical attack. This was later extended to address architectures for both chemical and biological defense of airports.

The principal collaborator for subway program was the Washington Metropolitan Area Transit Authority (WMATA). The major areas of technical focus included a detection testbed to evaluate chemical detectors in the subway environment, characterization of station airflow characteristics to provide information on detector siting and operational response options, and development of a decision support and crisis management system. The detection testbed began operations in May 2000. The first demonstration of the integrated system for a single station was conducted on December 5, 2001. Subsequently, development of the complete system was accelerated to create a system-wide operational capability for WMATA [3].

The airport work has been done in collaboration with the San Francisco International Airport. While the airport activities comprised a small fraction of the PROTECT program, use of systems and tools developed for the subway program allowed rapid contributions to be made. In 2003, the Protective and Responsive Options for Airport Counter-Terrorism

(PROACT) program was created to focus specifically on defensive options for large facilities served by air handling and conditioning systems. The PROACT program is also investigating the integration of biological detection systems employing both detect-to-warn and detect-to-treat response options.

4.2 Technical Contributions

The PROTECT program provided the basis for the first domestic, public, continuously operable, chemical defense system in the U.S. In the process, several significant technical advances were accomplished. These include the following:

- Detector Modification and Qualification – Commercial detection systems were modified and qualified for continuous operations in a subway environment. Modifications included several features needed to increase reliability of pumps and improve filtration to operate in the relatively dirty subway environment, as well as development of techniques that allow determination of agent concentration information (in addition to threshold alarms) to support predictive modeling of agent cloud transport by the decision support system.
- Rapid Site and Response Characterization – Experimental procedures utilizing smoke aerosol and tracer gas releases to determine air flow characteristics within subway stations and airport terminals were developed. Such information is required to determine the number and sitting of detectors and to identify response actions (e.g., subway train movements, air handler controls, station evacuations) that will minimize personnel exposure to contaminated air
- Decision and Information System – An information and decision support system (the Chem.-Bio Emergency Management Information System or CB-EMIS) that incorporates specific site information with real-time transport models permits system operators to assess the current and predicted status of an attack and to confirm control actions for reducing the spread of the agent cloud.
- Communications and Control Infrastructure – A secure, fault-tolerant networking system was demonstrated that interfaced with diverse detectors and other information sources and that linked to the central decision support computer and software. Scalability was a key consideration in the design of the information system and networking infrastructure.

4.3 Operational Experience and Impact on Key Participants

There were several operational areas in which the experience gained from the PROTECT program is of particular interest. They include the following:

- False Alarm Protocols - The problem of differentiating false alarms from real attacks is particularly important for major infrastructure elements such as subway systems. Intermittent false positive alarms cannot be avoided with currently available detectors. However, several techniques can be employed to validate alarms. If several detectors are included in an area, the report from multiple detectors can be used to validate an initial alarm, although with a penalty in system response times. Initial detector alarms can also be used to alert system operators to review closed circuit video monitors of the area to determine whether immediate effects on individuals are apparent. This is a high confidence confirmation strategy although it does not apply for delayed-acting agents. Several detectors utilizing different detection principles can also be deployed within each detection module to provide independent reports at each detection location. Experience in the detection testbed aspect of the PROTECT program proved that current detector false alarm performance, supported by an appropriate choice of the strategies outlined above, were sufficient to provide the very low system false alarm rates required for transition to a fully operational status.

– - Rapid Transition to Operational Status – Following extended operation of the detection testbed and an initial demonstration of full system performance for a single subway station in December 2001, the rapid expansion to operational status limited the ability to transition PROTECT from national laboratory ownership to commercial suppliers and subsequently to iterate the design prior to the initial large scale deployment. The previously planned PROTECT commercialization program, whereby the lessons learned from demonstration activities would be transferred to the private sector, was bypassed due to the urgency caused by national concerns over the threat of WMD attacks.

– - Testing of Advanced Technology – The detection testbed and user cooperation permitted the testing of advanced sensors that will support upgrades in the future system. This has also enabled evolution of the architecture toward concepts of detect-to-warn biodefense in the PROACT program.

5. EXAMPLE 2 - TECHNICAL AND OPERATION FEATURES OF BASIS

5.1 Summary

The BASIS system was designed to detect and locate an aerosol release of biothreat organisms quickly and accurately enough to implement an effective response. In most cases, the prognosis for individuals infected in such an attack depends on how rapidly antibiotics, vaccines, or other medical interventions can be administered [4]. The ability to make a timely detection and characterization of an attack can save many lives. The BASIS system collects aerosol samples at specified time intervals in locations chosen to best characterize an attack. These samples are then evaluated in a mobile field laboratory utilizing DNA-based signatures that permit identification of potentially lethal bacteria and viruses within hours. The system sensitivity is sufficiently high to detect even the tiny amounts of pathogen in a well-dispersed release cloud.

The BASIS system includes three primary elements. The aerosol collector system continuously collects, time-stamps, and stores samples. The mobile field laboratory analyzes the numerous samples from the array of collectors. The operations and management software tracks the samples through all stages of the process and provides real time control of the overall system.

The BASIS program was initiated in 2000 and was first deployed in the month following the September 11 terrorist attacks. It was deployed at the 2002 Winter Olympics in Salt Lake City and operated for 35 days in sports venues, urban areas and transportation hubs. It has been subsequently deployed at numerous special events. The BioWatch system, a derivative of BASIS, is currently deployed in over 30 U.S. urban areas under the auspices of the U.S. Department of Homeland Security. The current BioWatch deployments utilize laboratories that comprise the federal Laboratory Response Network operated by the Centers for Disease Control [5] rather than the mobile laboratories that continue to be available for BASIS special event deployments.

5.2 Technical Contributions

BASIS was the first integrated system that permitted continuous air sampling followed by rapid laboratory analysis to determine the presence of pathogens that might be indicators of a bioterror attack. To accomplish this goal, notable advances in several areas were required. These include the following:

- - Bio Assay Development – A key capability that enables the robust performance of the BASIS architecture is the development of DNA-based assays that uniquely identify a pathogen. The BASIS laboratory utilizes DNA amplification via polymerase chain reaction (PCR) to identify specific microbes. While PCR provides a rapid process for detection and identification, its reliability depends critically upon identification of specific regions of the DNA or RNA that uniquely identify the pathogen. Previous assay development processes were cumbersome and did not permit examination of the large number of candidate signatures available in pathogen genomes. The BASIS team created an automated, signature-design and maintenance system that allowed computer-based screening of candidate signatures to identify those candidates most likely to be unique. The use of this process has enabled the generation of a large family of assays to support BASIS, the CDC Laboratory Response Network, and other users. Without these assays, the attainment of rapid analysis at low false alarm rates, as required for BASIS success, would have been impossible [6].
- - Sample Handling and Laboratory Automation – The collection and processing of the large number of samples associated with deployment of multiple collectors to cover the diverse activities associated with an event such as the Olympics requires advanced information systems for sample tracking, lab analysis reporting, and system status monitoring. Readily available filter-based aerosol collectors were used for BASIS, but were augmented by an advanced Sample Management System for filter tracking and the BASIS Operations Center to monitor systems operation [7]. Note that the handling of the filter samples must adhere to applicable U.S. chain-of-custody rules that apply to criminal evidence, in the event that the detections are associated with subsequent legal proceedings. This adds another set of assurance and control requirements on the system.

5.3 Operational Experience and Impact on Key Participants

The following are several areas associated with the operation of BASIS that have impacted the key users and other participants.

- Alarm Protocols – The specificity of the assays developed and validated for BASIS has resulted in no false positives in a very large (>700,000) assays completed to date [8]. However, since some of the pathogens of concern are endemic to the environments of special events, and because the PCR process can amplify very small amounts of DNA to detectable levels, it is possible to have confirmed detections that are not associated with bioterrorist releases. One example drawn from BioWatch, a system that employs the BASIS architecture, was the detection of fragments of F. tularensis, the bacterium found in rabbits and rodents that can spread to humans to cause tularemia, during sampling in October 2003 in Houston [9]. Such alarms represent detection of the very low levels of endemic pathogens at monitoring sites and must be differentiated from real bioterrorist releases by use of other operational techniques. Several options for validating alarms have been identified. These include additional environmental sampling to determine surface deposition levels, event reconstruction using computer modeling tools, and review of public health and veterinary surveillance inputs. Operational measures to engage local, state, and federal resources in the examination of potential alarms are a part of the current BioWatch program [10]. The capability to pursue joint planning that integrates realistic biodetection capabilities into the operational context of the public health and first responder communities is an important benefit of the direct user interactions created initially through the DDAP process.

- Government-owned Capabilities from DDAPs – The BASIS DDAP is an excellent example of the role of government-owned and operated capabilities as the basis for development and demonstration of advanced biodetection capabilities. The capability satisfies an ongoing need for enhanced protection at special events for which the federal government assumes significant security responsibilities. Furthermore, the limited duration and high emphasis placed on such events permits support of advanced systems by technical personnel and labor-intensive processes that might not be affordable for day-to-day security operations at other infrastructure facilities. The government-owned system can also provide a base of operational experience that supports decisions for broader deployment and multi-agency operation of the defensive system.
- Focus for Technology Development – The BASIS DDAP has provided a focus for technology development in a variety of areas. As noted earlier, advanced processes for deriving DNA-based PCR assays enabled the initial deployment. Subsequently, U.S. Department of Homeland Security programs in advanced bioassay development, laboratory automation, and detector development have targeted specific improvements of the baseline BASIS and BioWatch architectures as developmental goals [11]. In this case, BASIS has provided a robust architecture to guide technology requirements, operational interfaces, and implementation strategies.

– Springboard for Broader System Development and Employment - The BASIS program has successfully established broad-based confidence in its architectural concept, hence enabling the major investment and expansion decisions required for establishment of the BioWatch program. These decisions followed the successful BASIS deployments in late 2001 and early 2002, most notably at the 2002 Winter Olympics. The transition of laboratory responsibilities to the CDC Laboratory Response Network and the multi-agency support of post-alarm consequence management processes are further evidence of broad, multiple agency adoption of the architectures initiated by the BASIS DDAP [12].

6. SUMMARY OF CONCLUSIONS FOR DETECTION-BASED DEMONSTRATION PROGRAMS

The examples in this paper have illustrated the important role that demonstration programs can play in the development and adoption of detection-based systems for defense against WMD attacks. The two DDAPs reviewed here represented somewhat different approaches. The first (PROTECT) addressed key technical elements of a detect-to-warn, chemical defensive system that was intended for acquisition and operation by critical infrastructure owners. The second (BASIS) developed a government-owned capability to provide biological monitoring of special events. Despite the differences in mission and approach for the two systems, common elements characterize both DDAPs.

First, the involvement of users provided a focus on key system requirements needed to make the detection technologies workable in real environments. Of particular importance in both cases were operational protocols that permit effective response to system alarms.

Second, the demonstrations resulted in significant improvements in operational capabilities. In the case of PROTECT, the transition to operational status at WMATA and the subsequent initiation of the PROACT program demonstrate the success of the program in solution of near-term problems and identification of growth directions for future systems capabilities.

Third, the demonstration architectures provided a basis for future technology development and deployment.

Fourth, the process of engaging users and related agencies was vital in developing confidence in advanced defensive systems. Experience in the performance of new technologies and joint development of operational plans and tools to communicate these plans are essential. The demonstration environment provides an excellent approach for bringing technology developers and users together to address these key issues and for promoting commercialization of the demonstrated architectures.

REFERENCES

1. For a more complete introduction to the ACTD concept, see the discussion at http://www.acq.osd.mil/actd/intro.htm.
2. The Chemical and Biological National Security Program (termed CBNP) was initiated in FY97 in response to the growing threat of chemical or biological use against U.S. population or infrastructure. The program was subsequently transferred to the Department of Homeland Security. Background on the accomplishments of the program during its first five years is summarized in a series of annual reports, beginning in FY99. The most recent of that series is "Chemical and Biological National Security Program FY02 Annual Report," U.S. Department of Energy, National Nuclear Security Administration, Office of Nonproliferation Research and Engineering, February 2003.
3. The milestones in the evolution of PROTECT have been documented by the media and in the open literature. The CBNP Annual Reports referenced in an earlier endnote contain system descriptions and describe program progress. Media references that discuss program milestones include: 1.) "Metro seeks high-tech security," Washington Post, October 16, 2001; 2.) "Metro drill tests response to an attack," Washington Post, December 5, 2001; 3.) "Metro set to initiate chemical sensors," Washington Post, December 25, 2001.
4. Numerous studies have addressed the tradeoffs between detection and various response actions and their impact on casualties expected in a bioterrorism event. One available example for an anthrax attack is Wein et al, "Emergency Response to an Anthrax Attack," Proceedings of the National Academy of Sciences, April 1, 2003, pp 4346-4351. (Available at www.pnas.org/cgi/reprint/100/7/4346.pdf) Other studies on similar scenarios reach different conclusions due to the many uncertainties in the attack and response parameters.
5. For a summary of the Laboratory Response Network see the CDC fact sheet at http://www.bt.cdc.gov/lrn/factsheet.asp.
6. For a somewhat more complete but nontechnical summary of the assay development process supporting BASIS, see "On the Front Lines of Biodefense," Lawrence Livermore National Laboratory, Science and Technology Review, April 2004 (available at www.llnl.gov).
7. See "BASIS Counters Airborne Bioterrorism," Lawrence Livermore National Laboratory, Science and Technology Review, October 2003 (available at www.llnl.gov).
8. As of April 2004, Lawrence Livermore sources estimated that, in more than 700,000 uses of the assays, no analysis using the prescribed set of assays had ever experienced a

false positive, that is, concluded that pathogens were present when they were not. (See "On the Front Lines of Biodefense," Lawrence Livermore National Laboratory, Science and Technology Review, April 2004)

9. Discussion of this event is included in "On the Front Lines of Biodefense," Lawrence Livermore National Laboratory, Science and Technology Review, April 2004. Initial public agency response to the Houston incident is summarized in "Signs of Tularemia Detected in Houston Air," Center for Infectious Disease Research and Policy, CIDRAP News, October 10, 2003.
10. The developers of the BASIS and BioWatch systems have continued to work with users to identify joint activities for event characterization and consequence management following BioWatch alarms. The Incident Characterization Action Plan (ICAP) program is currently underway to address this problem (www.lanl.gov/orgs/d/d3/chembio.shtml). The plan is intended to provide a basis for a coordinated follow-up to all alarms generated by the BioWatch system.
11. A wide array of U.S. detection programs is targeting reduction in response times and operational costs within the BioWatch framework. These included concepts for autonomous networked detectors that can eliminate the costly sample collection and laboratory analysis functions of BioWatch by providing high quality, low false alarm, assays at the sampling location. These are represented by the Autonomous Pathogen Detection System (APDS) and successors developed by Lawrence Livermore National Laboratory. The Homeland Security Advanced Research Projects Agency (HSARPA) has also identified new technologies for development. (For information on current HSARPA initiatives in this area, see HSARPA Research Announcement RA 03-01, "Detection Systems for Biological and Chemical Countermeasures," September 23, 2003)
12. The establishment and growth of the BioWatch system has been well documented by the public media. For example, the following news references document the chronology of program decisions: 1.) "BioWatch program aims for nationwide detection of airborne pathogens," CIDRAP News, February 26, 2003 (www.cidrap.umn.edu); 2.) "Secret sensors scour air for bioterrorism," Boston Globe – Associated Press, July 21, 2003; 3.) "Government discloses details of nation's bioterror sensors," Ted Bridis, Associated Press, November 15, 2003; 4.) "Bush Oks directive to boost biodefense," Washington Post – Associated Press, April 28, 2004. A summary fact sheet is also available at www.dhs.gov.

VALIDATION TESTING FOR BIOLOGICAL THREAT ORGANISMS

T. L. HADFIELD
Midwest Research Institute, Kansas City, Missouri, USA

Abstract: Conducting validation studies of qualitative biothreat identification assays is a new field compared to validation of standard clinical laboratory techniques or bioanalytical methods. The technical challenges in biothreat assay testing include such things as (a) effects of various bacterial or viral concentration techniques; (b) residual matrix-associated amplification inhibitors; and, (c) differentiating live from dead cells. The determinations that must be made in conducting qualitative assay validation studies are: (1) specificity rate, (2) sensitivity rate, (3) false positive and false negative rates, (4) system suitability testing, (5) robustness, (6) linearity, (7) range, (8) precision and (9) limit of detection. The Association of Analytical Communities (AOAC) requires that each assay validation is completed with a Package Insert review ensuring validation parameters are accurately reflected. Upon completion of validation, a "Quality Policy Certification," will be issued for each assay and published in the AOAC journal. The focus of the project presented in this paper is to provide the Department of Defense (DoD) and Department of Homeland Security (DHS) with a well-executed plan leading to the successful validation of assays for biothreat agents. The project will hopefully show the need for a Validation Center which can provide thorough and well-designed validation studies for qualitative assays and instrumentation. Validation that qualifies the technique/instrument to sample type, outcome expectations (forensic or force protection) and reproducible detection limits with false positives, false negatives and confidence calculations would also be included. In addition to the main laboratory (Validation Center) executing the validation study, five additional laboratories would also be required to complete validation testing.

D. Morrison et al. (eds.), Defense against Bioterror: Detection Technologies, Implementation Strategies and Commercial Opportunities, 105–117.

1. INTRODUCTION

Validation testing of biological threat organisms presents a unique set of complications not experienced in validation of most other test methods and assays. The lack of a ready source of positive samples severely restricts the approaches available to validate the test methods. It may be easy to validate the assay for culture identification of the organism but virtually impossible to validate the test for detection of the organism in other types of samples such as air collections, clinical samples or water samples if the organism can not be found in those samples at a reasonable concentration. In such cases, validation of the assay becomes much more problematic. One can resort to spiking experiments or similar types of experiments but the developer must realize these experiments only approximate real sample testing. Testing of real sample matrix from a variety of sources is also an important aspect of validation testing. This will provide good data for specificity values. Understanding the limitations of the validation testing is an important factor for both the developer and the validation-testing group for successful evaluations to occur. If the developer plans to commercialise the assay, the validation group should work with the developer and any state and federal organizations to understand all the requirements of any licensing agency that are over and above the requirements for validation of the assay. Validation testing of a new instrument, method or biological assay requires significant thought and expense. Determination of how the device, method or biological assay is to be used will focus the testing requirements for validation. Statistical support is an essential component for designing the validation plan. An assessment of how many instruments, replicates of the method, determinations made by the assay for a particular substrate are all key aspects addressed by the statistician before the plan is written. The developer and the end user should be in agreement regarding the acceptance criteria for successful completion of the validation. Both parties must be reasonable and understand the limitations of the instrumentation, method and/or the assay. Unrealistic expectations result in failed validation tests. For biological assays, two major phases of validation occur after assay development and optimization of a test. The analytical phase confirms the developers "claims" for performance of the test with "neat" reagents. By itself, the analytical phase of the validation plan does not validate the test or the system. Areas addressed in this phase of testing include determination of instrument accuracy, precision, limit of detection, dynamic range, selectivity, analytic sensitivity, analytic specificity, linearity and reproducibility. As part of the development or analytical validation, positive and negative controls should be established and standards should be established for quantitation and or calibration. Standardization and quality control of the procedures for preparing the stock solutions of standards and controls must be determined. Use of the controls and standards should be standardized for routine operations. This phase of testing describes all the performance characteristics of the assay and the instrument being used. The second phase of validation investigates the overall performance of the assay. It begins with sample collection and goes through sample reporting. Many of the same categories of data are examined again in this phase of testing. In addition to the accuracy, precision, selectivity, sensitivity, specificity, linearity and reproducibility, this phase also examines assay robustness, instrument variation, technician variability, reagent stability, and system compatibility. This phase of testing also includes a multi-laboratory test component. For this component of the testing to be successful, well-written standard operating procedures and highly standardized operations must be in place. Quality control and quality assurance of the samples and methods are particularly important in this phase of the study to ensure everyone participating in the multicenter trial is testing equivalent samples. In addition, before the testing begins, all instruments should be calibrated and shown to perform equally. Another critical determination is what constitutes a correct result in an unknown sample. For instance, if the standard method is negative and the new method is positive (a situation often seen with PCR), how do you determine which test is correct. Development of unique positive controls may be one solution to determining if the result is real or a contamination problem. Compilation of the data and analysis by a statistician provides the foundation to determine if the validation was successful.

Documentation of the development and preliminary testing establishes a foundation for future work. Once the test/method is selected for additional development and validation, documentation must meet an equivalent of GMP standards to be a candidate for commercialisation. Standardized testing methods will achieve these goals and determine if the developer performance characteristics are, in fact, well established and suitable for commercialisation.

2. ANALYTICAL PHASE TESTING

The development and validation of a new method is an iterative process beginning with the initial design concept. Initial testing of a new method can be investigated in a research laboratory but as soon as a decision is made to move a new method to development for commercialization or "community" use, a higher standard of investigation is required. All methods being considered for validation should be developed using good manufacturing practices. This insures a good audit trail and documentation of all aspects of the development and validation process. The extensive record bank is frequently required by agencies approving the use of the test for a particular application (i.e. Food and Drug Administration). A quality assurance officer should be given responsibility for overseeing the quality of the studies and should not answer to any of the individuals or groups involved in performing the study. The quality assurance officer is responsible for verifying the quality of the process used during development and testing of the product. Audits are an integral component of the QA officers tasks to provide adequate confidence the process meets the requirements for quality system. A product team should be established when the product is selected for progression of development. The product team should consist of statisticians, the developer, a quality assurance officer, and others invested in the product. Their first order of business is to decide what the proposed application of the product will be and determine the minimal acceptance criteria for test method. This can be a daunting task, especially as it relates to biological threat studies. Once the application is determined, the next critical determination is defining minimum requirements equating to acceptance specifications for each step in the method validation. Ideally, the acceptance requirements will be agreed upon by the developer and end users of the product. Once the intended use and acceptance criteria are established, a detailed validation plan can be written. The statistician should work with the plan writer to establish the minimum number of tests for each experiment to yield data meeting the acceptance criteria. The criteria should allow some flexibility in the testing protocols. The plan should include remedial actions if an acceptance criterion is not met and all modifications and changes should be well documented. Performing a thorough method validation is a tedious and expensive process. Carefully designed, defensible experiments provide quality data demonstrating the validity of the method at the least cost.

3. VALIDATION TESTING CONSIDERATIONS

When writing the validation plan a description of the instrument/assay and its use should be stated. This will define if the method will be quantitative or qualitative. It will also define specimen collection methods, data capturing and recording methods, and schedule meetings among the developer and validation teams. It should also include the acceptance criteria for each of the steps in the validation protocol and will generate a set of standard operating instructions such that personnel "skilled in the art" can replicate the procedure. Potential remedial actions should be identified in the plan for recognized portions of the study where a "problem" might occur.

All of these considerations will impact the remaining steps in the validation plan. Subsequent to the execution of the validation protocol, the data is analyzed with results, conclusions and deviations presented in an official final validation report. Provided the pre-defined acceptance criteria are met, and the deviations (if any) do not affect the scientific interpretation of the data, the method is considered valid. A statement of the method's validity should be placed at the beginning of the final validation report, along with signatures and titles of all significant participants and reviewers. In the final analysis, the purpose of validating methods is to ensure the delivery of high quality data and a reproducible method. Once the final report is in hand, it will serve as the foundation for future testing of products related to the method i.e. reagent stability. Data should address a number of specific topics described below. The validation-testing plan should also define safety considerations and methods to ensure personnel are not exposed to hazardous material. This may require specialized laboratories or personal protective equipment. Validation testing of biological threat organisms may impose additional safety requirements beyond those normally anticipated when working with cultures of bacterial or viral agents. Evaluation of the test method requiring aerosols creates additional concerns about safety and the reproducibility of the aerosols. The impact of doing the work in biological safety cabinets, or requiring personnel to wear head to toe protective equipment may impact results obtained during validation testing and influence the way laboratorians are required to perform the assay at their duty stations. Also, use of live organisms may generate results different from use of DNA in development of a new nucleic acid method. Safety of personnel should always be included in the validation plan and the impact of doing testing in appropriate protective environments should be a top priority when preparing to initiate validation studies using biological threat organisms.

4. ACCURACY

Accuracy is the expression of agreement between an accepted value and the value determined by the assay. Accuracy can be determined by analyzing a sample of known concentration. The true value is compared to the measured value to determine the accuracy of the assay. Another means for determination of accuracy is to compare results of a new method with results from an existing alternate method known to be accurate [1,2]. Two additional approaches are based on recovery of known amounts of analyte spiked into sample matrix [3,4]. One method spikes analyte in blank matrices. Spiked samples are tested in triplicate over a range of 50-150% of the target concentration. The final way to assess accuracy is use of standard additions. This approach is used if it is not possible to prepare a blank sample matrix without the presence of the analyte. An example of accuracy criteria for an assay method is the mean recovery of 100 ± 2% at each concentration over the range of 80-120% of the target concentration. Another example is the method demonstrates recoveries of 70-120% of the accepted standard value (mean value) with a coefficient of variation (CV) of less than 20% for the measured recoveries at each spike level. For quantitative procedures, accuracy should be determined across the specified range of the procedure. Accuracy may be inferred once precision, linearity and specificity are established [9]. For qualitative analysis, accuracy is expressed as the false positive and false negative rates determined at the limit of detection concentration [5]. False negative rates are determined as follows: % false negatives=false negatives/total number of known positives x 100. False positive rate = false positives/total known negatives x 100. For quantitative analysis, accuracy is reported as the mean recovery at several levels across the quantitative range and is demonstrated by measuring the recovery of target analyte from fortified samples. Development of assays for biological threat organisms introduces some additional considerations. What units are going to be used, CFU or PFU for the bacterial/viral analytes. Are they appropriate for the method? Does it matter if the organism is alive or dead? Will the method of killing influence the results if killed cells are used? Accuracy determinations may be significantly diverged if one compares CFU with gene copies, how will the discrepancy be resolved.

5. DETECTION LIMIT

Detection limit of an analytical procedure is the lowest detectable amount of analyte in a sample but not necessarily the lowest quantifiable as an exact value8. The detection limit can also be defined as the lowest analyte concentration producing a consistent response detectable above the noise level of the system. It is important to test the method detection limit on different instruments in different laboratories to get a true estimate of the limit of detection (LOD). LOD is an approved way of estimating sensitivity [6,7]. The LOD may be defined as the minimum amount of analyte detectable in a sample at a defined level of confidence. LODs may vary between laboratories due to differences in the standards, instruments and technician performance, however, in a validation study, all laboratories should be using the same standards and each laboratory should demonstrate they can reproduce the "acceptance values" for the standards and methods being used. This approach essentially "qualifies" the technical staff performing the study before the study begins. Multicenter studies using a single defined set of standards give a more "universal" LOD value. The analytical LOD may vary from an operational LOD because of matrix effects. If matrix alters the analytical determined LOD, it should be established again using known quantities of analyte spiked into the matrix. The LOD should be stated in terms of a confidence interval, e.g. an analyst is 99.7% confident a sample containing the analyte at the declared LOD will be detected. LOD often varies from "neat" laboratory experiments where clean analyte is used to assess performance of detection and "real-world" samples where matrix influences the results of the LOD. Both values need to be established to understand the limitations of the assay and detection capabilities.

6. LINEARITY

Linearity is determined across a range of concentrations of analyte to determine the concentrations where linearity is lost. Linearity obtains test results directly proportional to the concentration of analyte in the sample. Linearity studies verify the sample solutions are in a concentration range where analyte response is linearly proportional to concentration. Linearity is evaluated by visual inspection of a plot of signals as a function of analyte concentration. Results are confirmed by calculation of a regression line by the method of least squares. Measurements using clean, standard preparations are performed to demonstrate detector linearity, while method linearity is determined during the accuracy study. A minimum of five concentrations is used to establish linearity. For each target concentration, standard solutions ranging from 50% to 150% of the target analyte concentration are prepared and tested. Validating over a wide range provides confidence the routine standard concentrations are well removed from nonlinear response concentrations. It also allows quantification of crude samples in support of process development. Acceptability of linearity data is often judged by examining the correlation coefficient and y-intercept of the linear regression line for the response versus concentration plot. Measurements using clean, standard preparations are performed to demonstrate detector linearity while method linearity is determined during the accuracy study. For linearity studies, the correlation coefficient, y-intercept, slope of the regression line and residual sum of squares should be determined. A plot of the data is included in the data set. Analysis of the deviation of the actual data points from the regression line is helpful for evaluating linearity.

7. PRECISION

Precision is the closeness of agreement between a series of measurements obtained from multiple sampling of the same homogenous preparation under prescribed conditions. Precision is considered at three levels; repeatability, intermediate precision and reproducibility [8,9]. Precision is usually expressed as the variance, standard deviation or coefficient of variation of a series of measurements. Repeatability (inter assay precision) expresses the precision under the same operating conditions over a short interval of time. Intermediate precision expresses within laboratory variations: different days, different analysts, and different equipment. Reproducibility expresses the precision between laboratories. Different sources of reagents and multiple lots of reagents should be included in this study.

Alternatively, the "round one, round two method to determine precision can be used. In round one an instrument precision study is performed. A minimum of 10 runs of one sample solution is made to test the performance of the instrument. For round one, an example of precision criteria for a particular assay method is the instrument relative standard deviation will be 1% and the intra-assay precision will be 2%. The second study is intra-assay precision; data are obtained by repeatedly analyzing aliquots of a sample, independently prepared according to the method procedure in one laboratory on one day. From these studies, the sample preparation procedure, the number of replicate samples prepared for evaluation and the number of runs required for each sample in the final method procedure are finalized.

Round two testing also includes assessing "ruggedness". Intermediate precision is obtained when the assay is performed by multiple analysts, using multiple instruments on multiple days in one laboratory. Included in the study is different sources or reagents and multiple lots of reagents from one source. Intermediate precision results are used to identify which factors contribute significant variability to the final result. Ruggedness testing allows documentation of analyst-to-analyst and/or lab-to-lab changes in a systemic manner. Ruggedness testing investigates the effects on either precision or accuracy of the method. For qualitative methods, ruggedness monitors false positive and false negative rates.

Reproducibility is determined by testing homogenous samples in multiple laboratories as part of interlaboratory studies. Reproducibility focuses on measuring bias in results rather than on determining differences in precision alone. Statistical equivalence us used as a measure of acceptable interlaboratory results. An alternative approach is the use of "analytical equivalence". This approach defines a range of acceptable results prior to the study and is used to judge the acceptability of the results obtained from different laboratories. The standard deviation, relative standard deviation (coefficient of variation) and confidence interval should be reported for each type of precision investigated. An example of reproducibility criteria for an assay method is the assay results obtained in multiple laboratories will be statistically equivalent or the mean results will be within 2% of the value obtained by the primary testing laboratory. Repeatability is assessed using a minimum of 9 determinations covering the specified range of the procedure (e.g., 3 concentrations/3 replicates of each) or a minimum of 6 determinations at 100% of the test concentration.

Typical studies include variations observed on different days, different analysts, different instruments, different reagent preparations and so on. The standard deviation, relative standard deviation (coefficient of variation) and confidence interval are reported for each type of precision investigated. CV values of 10 or less are generally acceptable.

8. QUANTIFICATION LIMIT

Quantification Limit is the lowest amount of analyte in a sample quantitatively determined with suitable precision and accuracy or it is the lowest level of analyte accurately and precisely measured [3]. The method for determination of the quantification limit should be presented and approved prior to the validation experiments. This limit is determined by reducing the analyte concentration until a level is reached where the precision of the method is unacceptable. An example of quantification limit criteria is the lowest concentration of analyte with a relative standard deviation (CV) of $\leq$ 20% when intra-assay precision studies are performed. Quantitation limits can also be determined by comparing signals from samples with known low concentrations of analyte to signals of blank samples and establishing a minimum concentration of analyte reliably quantified9. The quantitation limit is validated by testing a suitable number of samples prepared at the quantification limit.

9. RANGE

Range is the interval between the upper and lower concentrations of analyte in a sample demonstrating a suitable level of precision, accuracy and linearity [3,8]. Precision may change as a function of analyte level. Higher variability is expected as the analyte level approaches the detection limit and saturation limit of the method. The developer should decide on the allowable level of imprecision for the assay. Experimental studies will determine the concentrations where the imprecision becomes to great for the intended use of the method. The specified range is derived from linearity studies and depends on the intended application of the procedure [9]. An example of range criteria for an assay is the concentration interval over which linearity and accuracy are obtained per previously discussed criteria with a precision of no greater than 3% relative standard deviation.

10. ROBUSTNESS

Robustness of the assay is a measure of a tests capacity to remain unaffected by small but deliberate variations in method parameters and provides an indication of its reliability during normal usage [8]. Method parameters are evaluated one factor at a time or simultaneously as part of a factorial experiment. An example of robustness is how the changes in a parameter affect the outcome of the assay. If changes are within the limits producing acceptable results, they will be stated in the method procedure. The most efficient way to make these determinations is through designed experiments. Such experiments potentially include multifactorial changes to investigate first order effects. These studies identify variables with the largesteffects on results with a minimal number of experiments. The actual method validation ensures the final ranges are robust. If measurements are susceptible to variations in analytical conditions, the analytical conditions must suitably controlled or a precautionary statement is included in the procedure. Robustness experiments should define a series of system suitability parameters.

11. SPECIFICITY

Specificity testing is influenced by the type of method or assay being evaluated. Early in development specificity testing may be done with analytes similar to the analyte of interest. For example, in a PCR evaluation for anthrax, testing of other Bacillus species for cross reactivity is important but not the final evaluation. If the assay is not cross-reactive with pure cultures, testing can proceed to evaluation of the method to detect the analyte in matrix. The intended use of the method will influence the types of matrix to be evaluated. For example, if it is a clinical diagnostic test, blood, body fluids, tissues or other clinical samples will be included in the assessment for specificity. If the matrix is air samples collected on dry filter units, a series of filters testing negative for anthrax must be evaluated. Ideally, filters from a number of sites geographically distributed throughout the area of interest (i.e. city, state country, or world) should be evaluated. When assessing specificity, validation of the reagent components and method is carried out for each matrix for the intended use of the assay. The specificity value frequently changes in different matrices or during different seasons with a single matrix.

Specificity may also be thought of as the ability to unequivocally assess the analyte in the presence of matrix (also called selectivity). The response to the analyte in test mixtures and all potential sample components is compared with the response of a solution containing only the analyte. An example of specificity testing is a PCR assay amplifying a target sequence from corn DNA isolated from fresh tissue may differ from the specificity of the same method applied to DNA isolated from corn chips. Specificity of qualitative PCR is evaluated by testing a number of reactions on known positive and negative samples using test conditions approximating sample matrices and limit of detection sensitivity values.

12. STABILITY

Stability testing varies from allowing delays in routine testing of samples to stability of reagents over time under various storage conditions. Tests of analytes and reagents should be compared with freshly prepared standards over a defined time. For daily testing, the time frame is often 48 hours. If solutions are not stable over 48h, storage conditions or additives should be identified to improve stability. An example of acceptable stability criteria is a 2% change in sample result relative to freshly prepared sample results. Long-term stability can be evaluated by an accelerated protocol (elevated temperatures with weekly testing) or by holding the products at the desired temperature and testing monthly until the reagent fails.

13. SYSTEM SUITABILITY TESTING

This testing component includes all components from sample receipt to reporting of results. Tests are based on the concept that the equipment, electronics, analytical operations and samples analyzed constitute an integral system [9]. System suitability testing should be conducted in the early stages, along with robustness, and again toward the end of the validation process. In validation of a PCR test, a key component of system suitability testing is recognition of inhibition along with restoration of amplification. System suitability testing can be done with blanks from a matrix sample and spiked samples that are processed from beginning to end. This testing will also provide data to determine a false positive and negative rate and an approximation of the sensitivity and specificity of the test method.

Standardized testing of new methods is essential to validate the test method. Developer laboratories often perform abbreviated validation studies and feel their assay is validated. Frequently the assay has only been evaluated using "neat" samples and the evaluation of matrix effects has not been performed. Systematic evaluation of the test method by independent groups will assess the quality of the development and will potentially discover any potential flaws in the design and development of the assay. Standardized testing and evaluation in multiple laboratories confirms the robustness and specificity of the assay and establishes the initial sensitivity and specificity values for the intended use of the assay. Each different intended use must be evaluated separately in order for the assay to be validated for that use. The more complete the validation studies are the better the test will perform in laboratories outside the initial developer's laboratory. [10-12]

14. SUMMARY

In summary, validation testing of assays and methods for biological threat agents introduces new challenges not usually encountered in testing validation. The nature of the organisms requires consideration for safety in design, development, and execution of the test method. This influences technical staff performance which must be factored into the design, and execution of the validation test plan. Quality assurance becomes a significant component of the testing to assure all preparations are standardized and are functional at the time of use in each of the laboratories participating in the validation studies. Without adequate documentation and verification of the quality of the standards used in each laboratory the validation testing may be of little value. Validation of a new method is a complicated and time consuming task but a task that is required to assure the method is providing reliable, reproducible data. Validation of a new method will identify steps which must be tightly controlled to obtain reliable, reproducible data. The documentation of the steps and components of the validation are also required to assure quality of the product and provide an audit trail for defensibility of the study. The entire process should provide an approximation of expected system performance and will serve as a baseline to recognize when the system is out of control. Documentation and quality assurance audits provide proof of the study being done using standardized methods and quality standards. Quality assurance also provides evidence of the standardization of newly designed quality control standards, quantitative standards and calibrators used throughout the study. After performing each of the steps in the validation test plan, documenting the data collected at each step and, providing the acceptance criteria are met and deviations do not affect the scientific interpretation of the data, the method is considered valid.

REFERENCES

1. http://www.devicelink.com/mddi/archive/96/07/010.html
2. http://perso.wanadoo.fr/framille.relland.jose/p11s/fda/820/21cfr820-%20Quality%20system%20regulation.html
3. http://pubs.accs.org/hotartcl/ac/96/may/may.html
4. http://www.fda.gov/cder/guidance/pv/html
5. http://www,omerva-biolabs.com/files/vgm_valid_e.pdf
6. GP22-A Vol. 19 No 13, Continuous Quality Improvement: Essential Management Approaches; Approved Guideline, NCCLS.
7. http://www.regulatory.com/forum/article/test-val.html
8. http://pharmacos.eudra.org/F2/eudralex/vol-3/pdfs-en/3aq14aen.pdf
9. http://www.nihs.go.jp/drug/validation/q2bwww.html
10. http://www.aoac.org/testkits/app%20packet%203/Ops.pdf
11. http://www.aeicbiotech.org/guideline/pcr_valid_final.pdf
12. http://www.aoac.org/testkits/app%20packet%203Micguide.pdf

DEVELOPMENT OF BIOAEROSOL ALARMING DETECTOR

A. V. WUIJCKHUIJSE[1], C. KIENTZ[1], B. V. BAAR[1], O. KIEVIT[1], R. BUSKER[1], M. STOWERS[2], W. KLEEFSMAN[2] AND J. MARIJNISSEN[2]

[1]*TNO Prins Maurits Laboratory, P.O. Box 45, 2280 AA Rijswijk, The Netherlands*
[2]*Delft University of Technology, Faculty of Applied Sciences, Julianalaan 136, 2628 BL Delft, The Netherlands*

Abstract: Although banned by the BTWC in 1972, biological warfare agents continue to be a threat to military and public health. Countermeasures can only be effective if rapid detection and reliable identification techniques are in place. TNO Prins Maurits Laboratory, and Delft University of Technology and Bruker-Daltonik (Germany) are developing an bioaerosol alarm detector, based on fluorescence pre-selection, Aerosol Time-of-Flight Mass Spectrometry (ATOFMS) and Matrix-Assisted Laser Desorption/Ionization (MALDI). Using this combination, mass spectra were obtained from single biological aerosol particles.

1. INTRODUCTION

Biological warfare agents (BWA) have been around since the Middle Ages, when victims of the Black Death were flung over castle walls using catapults. BWA use has also been documented in the French and Indian War, when British troops distributed blankets infected by smallpox to Native Americans. Although the use of such weapons was banned by international law, through the Geneva Protocol of 1925 and the Biological and Toxin Weapons Convention in 1972, it is believed that the number of countries capable of producing such weapons is still increasing. The (suspected)

D. Morrison et al. (eds.), Defense against Bioterror: Detection Technologies, Implementation Strategies and Commercial Opportunities, 119–128.

development of bioweapons in states such as Iran, Libya, and North Korea is of special concern. Furthermore, recent events have shown that biological weapons are very attractive for terrorist organizations such as Aum Shinrikyo and Al Qaeda. A study by the World Health Organization (Huxsoll et al., 1989) estimates that 50 kg of Bacillus anthracis released upwind of a population center of 500 000 may result in up to 95 000 fatalities, with an additional 125 000 persons incapacitated. Countermeasures can only be effective if rapid reliable detection and identification techniques are in place. However, current technologies perform poorly when required to combine rapid detection with a good capacity to differentiate. As a result a warning system tends to act in one of two ways: it is quick to issue a warning but doesn't make clear the nature of the threat or it waits until the threat is understood before issuing the warning. Examples are the ASPEC (BIRAL), an instrument that uses shape information, and the UV-APS (TSI model 3312), which uses fluorescence to detect biological particles. Neither approach is desirable in an operational situation: neither enables the rapid implementation of the appropriate counter measures.

Microbiological identification techniques, such as PCR (polymerase chain reaction), typically take hours to yield results. The drawback of this particular method is that it involves a sampling step and wet chemistry to bring the DNA from an airborne bio-particle into contact with a DNA-probe. Nevertheless, modern instruments have reduced the total time needed for analysis to some 15 minutes.

Historically, mass spectrometry has provided a key method in the chemical detection of biological aerosols. In 1985, (Sinha et al.) developed a technology for the on-line analysis by particle beam mass spectrometry. Developments along the same line of development are laser based aerosol mass spectrometers, which attempt instantaneous detection of species-level differences between single cells (Gieray R.A. et al.(1997), Russell S.C. et al.(2004)).

Another versatile mass spectrometric technique for identifying biological compounds is MALDI-TOF-MS (Matrix Assisted Laser Desorption/Ionisation Time of Flight Mass Spectrometry). This is a generic analytical technique, suitable for toxins, viruses, bacteria and spores. However, MALDI-TOF-MS also involves sampling by deposition: the material has to be collected on a suitable substrate before it can be analyzed. Researchers at Johns Hopkins University (McLoughlin et al., 1999) have addressed this problem by combining their instrument with an aerosol impactor. Material is deposited on a video tape, which is periodically transported into the mass spectrometer for analysis. This configuration reduces the total analysis time to 10 to 15 minutes, which is still long for detection purposes.

Researchers at Delft University of Technology (DUT) and the TNO Prins Maurits Laboratory (TNO-PML) developed an alternative approach (Stowers et al., 2000 and 2004), which demonstrates that real-time aerosol time-of-flight mass spectrometry (ATOFMS) can be applied to bio-aerosols. This approach is also followed by Jackson, S.N. et al. (2002). The system is an experimental, laboratory scale set-up modified from an existing ATOFMS, which latter system had been developed for the analysis of inorganic and organic chemical aerosol particles (Marijnissen et al., 1988, McKeown et al., 1991, Prather et al.,1994, Kievit et al., 1996, Weiss, et al. 1997). The existing ATOFMS system was considerably modified with regard to the mass range and detectability, in order to produce selective information (particle size and mass spectrum) for single aerosol particles of peptides, small proteins and Bacillus subtilis spores (Stowers et al., 2000 and 2004).

Mass spectra from a single or a few individual aerosol particles can be obtained within a few seconds, enabling rapid identification of the aerosol material. Recently the system has been further improved by implementation of fluorescence pre-selection. Those particles that are potential biological in origin are then selected and analysed with the mass spectrometer. This biological pre-selection increases the warning's reliability and makes it an extremely fast and accurate method of detection.

2. EXPERIMENTAL

Bioaerosol was produced either by nebulizing aqueous solutions or by dispersing fine powders. For MALDI experiments, matrix material could be pre-mixed with the bioaerosol (in the case of aqueous solutions) or could be added on-line with an evaporation/condensation device which was fitted onto the entrance of the ATOFMS. The core system (Figure 1) consists of an aerosol inlet to the vacuum, and laser systems for particle sizing, pre-selection and ionization, and a linear time-of-flight mass spectrometer for ion mass analysis. The system detects particles based on light scattering emitted as the particles pass the two detection laser beams of the MBD-266 cw UV laser. With the time between scattering events related to particle size, the pulsed 308 nm excimer laser is triggered for ionisation.

The matrix of UV-absorbing material enables efficient ion formation, and the resulting mass spectra can be used to identify the aerosol material.

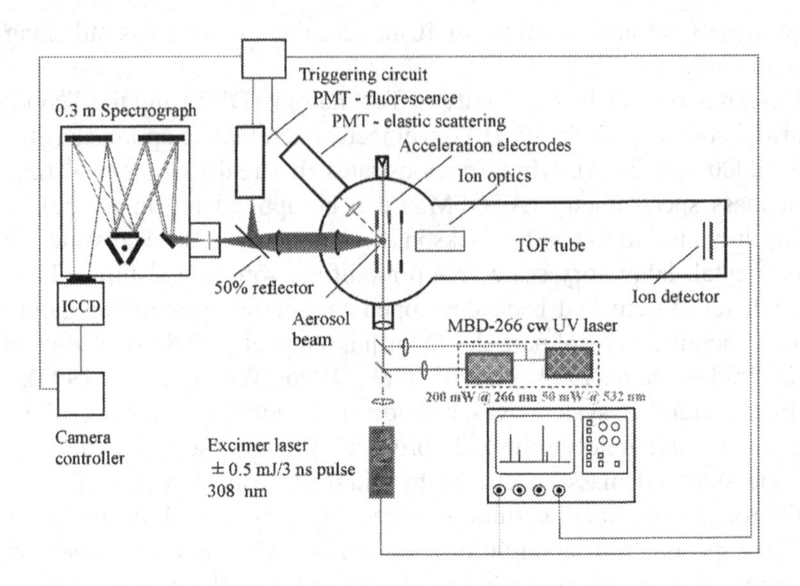

Figure 1. Set-up of the ATOFMS system

The aerosol mass spectrometer is designed to detect single particles in the size range of 0.5 to 20 μm in diameter. The concentration of these particles in the atmosphere is highly variable, but is commonly on the order of 104 particles per litre of air. The future NATO standard for a bacteria detector is 1 agent-containing particle per litre air (ACPLA) detected within 5 minutes. If all particles in this size range are analysed, this implies that, on average, 10.000 particles must be analysed for every agent-containing particle. Given the maximum rate of operation of the system, 100 Hz, only a few agent-containing particles would be analysed during any 5-minute period.

To address this situation, a mechanism is developed whereby the system reacts only to particles that are very likely to contain bacteria. Bacteria are known to emit fluorescence when excited with ultraviolet light; this distinguishes bacteria from virtually all other particles likely to be present in the atmosphere. The particle detection system has been designed to use this property to pre-select bacteria from a mixed aerosol. The system detects particles based on light scattering emitted as the particles pass two detection laser beams, with the time between scattering events related to particle size. In our system, one of these beams is composed of 266 nm light. As the particles pass this beam, emitted fluorescence in the range of 300 nm to 450 nm distinguishes the particles existing of bacteria from the larger background population of particles. By limiting the subsequent mass spectral analysis to fluorescing particles, our system performs an efficient pre-selection of the bacteria present in the sample.

3. RESULTS

To demonstrate the pre-selection mechanism, the system was operated using two optical configurations simultaneously. The first, referred to as "2 colour" detection, is described above. The second is the original (for the system) "1 colour" detection scheme, where both detection laser beams are of visible light as described earlier (Stowers et al., 2000). In this case, no fluorescence is detected and there is no discrimination between particles. For the results discussed below, the 266 nm beam was present during "1 colour" detection, but was not used for triggering. Therefore, fluorescence intensity is measured for all particles that are detected. We need only toggle a switch on the triggering circuit to change between the two detection modes. Therefore, we can conveniently change between ionising all suitably sized particles and ionising only suitably sized fluorescing particles. The detection system provides the aerodynamic size and the intensity of the fluorescence emitted from each particle. This information can be conveniently represented in a 2-dimensional histogram plot such as in Figure 2.

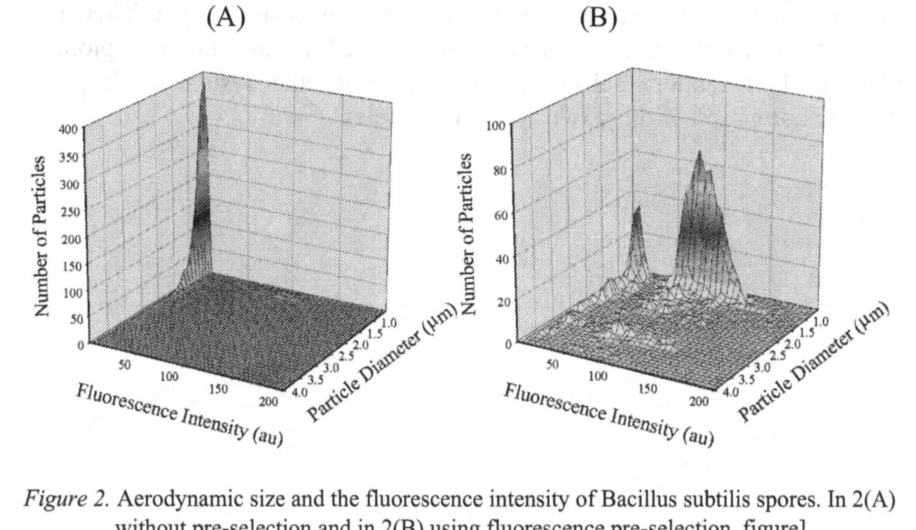

Figure 2. Aerodynamic size and the fluorescence intensity of Bacillus subtilis spores. In 2(A) without pre-selection and in 2(B) using fluorescence pre-selection. figure]

Figure 2 shows the utility of the pre-selection configuration for a biological sample. The aerosol for this experiment is generated from a sample of unwashed Bacillus subtilis spores. In 2(A) all suitably sized particles (that is, using "1 colour" detection) are selected by the system. In 2(B) the fluorescence pre-selection mechanism is used. In the left figure, the system analyses mostly small particles with little fluorescence. These particles are composed of salts and other material present in the original solution, but do not contain the spores of interest. The system is occupied most of the time by non-interesting particles. There is, however, a minority of particles, somewhat larger than 1 μm in diameter that do have significant fluorescence. In 2(B), using "2 colour" detection, fluorescence allows discrimination between particles. In this case, the selected particles are almost certainly spores. In this way, the system reserves its precious capacity for the particles that are most likely of interest for further analysis. In the following experiment, a model aerosol is generated from (1) an aqueous solution containing lead chloride (PbCl3) and (2) tryptophan particles generated by dispersing a dry powder. Lead chloride does not fluoresce, while tryptophan is the molecule that dominates the fluorescence emitted from bacteria excited at 266 nm. Lead chloride is characterised by the ions Pb+ (m/z = 207 Da) and PbCl+ (m/z = 242 Da) and by a cluster of ions containing two lead atoms at masses above 400 Da. The major ion peak in the tryptophan spectrum is the dehydroindole ion at a mass of 130 Da. Figure 3 (a) shows sequential mass spectra generated using "1 colour" detection. The system detects a mixture of particles. The first one in this sequence contains lead with its characteristic lead-containing ions, but the second one is a tryptophan particle, with its characteristic mass at 130 Da. Then five consecutive lead-containing particles are analysed, followed by a tryptophan particle. This random selection continues, with the system analysing a mixture composed of these two types of particles.

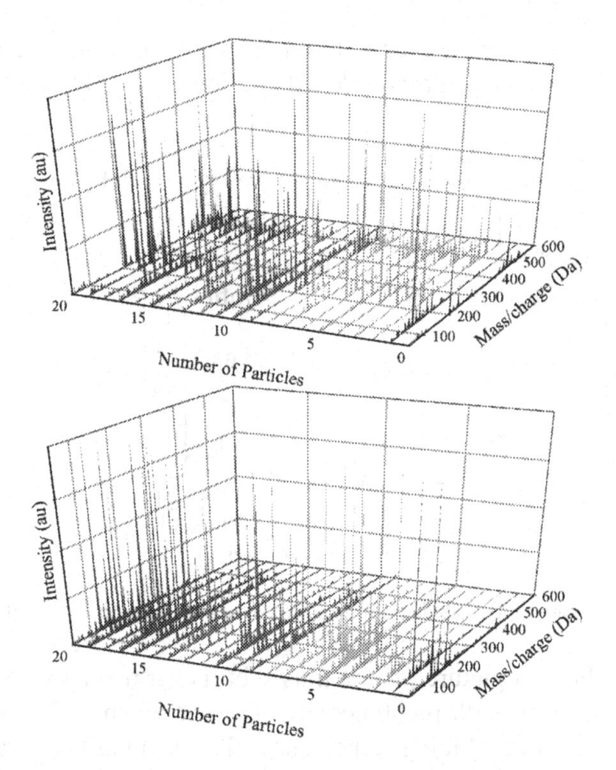

Figure 3. MS spectra of lead chloride and tryptophan originating from a lead chloride tryptophan aerosol mixture without pre-selection (A), and 3(B) only tryptophan spectra with usage of pre-selection.

Figure 3 (B) shows the result only a moment later after switching to "2 colour" detection when the system reacts only to the fluorescing tryptophan particles. The result is that the system generates a laser pulse only for the particles composed on tryptophan, thereby disregarding all non-fluorescing lead chloride particles. This experiment demonstrates that the Aerosol Time-of-Flight Mass Spectrometer system is capable of selecting fluorescing particles from a mixed aerosol. This capability allows for mass spectral analysis of bioaerosols without triggering the ionisation laser on non-biological particles. As a result of this modification, the mass spectrometer is far more likely to reserve its limited capacity for the bacteria particles that are of primary interest in this application.

An example of the subsequent mass spectrometric analysis by the Matrix Assisted Aerosol Time-of-Flight Mass Spectrometer is given by the analysis of sporal Bacillus subtilis. Bacillus subtilis is often applied as a simulant of Bacillus anthracis. The aerodynamic properties of these two bacteria are almost identical. Spores of Bacillus subtilis var niger were aerosolized and prior to analysis on-line coated with either picolinic acid or ferulic acid as matrix materials by an evaporation/condensation flow cell (Figure 4).

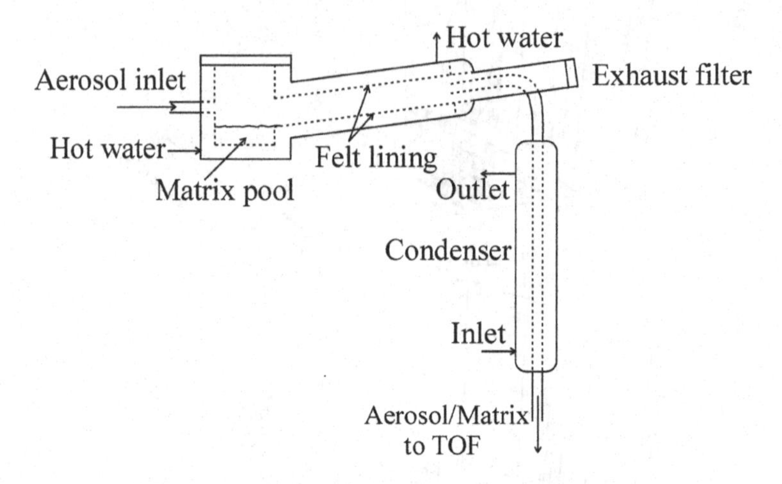

Figure 4. . Evaporation/Condensation cell to cover the aerosol particle with matrix.

Figure 5 shows a picture of Bacillus subtilis var niger spores before and after on-line coating with picolinic acid. The coated spore has increased in size and has a more shining appearance. The shining appearance of the coated spores is probably caused due to an increased conductivity of the sporal particle due to the condensed matrix material.

The spectrum (Figure 6) displays an earlier observed peak at 1,224 Da (Stowers et al., 2000),

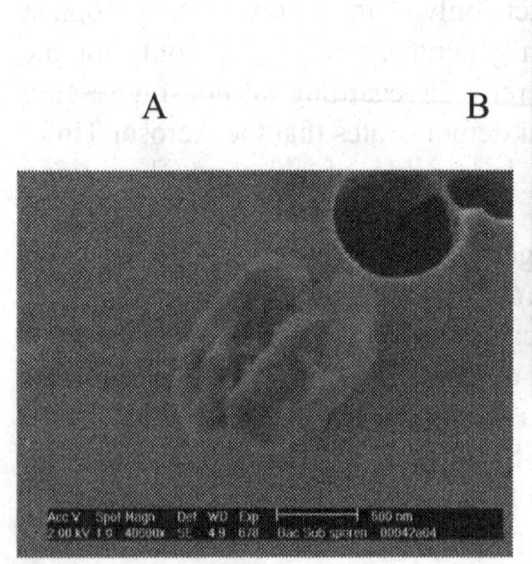

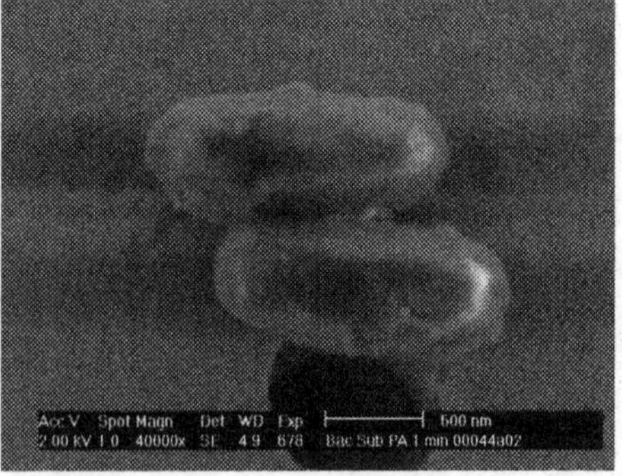

Figure 5. SEM-pictures of Bacillus subtilis var niger spores before (A) and after on-line coating (B) with picolinic acid.

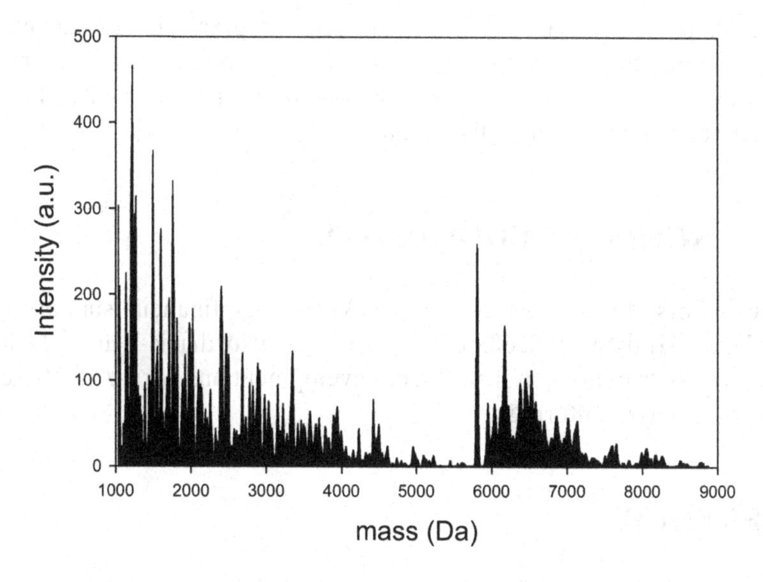

Figure 6. Aerosol mass spectrum of Bacillus subtilis var niger.

which is the mass of peptidoglycan, a compound of the cell wall. Peaks in the range up to 4,500 Da are seen, but no well resolved peaks are found in the range from 5,000 to 9,000 Da. Even though the quality of the signal cannot compete the signal of traditional MALDI mass spectrometry, identical information is obtained for analysis upon Bacillus subtilis var niger by Matrix Assisted Aerosol Time-of-Flight Mass Spectrometry.

The lack of distinct high mass signals may be due to the omission of an acid in the particle coating process as compared to liquid MALDI sample treatment. Addition of acid in the particle coating process is the objective of present further research, which might lead to spectra upon which identification of bacteria can based.

4. CONCLUSION

Fluorescence pre-selection was successfully implemented in the ATOFMS system, to distinguish biological aerosol particles from non-biological particles. Results from experiments with spores are promising, but more research is required to improve the quality of the mass spectra, to allow identification of bacteria.

The long-term goal of this research program is to develop a (trans)portable instrument, suitable for point detection of biological aerosols in the field. The transportable system will be built into a dedicated B-Fuchs reconnaissance vehicle for further evaluation.

5. ACKNOWLEDGEMENTS

The authors of this paper gratefully acknowledge financial support by the Netherlands Ministry of Defence. The authors also thank Bruker Daltonik GmbH, for contributions in instrument development and Mrs. A.I. Visser for help with bacterial culturing.

REFERENCES

Gieray R.A. et al.(1997). Real-Time Detection of Individual Airborne Bacteria, J. Microbio. Methods, 29, 191.

Jackson, S.N. et al. (2002). Matrix addition by condensation for Matrix-Assisted Laser Desorption/Ionization of collected aerosol particles, Anal. Chem., 74, 4841.

Kievit O. et al. (1996). The on-line chemical analysis of single particles using aerosol beams and time of flight mass spectroscopy, Chem. Eng. Comm. 151, 79-100.

Marijnissen J., Scarlett, B and Verheijen P. (1988). Proposed On-Line Aerosol Analysis Combining Size Determination, Laser-induced Fragmentation and Time-of-Flight Mass Spectroscopy. J. Aerosol Science, 19, 1307-1310.

McKeown, P. J. et al. (1991). On-Line Single-Particle Analysis by Laser Desorptin Mass Spectrometry. Anal. Chem., 63, 2069-2073.

McLoughlin M.P. et al. (1999). Development of a field-portable time-of-flight mass spectrometer system, Johns Hopkins APL Tech. Dig. 20, 326-334.

Prather et al. (1994). Real-Time Characterization of Individual Aerosol Particles using Time-of-Flight Mass Spectrometry. Anal. Chem., 66, 1403.

Russell S.C. et al.(2004). Toward understanding the ionization of biomarkers from micrometer particles by bio-aerosol mass spectrometry, J Am Soc Mass Spectrom,15, 900.

Sinha, M. P. (1984) Analysis of Individual Biological Particles by Mass Spectrometry. International Journal of Mass Spectrometry and Ion

Stowers M.A. et al. (2004). Fluorescence Preselection of Bioaerosol for Single Particle Mass Spectrometry, submitted to Appl. Optics.

Stowers M.A., Wuijckhuijse A.L. van, Marijnissen J.C.M., Scarlett B., Baar B.L.M. van, and Kientz Ch.E. (2000). Application of Matrix-Assisted Laser Desorption/Ionization to an On-line Aerosol Time of Flight Mass Spectrometer, Rapid Commun. Mass Spectrom. 14, 829-833.

Weiss M., Verheijen P, Marijnissen J.C.M. and Scarlett B. On the performance of an on-line time of flight mass spectrometer for aerosols. J. Aerosol Sci. 28 (1997), 159,171.

BIODETECTION USING MICRO-PHYSIOMETRY TOOLS BASED ON ELECTROKINETIC PHENOMENA

R. PETHIG
School of Informatics, University of Wales, Bangor, Gwynedd, LL57 1UT, UK, and Aura BioSystems, Inc, San Jose, CA 95131-1323, USA

Abstract: Cell Physiometry tools will be described, having applications in cell diagnostics and cell separations, that can be directly exploited in biodetection technologies. These tools employ dielectrophoresis (DEP), using microelectrode arrays energized by broad band frequency generators. Biochemical labels, beads, dyes or other markers and tags are not required. Extensive and validated studies have shown that different cell types, cells at various stages of maturation or proliferation, and cells exposed to toxic agents, exhibit characteristic DEP signatures associated with their distinctive morphologies and cellular structures. The microelectrode array, sample chamber and fluidics are located with on-board electronics on a microscope stage. The electronics is used to control voltage signals to the electrode arrays, and images of the DEP responses of the cells are captured using an image processing tool. Further details are provided elsewhere. The technology is inherently fast and inexpensive. Particles, such as cells and bacteria, are characterized or sorted as a flux, introducing a high degree of parallelism as opposed to the serial analysis of events using conventional cell sorting instruments.

1. INTRODUCTION

The 'Cell Physiometry' tools described here employ the electrokinetic techniques of dielectrophoresis and electrorotation. An understanding of the theory and applications of these two phenomena can be gained from

D. Morrison et al. (eds.), Defense against Bioterror: Detection Technologies, Implementation Strategies and Commercial Opportunities, 129–142.

references given [1-9] and the works cited therein. In this paper we provide an overview of the two techniques and propose how they may be applied as biodetectors in biothreat defense.

Dielectrophoresis manifests itself as a movement of a particle when it is electrically polarized in a non-uniform electric field. The effect differs from electrophoresis because the particle need not carry its own intrinsic electric charge, and alternating current electric fields rather than D.C. fields are employed. The non-uniform fields are usually generated by microelectrode arrays energized by broad band frequency generators. An example of a simple electrode geometry is the planar quadrupole arrangement shown in Fig.1, which can readily be constructed and miniaturized using modest photolithographic facilities.

In electrorotation measurements, a particle is subjected to a uniform rotating electric field that causes the particle to rotate. As shown in Fig.1, the simple quadrupole geometry allows us to generate non-uniform fields for dielectrophoresis and also rotating fields.

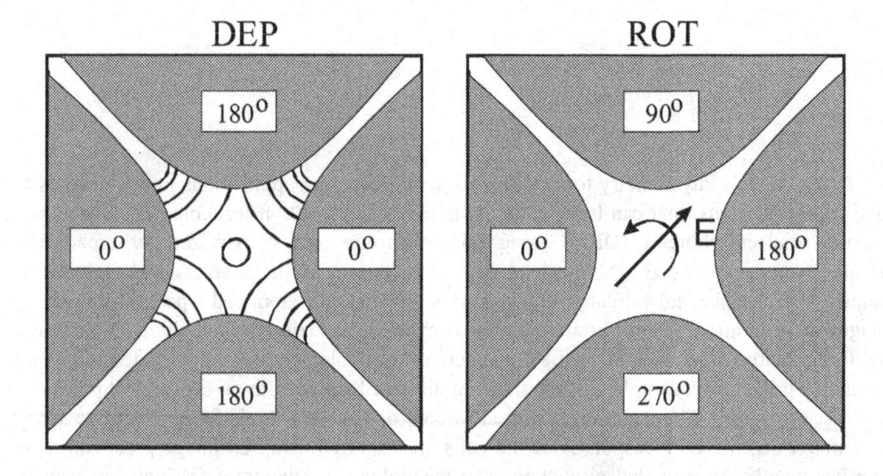

Figure 1 Depending on the phasing of sinusoidal voltage excitation, quadrupole planar electrodes can be used for either dielectrophoresis (DEP) or electrorotation (ROT) measurements. The DEP mode creates a highly non-uniform electric field (depicted schematically by the curved lines), whereas with the ROT mode a rotating uniform field is generated.

Extensive and validated studies have shown that different types of bioparticles (e.g., cells, bacteria, parasites) at various stages of maturation, proliferation or viability, exhibit characteristic dielectrophoretic and electrorotation signatures associated with their distinctive morphologies, structures and physiological state [1, 3-5, 8].

1.1 Dielectrophoresis (DEP) and Electrorotation (ROT) Applied to Bioparticles

A particle subjected to an electric field becomes electrically polarized, and this polarization can be considered to take the form of an induced dipole moment. The DEP force and ROT torque acting on a particle can then be considered to arise from the interaction of this induced dipole moment with the applied field phasor. For a spherical particle of radius r suspended in a medium of absolute permittivity ε_m a general expression for the time-averaged DEP force F acting on the particle is given by:

$$F = 2\pi\varepsilon_m r^3 \{\mathrm{Re}(m)\nabla E^2 + \mathrm{Im}(m)\Sigma E^2 \nabla\phi\} \quad (1)$$

∇E^2 is the gradient of the square of the applied field E, and $\Sigma E^2 \nabla\phi$ represents a summation involving the magnitude and phase ϕ of each field component in a Cartesian coordinate frame. The dependence on the *square* of the field emphasizes that the force is independent of the polarity of the applied field, so that either A.C. or D.C. fields can be used. The range of frequencies that can be employed is large, extending from around 100 Hz or less, up through radio wave frequencies and well into the microwave region (100 MHz and above). The polarizability factor m, given by the so-called Clausius-Mossotti factor, determines the magnitude of the induced dipole moment and is a function of the frequency of the applied field, and the conductivity and permittivity of the particle and its suspending medium. Re(m) and Im(m) relate to the real (in-phase) and imaginary (out-of-phase) components of this dipole moment, respectively. Re(m) is bounded by values of +1 and -0.5, and Im(m) falls within the bounds of +0.75 and -0.75.

For the usual DEP case, where the electrodes are energized with the two phases of 0° and 180°, a stationary field is generated with no spatially varying phase. In this case the factor $\Sigma E^2 \nabla\phi$ in equation (1) is zero. The resulting DEP force acts to either attract the particle to the electrodes (positive DEP) or to repel it from the electrode edges (negative DEP), depending on the sign of Re(m). Examples of this for yeast cells are shown in Figs.2. For the case $\Sigma E^2 \nabla\phi \neq 0$ we have traveling wave dielectrophoresis [8].

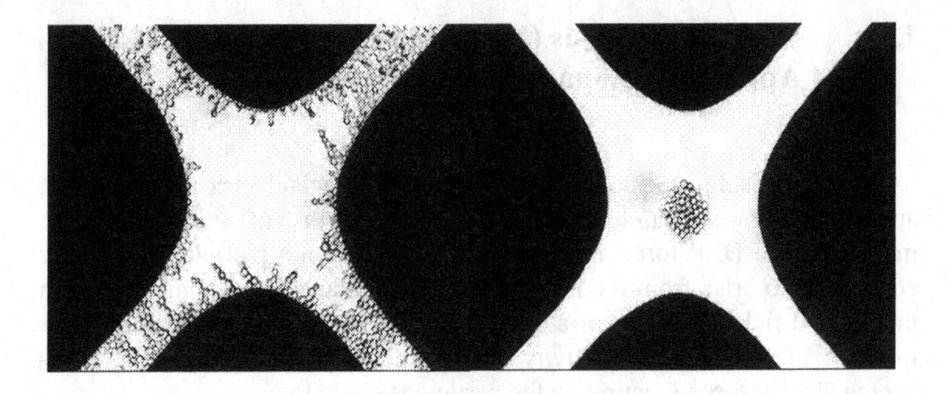

Figure 2. Left: Yeast cells collecting at electrode edges under the influence of positive dielectrophoresis. Right: Yeast cells repelled away from electrode edges, into a field cage, under the action of negative dielectrophoresis. (The only difference in experimental conditions between these two images is the frequency of excitation of the quadrupolar electrodes)

The ability to attract or repel particles from electrodes is an important aspect of DEP, and finds particular application in the selective separation of bioparticles. This can be illustrated using the DEP frequency response plots for the Gram positive bacteria *Micrococcus lysodeikticus* and Gram negative *Escherichia coli* shown in Fig.3.

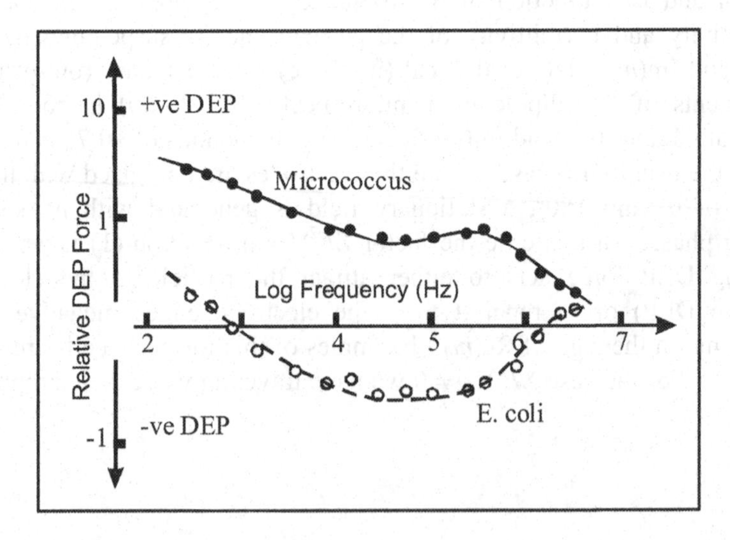

Figure 3. The DEP spectra of Micrococcus lysodeikticus and E. coli when suspended in a 280 mM mannitol solution of conductivity 55 mS/m.

The different DEP behaviour shown in Fig.3 arises because Gram positive and Gram negative bacteria have different cell wall and membrane structures. This leads to their ready discrimination by DEP as shown, for example, in Fig.4.

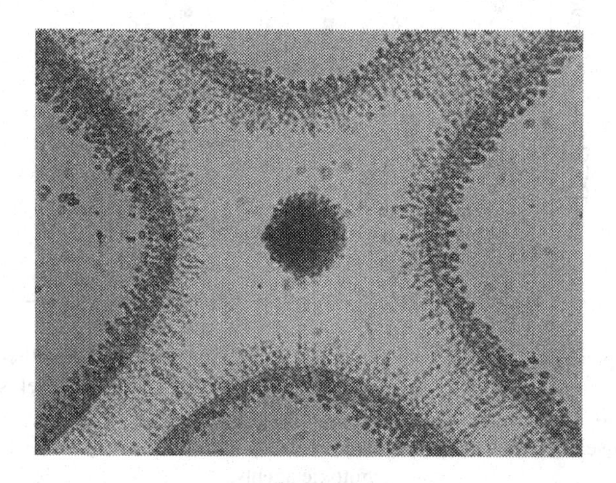

Figure 4. Separation of M. lysodeikticus (experiencing positive DEP) and E. coli (experiencing negative DEP) in a quadrupole electrode system after application of a 100 kHz signal. The suspending medium was 280 mM mannitol of conductivity 55 mS/m.

Gram-positive bacteria synthesize a uniform peptidoglycan-structured cell wall, whilst Gram-negative cells have a more complicated wall structure involving lipids and proteins. The cell walls of Gram-positive bacteria are also characterized by the incorporation of covalently bound teichuronic acids and proteins, forming open networks with high charge densities. These charged groups and the absence of an outer lipid membrane leads to Gram positive bacteria exhibiting higher polarizabilities than Gram negative ones. The DEP separation shown in Fig.4 was achieved by suspending a mixture of Gram positive *M. lysodeikticus* and Gram negative *E. coli* in a solution whose conductivity had been adjusted to a value roughly mid-way between the effective particle conductivities of these two bacteria types. Using this approach, it is also possible to separate different Gram positive (or Gram negative) species.

The DEP frequency spectrum typically observed for a mammalian cell (e.g., blood cell, cancer cell) is shown in Figure 5.

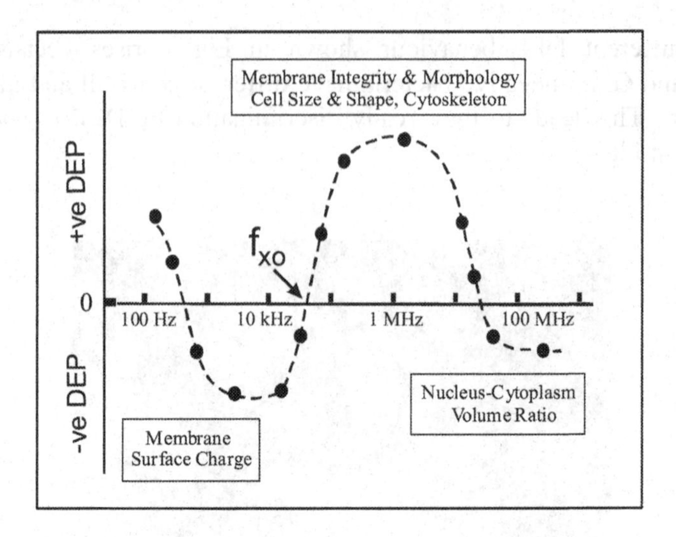

Figure 5. A typical DEP characteristic is shown for a mammalian cell, together with the main controlling physiometric parameters. Spectra vary according to cell type, cell state, and with the dielectric properties of the suspending medium. Measurement of the cross-over frequency fxo can be used to formulate protocols for cell separations and to monitor the effects of cytotoxic agents.

At low frequencies the electrical double-layer associated with intrinsic charges on the membrane dominate the polarizability of a cell, and positive DEP is usually observed. As the frequency is increased, a viable cell suspended in a weak electrolyte appears as an electrically insulating particle. The term *viable* implies the cell has an intact membrane that acts as a barrier to passive ion flow. The imposed electric field, and resulting ionic conduction flow, will therefore skirt around the cell membrane and seek more conductive paths in the surrounding electrolyte. Induced charges will appear at the cell surface to produce a dipole moment opposing the applied field, and so the cell will exhibit negative DEP. At higher frequencies the membrane resistance is electrically shorted-out by the membrane capacitance, and the cell appears more polarizable than the surrounding electrolyte. The field will penetrate the cell interior, the induced dipole moment will be oriented in the same sense as the field, and the cell will exhibit positive DEP.

A characteristic frequency, known as the DEP cross-over frequency f_{xo} is shown in Fig.5. This is determined as the frequency where there is a transition, at radio frequencies, between positive and negative DEP. Basically, at f_{xo}, we have 'tuned' to the situation where the dielectric properties of the cell exactly match those of the suspending medium. A very small change in either the cell state or its physico-chemical properties can result in a measurable change of f_{xo}. In particular, a significant change will occur if a chemical agent acts on the cell so as to change its membrane morphology (increasing the degree of membrane folding, microvilli or blebbing, for example) or degrades the membrane so that it no longer acts as an insulting barrier to passive ion flow.

When the electrodes are energized with quadrature phases (0°, 90°, 180° and 270°) shown in Fig.1, a rotating field is generated. A particle suspended between the electrodes will experience a rotational torque Γ given by:

$$\Gamma = -4\pi\varepsilon_m r^3 \,\mathrm{Im}(m) E^2 \tag{2}$$

The sense and rate of the induced particle rotation depend on the magnitude and sign of Im(*m*). Co-field or anti field rotation corresponds to negative or positive values of Im(*m*), respectively. ROT spectra are shown in Fig.6 for the intestinal parasite *Giardia intestinalis*, grouped according to their viability using a fluorogenic vital dye assay and morphological indicators.

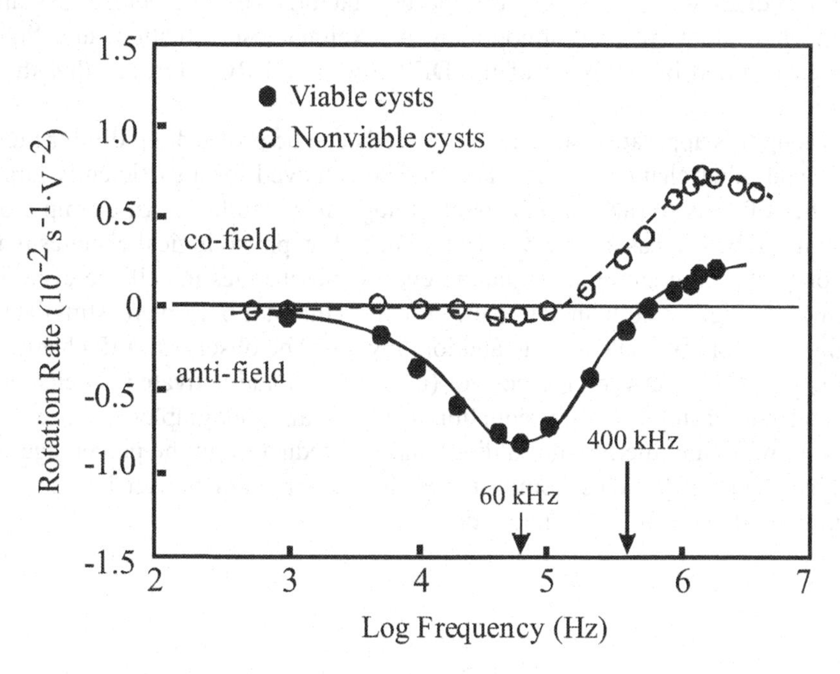

Figure 6. ROT spectra of viable and nonviable Giardia intestinalis cysts, assayed according to inclusion or exclusion of the vital dye propidium iodide and morphology (from Dalton et al [4]).

Inspection of Fig.6 shows that there is a frequency window, centered around 400 kHz, within which viable and nonviable cysts rotate in opposite directions. Also, a significant difference exists in the magnitudes of the anti-field rotation rates around 60 kHz. Automated image processing techniques are available to detect these characteristics.

2. INSTRUMENTATION AND VALIDATION STUDIES

We have developed a 'Cell Physiometry' instrument that can be used for both cell characterization as well as cell sorting. This versatility is achieved through specific programming of the electric fields imposed on the cells in the chamber, as well as by the design of the chamber and associated fluidics. The design of the microelectrodes used to generate the required non-uniform radio frequency fields is also an important factor. The microelectrodes are fabricated either by photolithography or excimer laser ablation, and their design is aided by analyses of the DEP forces and ROT torques that they generate.

Using this apparatus, suspensions of cells can be profiled rapidly for their DEP and other characteristics. This can be achieved for a sufficiently large number of cells to take into account biological variability. An example of how it can be used as a noninvasive tool to follow physiological changes that accompany transmembrane signaling events and changes in cell life cycle is shown in Fig.7. Human leukemic T cells (Jurkat E6-1) were stimulated using phorbol myristate acetate and ionomycin. The observed DEP changes, shown in Fig.7, between the control (unactivated) and activated T cells can be understood in terms of a reduction of membrane topography (e.g., extent of microvillii and membrane folding) and to a reduction in the percentage of S phase cells [5]. This demonstrates that the instrument can be used to analyze cell cycle population kinetics.

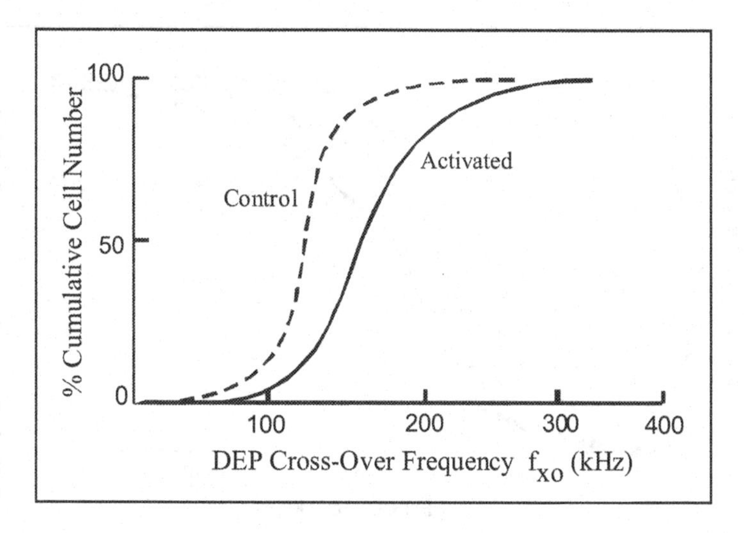

Figure 7. Cumulative DEP cross-over frequency (fxo) values for 1,160 control cells and 1,213 activated human leukemic T cells [5].

Another example of the sensitivity of the technique is its use as a rapid and low cost assay to detect and quantify apoptosis in cell populations. Apoptosis is a process where cells undergo gene-directed self-destruction. For example, the suicide of activated cycling T cells upon encountering autoantigens is likely to be one of the mechanisms that our immune systems use to prevent autoimmunity. The study of programmed cell death is a topic of intense research interest, and has led to new strategies for drug discovery and cancer therapy. Figs. 8 & 9 show the changes in the DEP cross-over frequency and cell radius for human T lymphocytes (Jurkat E6-1 cells) following treatment with the apoptosis-inducing chemical etoposide. Analysis of this data reveals the existence of several physiometric events. Within 2 hours of etoposide treatment the majority of cells become slightly larger, and a number of smaller cells (apoptotic cells) appear. Another subpopulation of cells of increased size also appears, with characteristics of dying (necrotic?) cells. The data also shows that the number of apoptotic cells, as validated by the Annexin V assay to detect externalized phosphatidylserine, increases with time. Also, close analysis of the DEP cross-over frequency values indicates that a reduction (smoothing) of the membrane topographical features of the apoptotic cells has also occurred. The observed shift with time of the DEP cross-over frequency to higher frequencies for the dying cells is consistent with their plasma membranes becoming increasingly rough and also physically degraded (i.e., loss of ability to act as a barrier to passive ion transport).

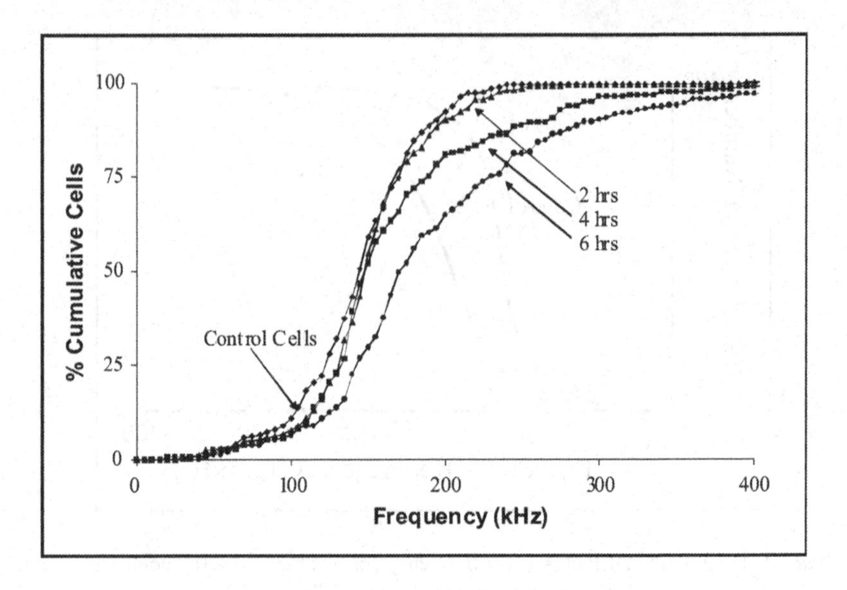

Figure 8. Change of the DEP cross-over frequency (fxo shown in Fig.5) for T lymphocytes with time after inducement of apoptosis [10].

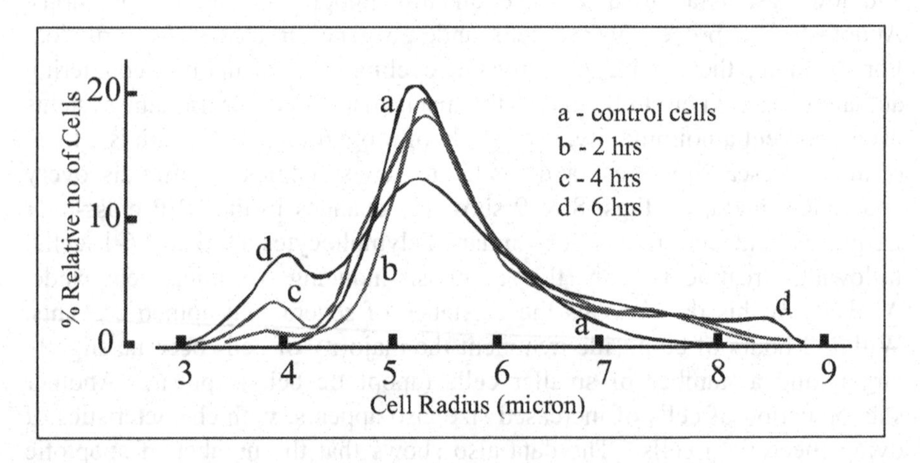

Figure 9. Change of cell radius of a population of T lymphocytes with time after inducement of apoptosis.

The results shown in Figs. 7-9 demonstrate that subtle changes in the physiological state of cells, induced by chemical agents, can be detected by monitoring the DEP behaviour of the cells.

3. POTENTIAL BIOTHREAT APPLICATIONS

Two possible approaches exist for using this technology as devices to aid biothreat defence:

Use DEP to isolate or concentrate target toxic species as a sample preparation procedure to increase the sensitivity of PCR analysis (for example), or

Monitor the DEP or ROT behavior of selected micro-organisms to detect the presence of toxins.

Cell Separation and/or Concentration

The ability of DEP to selectively separate cells and bacteria is described elsewhere [1, 3, 8]. The basis for efficient separation is a distinct difference of the DEP cross-over frequency of the target bioparticle from that of other (often more numerous) particles in a mixed suspension. The data shown in Fig. 10 demonstrate that the separation of bacteria from blood can be achieved using DEP.

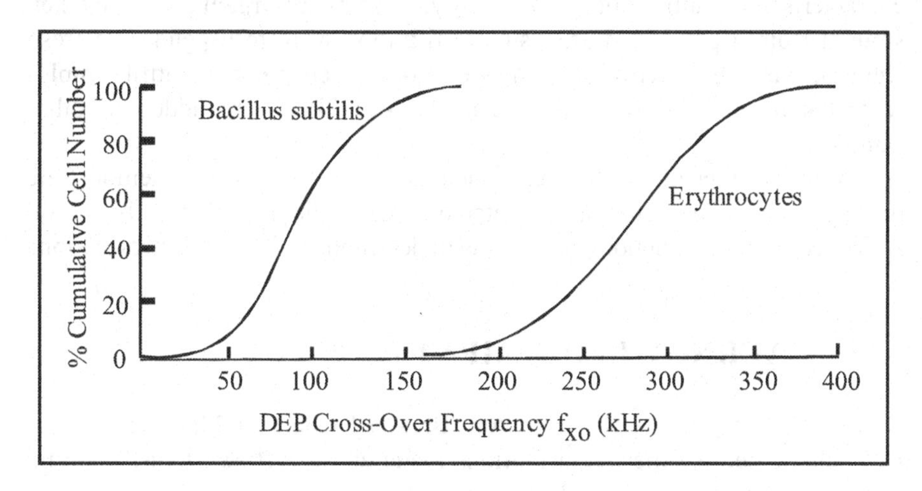

Figure 10. Cumulative DEP cross-over frequency (fxo) values for Bacillus subtilis and red blood cells (medium conductivity 40 mS/m)

Work could usefully be performed to develop DEP-based protocols to detect ,isolate and concentrate species such as anthrax from biological fluids (or dust), for example.

3.1 Biodetection using Microorganisms

Neurotoxins act specifically on nerve cells, typically by interacting with ion channels and membrane proteins. For example, some classes of neurotoxin block or prevent the closing of sodium, potassium or calcium channels, whilst others target the operation of acetylcholine or glutamate receptors.

It is possible to isolate neurons and to subject them to study by DEP and ROT, but other approaches might be less difficult and just as effective in terms of developing biodetection assays for toxins. For example, the torque generated by bacterial flagellar motors has been studied using DEP and ROT measurements, making use of equations 1 & 2 [11 - 13]. An example of how this approach can be extended is shown in Figs.11 and 12.

Figure 11 depicts the experimental procedure that was used to study how the motility of a common algae was affected by exposure to the well known free radical compound DPPH (diphenyl-picryl hydrazyl) This chemical causes oxidative damage to membranes and is commonly used in assays to characterise new anti-oxidants. Basically, these micro-organisms were taken straight from a pond, and after simple filtration were resuspended in fresh water between the electrodes of an electrorotation chamber. Control samples were taken, and different concentrations of DPPH were added to other samples.

Preliminary results of this experiment are shown below, and demonstrate one approach to developing biodetectors for biothreat defence, based on A.C. electrokinetic phenomena such as dielectrophoresis and electrorotation:

4. ACKNOWLEDGEMENTS

I thank Cathy Carswell-Crumpton, Brenda Kusler, Dr Mark Talary and Dr Richard Lee for their contributions made at Aura BioSystems, Inc, and also Drs Julian Burt and Andy Goater of the University of Wales, Bangor, for valuable discussions.

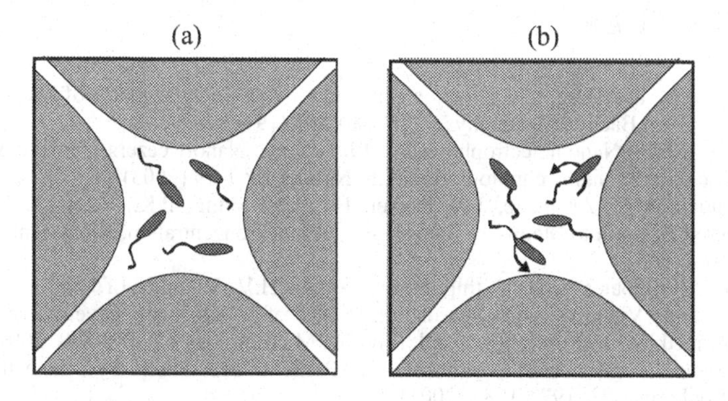

Figure 11. Schematic outline of an experiment to monitor the action of a toxic agent (DPPH) in reducing the operation of flagellar motors of Euglena gracilis suspended between ROT electrodes. The time taken between (a) when the micro-organisms were first exposed to a rotating field, and (b) when 50% of the organisms were no longer able to swim against the field, was noted for each DPPH concentration.

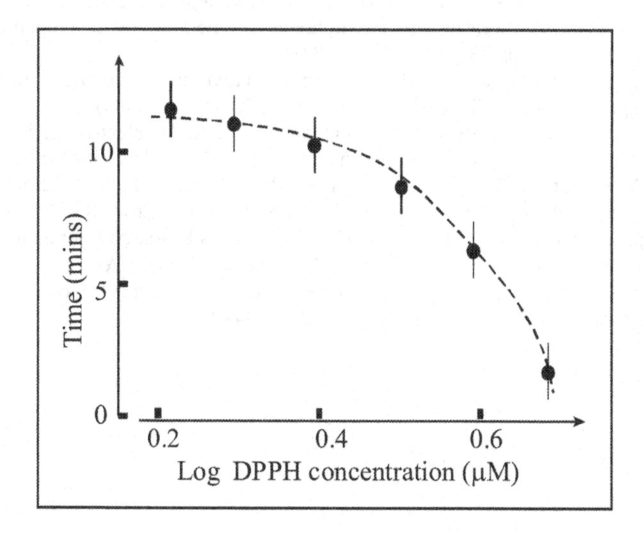

Figure 12. Plot of the time interval between exposure of Euglena gracilis to DPPH and the observation that 50% of the algae succumb to the rotational torque generated by a rotating field.

REFERENCES

1. Berry, R.M. and Berg, H.C., Torque generated by the flagellar motor of Escherichia coli while Driven Backward, Biophys. J. 76: 580-587 (1999)
2. Burke, P.J. Nanodielectrophoresis: Electronic Nanotweezers, Encyclopedia of Nanoscience & Nanotechnology (Ed. H.S. Nalwa) 10: 1-19 (2003)
3. Chernova, A.A., Armitage, J.P., Packer, H.L. and Maini, P.K., Response Kinetics of Tethered Bacteria to Stepwise Changes in Nutrient Concentration, Biosystems 71: 51-59 (2003)
4. Dalton, C., Goater, A.D., Pethig, R. and Smith, H.V. Viability of Giardia interstinalis Cysts and Viability and Sporulation State of Cyclospora cayetanensis Oocysts determined by Electrorotation, Appl. Environ. Microbiology 67: 586-590 (2001)
5. Gascoyne, P.R.C. and Vykoukal, J. Particle separation by dielectrophoresis, Electrophoresis 23: 1973-1983 (2001)
6. Jones, T.B. Basic Theory of Dielectrophoresis and Electrorotation, IEEE Med. Biol. Magazine 22: 33-42 (2003)
7. Müller, T., Pfennig, A., Klein, P., Gradl, G., Jäger, M. and Schnelle, T., The Potential of Dielectrophoresis for Single-Cell Expriments, IEEE Med. Biol. Magazine 22: 51-61 (2003)
8. Pethig, R. Dielectrophoresis: Using Inhomogeneous AC Electrical Fields to Separate and Manipulate Cells, Critical Reviews in Biotechnology 16: 331-348 (1996)
9. Pethig, R., Bressler, V., Carswell-Crumpton, C., Chen, Y., Foster-Haje, L., Garcia-Ojeda, M.E., Lee, R.S., Lock, G.M.,Talary, M.S. and Tate, K.M. Dielectrophoretic studies of the activation of human T lymphocytes using a newly developed cell profiling system, Electrophoresis 23: 2057 - 2063 (2002)
10. Pethig, R., Talary M.S. and Lee, R.S. Enhancing Traveling-Wave Dielectrophoresis with Signal Superposition, IEEE Med. Biol. Magazine 22: 43-50 (2003)
11. Ramos, A., Morgan, H., Green, N.G. and Castellanos, A. AC electrokinetics: A review of forces in microelectrode structures, J. Phys. D: Appl. Phys. 31: 2338-2353 (1998)
12. Talary, M.S., Carswell-Crumpton, C., Lee, R.S., Kusler, B., and Pethig, R., Micro-Physiometry Tools for Cell Analysis and Cell Separations, Proc. μTAS 2003 (Eds. M A Northrup, K F Jensen & D J Harrison) The Transducers Research Foundation (2003)
13. Washizu, M., Kurahashi, Y., Iochi, H., Kurosawa, O., Aizawa, S., Kudo, S., Magariyama, Y. and Hotani, H., Dielectrophoretic Measurement of Bacterial Motor Characteristics, IEEE Trans. Ind. Appl. 29: 286-294 (1993)

ELECTRO-OPTICAL TECHNIQUE FOR DETECTION AND IDENTIFICATION OF BIOLOGICAL AGENTS

VICTOR BUNIN
State Research Center for Applied Microbiology, 142253 Ibikebsj Nisci Reg., Russia

Abstract: A novel, portable Electro-Optical (EO) Biosensor (CELAN) for the detection of extremely low concentrations of microorganisms in real-time (<25 minutes) with immediate on-the-spot interpretation of the results has been developed. The EO biosensor consists of three major subsystems: (1) a fluid-handling subsystem, (2) electro-optical device and (3) computer hardware and software with a graphical user menu. The method is based on measurement of suspended cells polarizability before and after their interaction with selective labels. Monoclonal antibodies, phages, and gold particles have been used as markers. Live vaccine strains of *Bacillus anthracis*, strain STI, and *Francisella tularensis* were used as test pathogen agents. Mathematical data processing, includes accumulation of the data, their filtration and comparison of two functions and Frequency Dependence of Anisotropy Polarisability (FDPA) before and after biospecific cell-antibody interactions have been optimized

1. INTRODUCTION

Current scientific technologies offer a wide range of methods for selective detection and identification of microorganisms. However, most of them require complicated chemical, biochemical and immunological manipulations or are time-consuming. Physical methods including optical techniques are more convenient, but still far from the sensitivity and selectivity required. Our contribution to this subject is directed towards developing an electro-optical approach, which relies on the measurement of AC electrokinetic effects. The theoretical and experimental aspects of AC electrokinetic effects have been developed in several laboratories. The AC

D. Morrison et al. (eds.), Defense against Bioterror: Detection Technologies, Implementation Strategies and Commercial Opportunities, 143–146.

electrokinetic effect depends on dielectric properties of bioparticles and their suspending media, and on the frequency of applied electrical field. When a bioparticle (bacterium or virus) is exposed to an external electric field, it becomes electrically polarized. As a result of this the action of an applied electric field induces electrical charges to appear at the boundary between the particle and the surrounding medium. These field-induced charges provide a large electric dipole moment of the bioparticle, which arises from the induced charges that accumulate at the interface of the bioparticle. In general, permanent dipoles are randomly oriented, but if an external electric field is applied, they will be reorienting statistically. Induced dipoles will have the direction of the applied field. If a direct current (D.C.) electric field of uniform intensity is replaced by one of alternating current (A.C.), then because of the particle's inertia the electrophoretic effect becomes small for frequencies above around 1 kHz. However, dipole moments associated with Maxwell-Wagner interfacial polarisations can exert their influence up to frequencies of 50 MHz and beyond. The cell polarization and temporal orientation of dipoles, with respect to the applied field, depends on the dielectric properties of the suspending medium and the bioparticles.

The electro-optical technique includes a number of interrelated processes [1-5]: 1) action of electric field on the suspended bioparticles; 2) generation of induced charges on the boundaries of cell structures and cell surface; 3) creation of the driving torque; 4) cell transition into oriented state, which leads to the anisotropy of optical properties of a cell suspension. Such effects induce the fluctuations of light scattering and relevant variations in optical density. The changes of optical density can be detected by optical techniques, thus allowing the rapid and accurate detection of target bacteria in aqueous solutions. Such effects are birefringence, dichroism, induced changes of light scattering and relevant variations in optical density [6-8].

2. SELECTIVE ANALYSIS OF CELLS BY ELECTRO-OPTICAL TECHNIQUE

The method is based on measuring electrophysical parameters of cells, in particular Frequency Dependence of Anisotropy Polarisability (FDAP), when exposed to a selective factor [9]:

- antibodies
- selective nutrient medium
- lysogenic selective factor

The focus of this investigation was directed on electro-optic analysis of interaction between microorganism and monoclonal (mAb) or polyclonal (pAb) antibodies. Due to polyvalent proteins generated on the cell surface, the double layer shrinks resulting in variations in frequency dispersion of anisotropy of polarizability at frequencies to hundreds of kilohertz. In electro-optical approach, the interaction between antigen (Ag) and mAb

resulted in variations of the signal between 50 kHz and 1000 kHz. Based on theory of light scattering in suspensions, changes of cell orientation reflect in the integral signal, when the size of an object exceeds the length of the light wave. This method is useful for handling cells of more than 0.5 μm in size. The upper limit of cell sizes is 15-20 μm and may be explained by low speed of re-orientation of cells due to the factor affecting their sedimentation. Determination of a threshold of sensitivity of this method for cells monoculture is based on technical parameters of the photometric system, as well as of parameters of light scattering for model of abnormal diffraction of light.

Practical investigation was focused on :

- electro-optic analysis of the process of interaction between cells and antibody:
- determination of ratios of concentrations of cells and label;
- influence of different physical and chemical factors on the process;
- intensification of the electro-optical signal (response) by using "sandwich" type structures.

3. DESIGN AND DEVELOPMENT OF ELECTRO-OPTICAL ANALYZER AND ITS SOFTWARE

The Electro-Optical Analyzer (ELUS EO) developed at the State Research Center for Applied Microbiology, Obolensk, Russia was used as a basic instrument for electro-optical measurements. The Analyzer consists of the following modules: unit for sample preparation, mixer, AC field generator, EO-flow cell, micro-controller of liquid streams transactions, thermal system, operator interface, and image processor. The sample preparation unit includes a hydraulic system, initial ingredients store containers, reaction vessel, filter module, and a peristaltic pump for sampling. Vacuum is used as a driving force. The Analyzer can be used for determination of electro-physical and morphometric parameters of the microorganisms. It operates at multiple software programmable frequencies from 0.4 kHz to 20 MHz. The software program is used for collection, calculation and data processing.

The orientational spectra (OS) of the cells were measured with an EO ELUS analyzer at a wavelength of 670 nm. An OS is given as the frequency dependence of the difference (□*D*) between the suspension-optical-density values (D_a and D_b) measured during the propagation of a beam of non-polarized light along (D_a) and across (D_b) the orienting-field direction. This difference was normalized to the optical density value (*D)* measured for cells at random orientation.

4. CONCLUSION

An Electro-optical technique for the analysis of biospecific interaction between cells and mAb was developed. Variations of different parameters that influence the results of the interaction, was investigated. Optimal composition of its parameters was selected. The analyzer CELAN with automatic system of long-time measuring and software for control and data analysis was tested with different kinds of microorganisms.

REFERENCES

1. Bottcher, G.P.F. Theory of electrical polarizability; Acad. Press.: New York., 1982; V.1, P. 480.
2. Bunin, V.D., Voloshin, A.G. Determination of cell structures, electrophysical parametersand cell population heterogeneity, J. of colloid and interphase Sci. 1996, V. 180, P. 122-126.
3. Kerker, M. The scattering of light; Acad. Press.: London, 1969; P. 347.
4. Landau, L.D., Liphshitz, E.M. Electrodynamics of Continious medium; Gostechizdat : Moscow, 1957; P. 43.
5. Miroshnikov, A.I., Fomchenkov, V.M. Electrophysical analysis and separation of cells; Nauka : Moscow, 1986; P. 184.
6. Shchyogolev, Yu., Khlebtsov, N G., Bunin, V.D., Sirota, A.I., Bogatyrev, V.A. Inverse Problems of spectroturbidimetry of biological disperse systems with random and ordered particle orientation. Proc. SPIE. 1994, V. 2082, P.167-176.
7. Stoylov, S. P. Colloid electro-optics: theory and applications; Academic Press: London, 1991; P. 156.
8. Styopin, A.A. Calculation of Polarizability Parameters. J of Exp. & Theor. phys. 1972, V 8, P. 1230-1238.
9. Van de Hulst, H.C. Light Scattering by Small Particles; Wiley : New York, 1957; P. 321

DETECTION OF MICROBIAL CELLS WITH ELECTROOPTICAL ANALYSIS

O. V. IGNATOV[b]*, O. I. GULIY[b], V. D. BUNIN[a], A. G. VOLOSHIN[a], D. O'NEIL[c,d], and D. IVNITSKI[d]
a - Institute of Applied Microbiology, Obolensk, Moscow Region, Russian Federation;
b - Institute of Biochemistry & Physiology of Plants & Microorganisms,
Russian Academy of Sciences, Saratov, Russian Federation;
c - CRADA International, Inc., Norcross, USA;
d – New Mexico Institute of Mining and Technology, Albuquerque, NM 87106, USA

Abstract: An electro-optical approach has been used for studies of microbial cells and some biological agents (antibodies and phages) binding. The electro-optical analyzer (ELUS EO), which has been developed at the State Research Center for Applied Microbiology, Obolensk, Russia, was used as the basic instrument for electro-optical measurements. Since the AC electrokinetic effects depend on dielectric properties of bioparticles, their composition, morphology, phenotype, the medium, and the frequency of applied electrical field, the electro-orientational spectra were used for discrimination of bacteria before and after selective binding with antibodies. It has been shown in the model systems that the biospecific interactions of *Listeria monocytogenes* cells as target cells with anti-*Listeria* antibody in the presence of *E. coli* K-12, and *A. brasilense* Sp7 change the electro-optical signals significantly. Thus, the determination of the presence of particular bacteria within a mixed sample may be achieved by selection and matching of antibodies specific to individual bacterium types and by comparing spectra of bacterium in the presence and in the absence of specific antibodies. The same principles were used for the investigations of bacteria – phage interactions. Biospecific binding was used between *Escherichia coli* XL-1 cells and the phage M13K07. The phage-cell interaction includes the following stages: phage adsorption on the cell surface, entry of viral DNA into the bacterial cell, amplification of phage within infected host and phage ejection from the cell. In this work, we used M13K07, a filamentous phage of the family *Inoviridae*. Preliminary study had shown

D. Morrison et al. (eds.), Defense against Bioterror: Detection Technologies, Implementation Strategies and Commercial Opportunities, 147–163.

that combination of the EO-approach with a phage as a recognition element has an excellent potential for mediator-less detection of phage-bacteria complex formation. The interaction of *E. coli* with phage M13K07 induces a strong and specific electro-optical signal as a result of substantial changes of the EO properties of the *E. coli* XL-1 suspension infected by the phage M13K07. Integration of the electro-optical approach with a phage has the following advantages: 1) bacteria from biological samples need not be purified, 2) the infection of phage to bacteria is specific, 3) exogenous substrates and mediators are not required for detection, and 4) it is suitable for any phage-bacterium system when bacteria-specific phages are available.

1. INTRODUCTION

Detection and identification of pathogenic bacteria in the environment present multiple challenges [1, 2]. Firstly, pathogen agents are effective in extremely low doses and, therefore, biodetection systems need to exhibit high sensitivity. Secondly, the complex and rapidly changing environmental background and the fact that many pathogenic organisms differ little from normal flora requires detection systems to exhibit a high degree of selectivity. Detection systems must be able to discriminate biological agents from other harmless biological and non-biological material present in the environment. Thirdly, another challenge that needs to be addressed is speed. Unlike chemical agents, many living biological agents can reproduce, multiply inside the host, and are passed from person to person. Therefore, the rapid detection and identification of biological agents are crucial.

The electro-optical technique includes a number of interrelated processes [3]: 1) action of electric field on the suspended bioparticles; 2) generation of induced charges on the boundaries of cell structures and cell surface; 3) creation of the driving torque; and, 4) cell transition into oriented state, which leads to the anisotropy of optical properties of a cell suspension. Such effect induced the fluctuations of light scattering and relevant variations in optical density. The changes of optical density can be detected by our optical technique, thus allowing the rapid and accurate detection of target bacteria in aqueous solutions.

The current focus of our research is to apply the electro-optical analyzer for direct detection and discrimination of bacteria based on biospecific binding. More specifically, the purpose is to demonstrate the potential of using of the electro-optical technique in combination with bacteriophage amplification for direct monitoring of *E. coli*. Bacteriophage are virus particles that generally attach to and infect a narrow range of host cells [4-7]. Infection of bacteria by a bacteriophage starts by recognition of the host through binding to an outer membrane receptor. Like human viruses, they inject their genetic material into the bacterial cell, replicate by the hundreds per cell, and then burst out before moving on to the next host cell. In the case of tailed phage, this binding triggers conformational changes that are transmitted along the tail to the capsid, allowing its opening and the release of the viral genome, which causes a change in the dielectric properties of cells. Advantages of the phage technology include its simplicity, ease-of-use, low cost, safety and immunogenicity (for vaccine studies). The bacteriophage specificity has been demonstrated at both the species and strain level in the literature. Specific bacteriophages are very good indicators for determining the species and type of bacteria. That is why they have found a wide application in medical practice for the fast identification of viable bacteria [5, 7, and 8].

The purpose of the paper is to demonstrate the potential of using the electro-orientation technique for monitoring *Listeria monocytogenes*-antibody reaction and monitoring reaction between *Escherichia coli* XL-1 cells and bacteriophage M13K07.

2. EXPERIMENTAL

2.1 Microorganisms

The *Listeria monocytogenes* vaccine strain was obtained from Institute of Applied Microbiology, Obolensk, Moscow region, Russian Federation. *Escherichia coli* XL-1, *E. coli* K-12, and *Azospirillum brasilense* Sp7 were obtained from the culture collection at the Institute of Biochemistry and Physiology of Plants and Microorganisms RAS, Saratov.

2.2 Culture Conditions

E. coli XL-1, *E. coli* K-12, and *A. brasilense* Sp7 were grown in liquid medium containing (g l-1): NaCl, 10; yeast extract (FLUKA, Switzerland), 5; peptone (FLUKA, Switzerland), 5. The cultures were shaken (160 rpm) aerobically for 24 h at a constant temperature of 30°C. The cells grown were used for EO studies.

2.3 Cell Preparation

Cells to be used in the analysis were washed three times by centrifugation at 2800 x *g* for 5 min and re-suspended in distilled water to an optical density OD_{665} of 0.4–0.5. To remove cellular aggregates, we re-centrifuged the cell suspension at 110 x *g*, 1 min, and further work was carried out using the suspension that remained in the supernatant liquid. Cell suspensions were prepared so that the OD_{665} for each type of cell suspension was within the range of 0.4–0.42.

2.4 Measurement of Cellular Orientational Spectra

The electro-optical analyzer (ELUS EO) (Fig. 1) has been developed at the State Research Center for Applied Microbiology, Obolensk, Russia, and was used as the basic instrument for electro-optical measurements. The orientation spectra (OS) of the cells were measured at a wavelength of 665 nm, as described earlier [3]. The measurements were performed with a discrete set of frequencies of the orienting electric field (10, 100, 250, 500, 1000, and 2000 kHz).

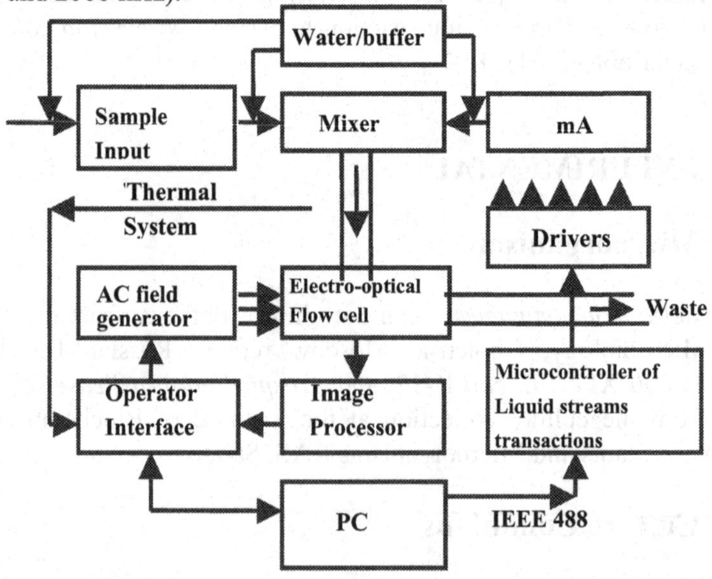

Figure 1 Schematic layout of the electro-optical analyzer.

2.5 Listeria monocytogenes - Antibody Binding

The monoclonal antibodies to *Listeria monocytogenes* were received from Hytest (Finland).

A stock solution containing $8.2*10^8$ cells/ml *L. monocytogenes* microorganisms was employed (conductivity of the dispersion medium, 1.6-

2.0 μS/m). The concentrations of organisms were checked by the standard technique of light microscopy. The cell suspension was separately complexed with anti-*Listeria monocytogenes* antibodies. For the assay, 10^8 cells/ml were incubated with different concentration of antibodies in 20.0 mM Tris-HCl buffer (pH 8.2) at 37°C for 30 min and then cells were prepared for EO measurements as described below.

2.6 Monitoring of the Interaction of *E. coli* with the Phage M13K07

The principle of analysis of the interaction of *E. coli* with the phage M13K07 is based on registration of changes of optical parameters of bacteria using an image processing technique. The assay includes measurement of the EO-spectrum of *E. coli* in the absence and in the presence of a phage. The phage-cell interaction includes the following stages: phage adsorption on the cell surface, entry of viral DNA into the bacterial cell, amplification of phage within infected host and phage ejection from the cell. In this work, we used M13K07, a filamentous phage of the family *Inoviridae*. M13 K07, a phage confers resistance commercial preparation manufactured by Stratagene (Sweden), was constructed on the basic of the wide-type phage M13 [9]. The phage has specificity toward *E. coli* XL-1. For transfection, *E. coli* XL-1 (a separate colony) was transferred from a plate containing agar-supplemented LB medium with 12.5 μg/ml tetracycline to a plate containing 2 ml of LB medium. The culture was incubated overnight with constant aeration at 35^{0}C; then, 1/10th of the overnight culture was transferred to a fresh medium of the same composition and was grown to exponential phase with aeration at 37^{0}C. When cells reached early log phase (OD_{660}=0.5-0.6, corresponding to 7.7×10^8 cells/ml), the aeration was stopped for 30–40 min in order that the cells restore their F-pili, and the suspension was incubated in a thermostat at 37^{0}C. The concentrations of organisms present were checked by standard techniques with the help of light microscopy. Twenty phages per bacterium were used for infection. Upon direct addition of the phages, the culture was incubated at 37^{0}C in a thermostat without shaking, in order for phage particles to sorb at the surface of the pili. After that the cells were prepared for EO measurements as described earlier. As a result of microorganism-phage complex formation, the optical properties of cell suspension, particularly light scattering, are changed and lead to changes in optical density.

3. RESULTS AND DISCUSSION

For identification of bacteria with the help of electro-optical techniques, the electro-optical approach should be integrated with immunoassay or phage technologies. Figure 2 demonstrates the variation in amplitudes of electro-optical signals for *L. monocytogenes* cells as a result of formation of *Listeria monocytogenes*-monoclonal antibodies biospecific complexes. The electro-orientation properties, e.g. the rate of *L. monocytogenes* cell orientation observed at a given field orientation frequency is altered sufficiently. During biospecific interaction an antibody is bound to the microorganism causing a change in the dielectric properties of the microorganism-antibody complex and the electrooptic signal reaches its maximum at 100-200 kHz. The experimental data of EO signals at different frequencies have indicated that EO spectral analysis should be conducted preferably at low frequencies (10-500 kHz) and the time of the measurement should be between 20-25 minutes.

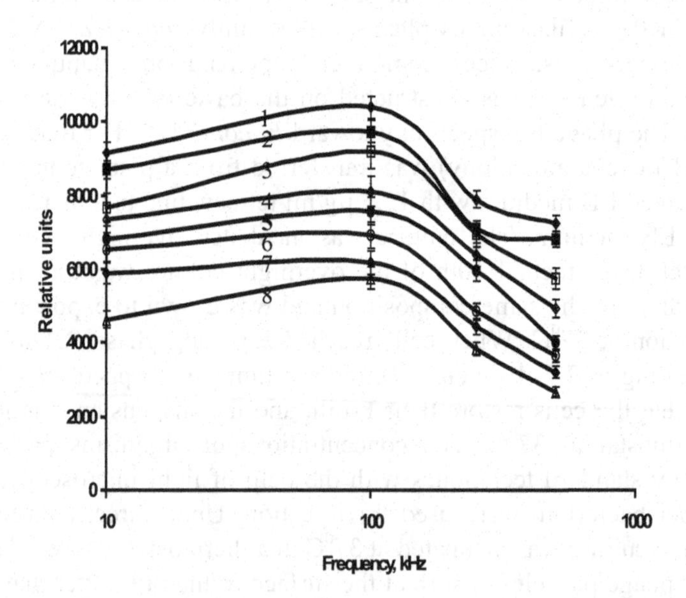

Figure 2 The electroorientation spectra of L. monocytogenes cells suspended in distilled water of conductivity 1.8 μS/m, obtained after incubation with different concentration of monoclonal anti-L. monocytogenes antibodies. (1) - control- without mAb; (2) - 0.475 μg/ml mAb; (3) - 1.425 μg/ml mAb; (4) - 1.9 μg/ml mAb; (5) - 2.3 μg/ml mAb; (6) - 4.75 μg/ml mAb; (7) - 7.1 μg/ml mAb; (8) - 9.5 μg/ml mAb.

One of the possible potential unfavourable cases of pathogen monitoring in the environment might be the case when targeted and debris cells present in suspension are of equal in sizes. Therefore, we investigated the possibility of registration of changes of EO-spectra of the *L. monocytogenes* in a mixed culture suspension containing *E. coli* K-12, and *A. brasilense* Sp7 as a result of specific interaction with monoclonal anti-*Listeria monocytogenes* antibodies. We found that the changes in the electro-optical signal *L. monocytogenes*-anti-*Listeria* antibody in the mixture of *E. coli* K-12, and *A. brasilense* Sp7 cells were significant (Figure 3). At the same time, control experiments in the absence of *L. monocytogenes* cells have shown that there are no non-specific changes in electro-optical properties in cell suspensions of *E. coli* K-12 and *A. brasilense* Sp7 in the presence of anti-*Listeria* antibodies. Thus, the determination of the presence of particular bacteria within a mixed sample is achievable by usage of antibodies specific to individual bacterium types and by comparing spectra of a bacterium in the presence and in the absence of an antibody. The results indicate that electro-orientational spectra can be used for discrimination of bacteria before and after interaction with selective antibodies.

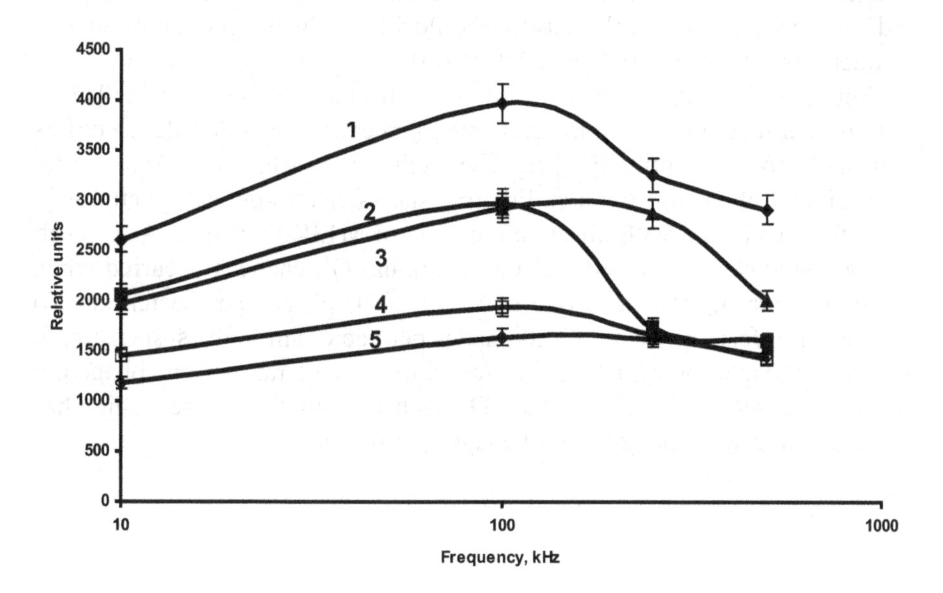

Figure 3. Electro-optical properties of mixed cells (A. brasilense, E. coli K-12, Listeria monocytogenes) suspended in distilled water with conductivity 1.8 μS/m, obtained after incubation with different concentration of monoclonal anti-L.monocytogenes antibodies. (1) - control without mAb; (2) - 0.475 μg/ml mAb; (3)- 1.425 μg/ml mAb; (4) - 1.9 μg/ml mAb; (5)- 2.3 μg/ml mAb.

The current focus of our research group is to apply the electro-optical analyzer for direct detection and discrimination of bacteria based on biospecific binding. More specifically, the purpose is to demonstrate the potential of using of the electro-optical technique in combination with bacteriophage amplification for direct monitoring of *E. coli.*

As is known, the infection of *E. coli* male cells by the bacteriophages M13, fd, or fl begins with the interaction of the minor phage capsid protein g3p (gene 3 protein) with bacterial F-pili (when is the primary phage receptor) and subsequently with the integral membrane protein TolA [10]. Phage action on a bacterial cell may follow different paths: a lytic reaction, lysis from the outside, and a lysogenic reaction. Whatever the path, due to entry of viral DNA into *E. coli* cell cytoplasm, with the capsid protein (g8p) integrated into the inner cytoplasmic membrane, the phage infection leads to considerable changes of the electrophysical parameters of the cells including changes of their dielectric properties and various kinds of damage of the intracellular structures [11-18,19]. Therefore, discrimination of microorganisms might be done based on the measurement of the electro-optical spectra of the microorganisms in the absence and in the presence of phage. The experimental data of EO signals at different frequencies have indicated that EO spectral analysis should be conducted preferably at low frequencies (250, 500, 1000, and 2000 kHz).

For controlling phage transfection to the bacteria, we grew the cells in a LB nutrient medium containing kanamycin, because phage M13K07 confers resistance to this antibiotic [9]. The cells grew well with kanamycin, indicative of phage transfection. Figure 4 shows electro-optical spectra of *E. coli* after incubation with different amounts of M13K07 phages. As a result of these studies, we showed that the maximum OS changes occurred when the cells were infected at a rate of twenty (20) phages per bacterium. To record cell infection, we added 20 phages per bacterium in the suspension in subsequent experiments. In fact, we found that even a ratio of one phage per bacteria leads to the changes of the EO-spectrum. But, the best resolution has been observed with the ratio of 20 phages per bacterium.

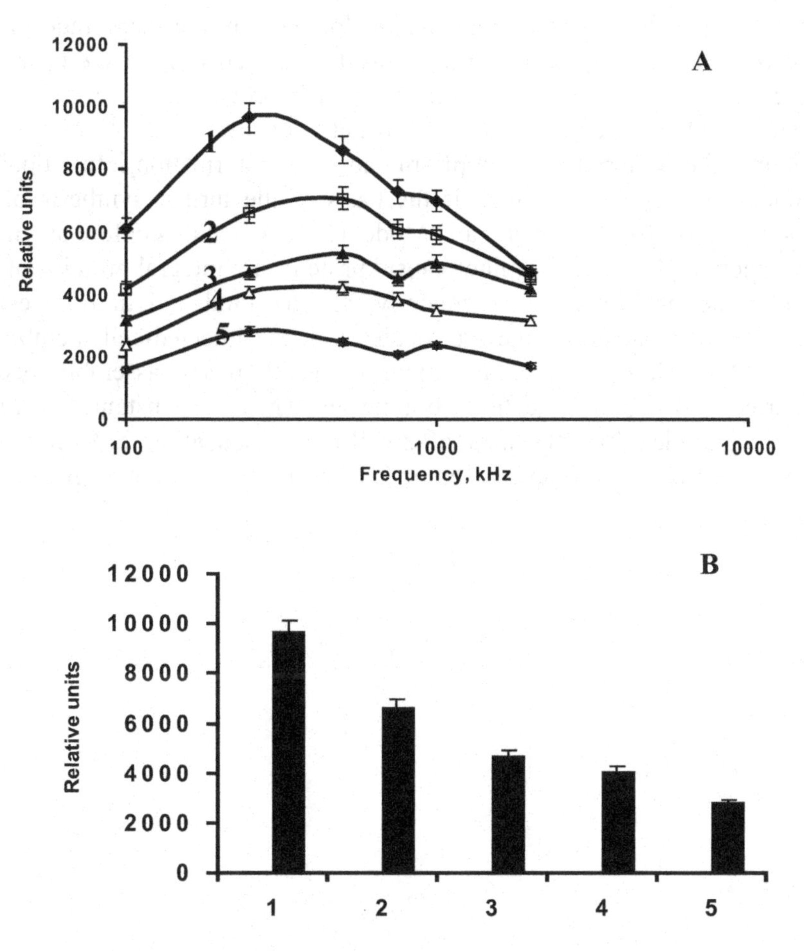

Figure 4. (A) The electroorientation spectra and (B) relative units at 250 kHz frequency of E.coli XL-1 cells suspended in distilled water (conductivity 1.8 μS/m), obtained after incubation with a range of M13K07 phages. (1) - control- without phages; (2) – 1 phage per bacterium; (3) - 5 phages per bacterium; (4) - 10 phages per bacterium; (5) - 20 phage per bacterium.

The next parameter that was studied was the effect of the duration of infection of *E. coli* XL-1 cell suspension by phage M13K07 on the electro-optical signal. The data are presented in Figure 5. We found that considerable changes in the magnitude of the EO signal are detected in the first 10 min after the phage injection. The reason might possibly be bacteria-phage complex formation. These results are in harmony with the data in the literature, because the translocation of phage DNA occurs with the participation of the bacterial protein TolQ, TolR, and TolA after the pili supposedly are retracted, thereby transferring the phage to the bacterial surface [20]. After cell-phage incubation for 60 min, the magnitude of the cell-suspension EO signal increased considerably. This may have been due to the entry of DNA into the cell and to the processes occurring in the cytoplasm. One major process is the penetration of single-stranded phage DNA into the bacterial-cell cytoplasm and its transformation into a double-stranded, replicate form, which is the basis of the further synthesis of all phage proteins [20]. Endemann and Model [10] showed that before assembly of the bacteriophage, all its minor capsid proteins are integral proteins of the *E. coli* inner membrane. Consequently, all structural and morphogenetic proteins of the bacteriophage are localized in the infected-cell membrane, compatible with the model according to which phage assembly occurs concurrently with phage ejection from the cell. Thus, the substantial increase in the magnitude of the EO signal after cell-phage incubation for 90 min may be explained by the possible assembly of the phages and their egress from the cell.

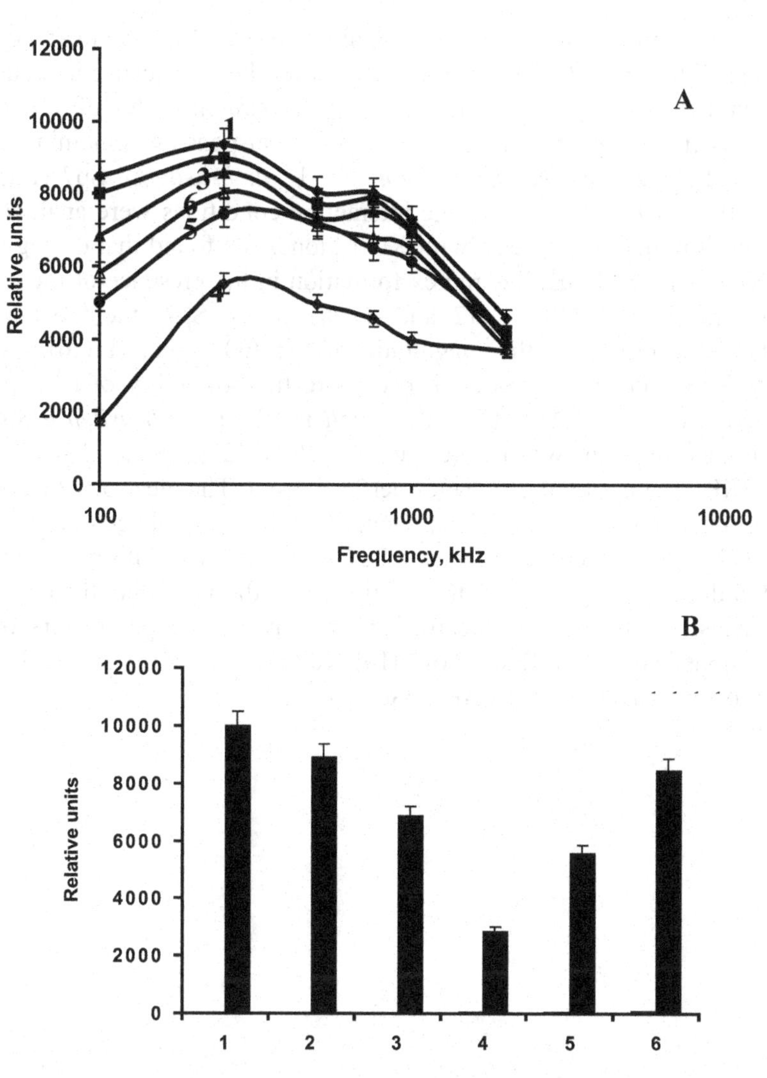

Figure 5. (A) The electroorientation spectra and (B) relative units at 250 kHz frequency of E.coli XL-1 cells suspended in distilled water (conductivity, 1.8 μS/m), obtained after incubation with 20 M13K07 phages per bacterium: (1) - control- without phages; (2) – after 1 min; (3) - after 10 min; (4) - after 30 min; (5) - after 60 min; (6) – after 90 min.

Since many pathogenic organisms differ little from normal flora, it important to test the specificity of the electro-optical system in the presence of interfering factors, first and foremost in the presence of foreign microflora. Therefore, in our next experiments the electro-optical measurements were done in the absence and in the presence of foreign microflora that would not be infected by phage M13K07 (Fig. 6). As control, we used cells of *E. coli* K-12 and *Azospirillum brasilense* Sp7. *E. coli* K-

12 was chosen because it is a parent strain of *E. coli* XL-1 and can possibly be infected by M13K07. *A. brasilense* Sp7 was chosen because it occupies a different taxonomic position and has a cell size similar to that of *E. coli* XL-1. To this end, the phage was added to a mixed suspension containing *E. coli* XL-1 and K-12 (Fig.6A), and *E. coli* XL-1 *A. brasilense* Sp7 (Fig. 6B) (OD_{665} 0.42–0.44). The cell-phage interaction conditions were analogous to those used in the experiments with XL-1 alone. We found that during the *E. coli* XL-1-phage M13K07 complex formation in the presence of the foreign microflora (both *E. coli* K-12 and *A. brasilense* Sp7) there occurred a substantial decrease in the magnitude of the EO signal (Fig.6). Control experiments were run in parallel to explore the possibility of non-specific interaction of phage M13K07 with *E. coli* K-12 and *A. brasilense* Sp7. To test this assumption, we infected an *E. coli* K-12 suspension with phage MM13KK07 at a rate of 20 phages per bacterium. The electro-optical signal did not change on addition of the phage; therefore, the phage did not infect K-12 (Fig. 7A). When *A. brasilense* Sp7 was incubated with the phage, the cell OS did not change either; that is, the phage did not infect the cells (Fig. 7B). Thus, we found that specific changes in the EO parameters of cell suspensions under the influence of M13KK07 occur only in *E. coli* XL-1 and do not occur in K-12 and *A. brasilense* Sp7.

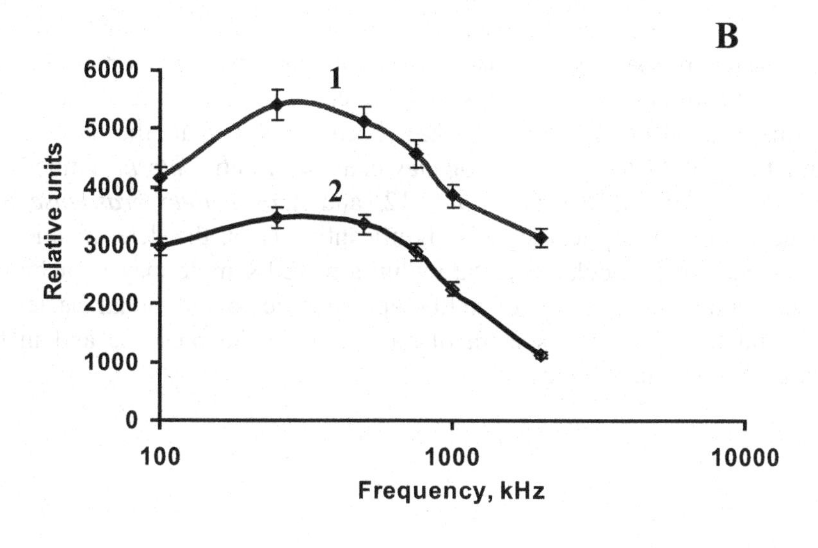

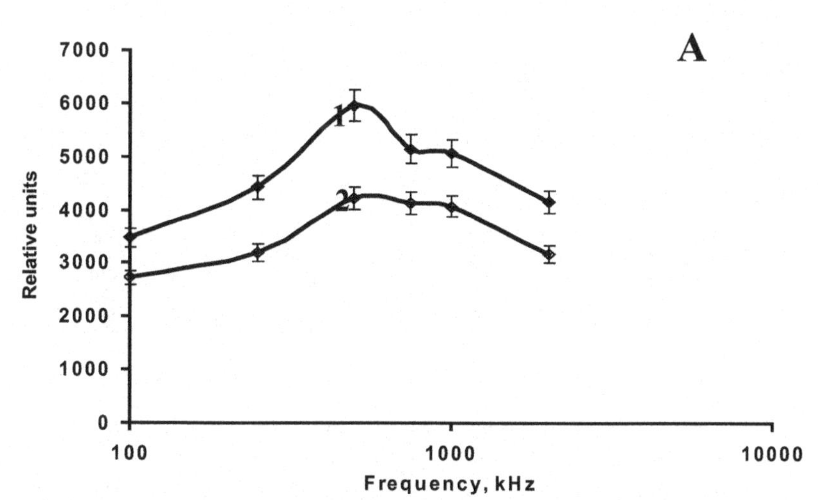

Figure 6. The electroorientation spectra of mixed cells (E.coli XL-1, E.coli K-12) (A) and (E.coli XL-1, A. brasilense Sp7) (B) suspended in distilled water (conductivity, 1.8 mS/m), obtained after incubation with 20 M13K07 phages. (1) - control- without phages; (2) – 20 phages per bacterium.

4. CONCLUSIONS

Electro-orientational spectral analysis of cell suspensions may be used for discrimination of different types of bacteria with the help of selective binding agents (antibodies, phages).

During biospecific interactions an antibody is bound to the microorganism causing a change in the dielectric properties of the microorganism-antibody complex and the electro-optical signal reaches its maximum at 100-200 kHz. It has been shown that the biospecific interactions of *Listeria monocytogenes* cells with anti-*Listeria* antibody in the presence of *Escherichia coli* K-12, and *Azospirillum brasilense* Sp7 change the electro-optical signals significantly. Thus, the determination of the presence of particular bacteria within a mixed sample may be achieved by selection and matching of antibodies specific to individual bacterium types and by comparing spectra of bacterium in the presence and in the absence of specific antibody.

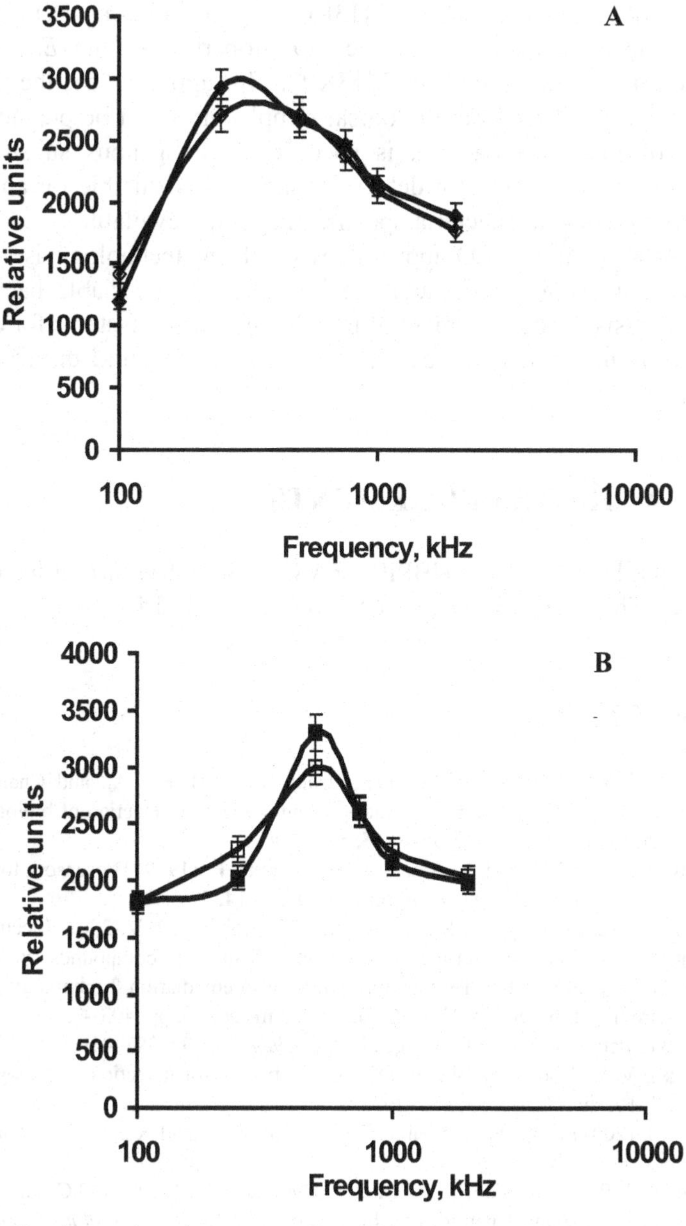

Figure 7. The electroorientation spectra of E.coli K-12 (A), and A.brasilense Sp7 (B) cells suspended in distilled water (conductivity, 1.8 mS/m), obtained after incubation with 20 M13K07 phages. (1) - control- without phages; (2) – 20 phages per bacterium.

This preliminary study has shown that the combination of the EO-approach with phage as a recognition element has an excellent potential for mediator-less detection of phage-bacteria complex formation. The interaction of *E. coli* with phage M13K07 induces a strong EO-signal as a result of substantial changes of the EO properties of the *E. coli* XL-1 suspension infected by the phage M13K07. This approach has the following advantages: 1) bacteria from biological samples need not be purified, 2) the infection of phage to bacteria is specific, 3) exogenous substrates and mediators are not required for detection, and 4) it is suitable for any phage-bacterium system when bacteria-specific phages are available.

Combination of the EO-approach with phage technology is a generic technique that enables rapid and specific detection of viable bacteria and might be a basis for development of portable biosensor systems for detection of pathogens in medical research, food processing and environmental analysis.

5. ACKNOWLEDGEMENTS

We thank D.N. Tychinin (IBPPM RAS, Saratov, Russia) for his technical assistance. This work was supported by ISTC grant 615.

REFERENCES

1. Iqbal, S.S., Mayo, M.W., Bruno, J.G., Bronk, B.V., Batt, C.A., and Chambers, J.P. (2000) A review of molecular recognition technologies for detection of biological threat agents. *Biosensors & Bioelectronics*. **15,** 549–578.
2. Ivnitski, D., Abdel-Hamid, I., Atanasov, P., Wilkins, E. (1999) Biosensors for detection of pathogenic bacteri. *Biosensors & Bioelectronics*. **14,** 599–624.
3. Ignatov, O.V., Shchyogolev, S .Yu., Bunin, V.D., Ignatov, V.V. 2002. Electro-physical properties microbial cells during aerobic metabolism toxic compounds. In V.P. Singh and R.D. Stapleton (eds.), Biotransformations: Bioremediation Technology for Health and Environment Protection, Vol. 36. Elsevier, Amsterdam, pp. 403–425.
4. Smith, G., Petrenko, V. (1997) Phage display. *Chem Rev.* **97**, 391–410.
5. Petrenko, V.A., Vodyanoy, V.J. (2003) Phage display for detection of biological threat agents. *J. Microbiol. Met.* **53**, 253–262.
6. Benhar, I. (2001) Biotechnological applications of phage and cell display. *Biotechnology Advances*. **19**, 1–33.
7. Goldman, E.R., Pazirandeh, M.P., Mauro, J.M., King, K.D., Frey, J.C, Anderson G.P. (2000) Phage-displayed peptides as biosensor reagents. *J. Molecular Recognition* **13,** 382–387.
8. Chatterjee, S., Mitra, M., Gupta, S. (2000) A high yielding mutant of mycobacteriophage L1 and its application as a diagnostic tool. *FEMS Microbiology Letters* **188**, 47–53.
9. Vieira, J., and Messing, J. (1987) *in* Production of single-stranded plasmid DNA. Methods in Enzymology **153,** pp. 3-11.

10. Endemann H., and Model, P. (1995) Location of filamentous phage minor coat proteins in phage and in infected cells. *J Mol Biol.* **250,** 496 – 506.
11. Bunin, V.D., and Voloshin, A.G. (1996) Determination of cell structures, electrophysical parameters, and cell population heterogeneity. *J. Colloid Interface Sci.* **180**, 122–126.
12. Bunin, V.D., Voloshin, A.G., Bunin, Z.F., Shmelev, V.A. (1996) Electrophysical monitoring of culture process of recombinant *Escherichia coli* strains. *Biotechnol.Bioenginer.* **51**, 720–724.
13. Ignatov, O.V., Shchyogolev, S .Yu., Bunin, V.D., Ignatov, V.V. (2001) Electro-physical properties microbial cells during aerobic metabolism toxic compounds *in* Biotransformations: Bioremediation Technology for Health and Environment Protection. **36**. pp. 403-425, Edited by: Ved Pal Singh & Raymond D. Stapleton, Elsevier Science B.V. The Netherlands.
14. Ignatov, O.V., Guliy, O.I., Shchyogolev, S.Yu., Bunin, V.D., Ignatov, V.V. (2002) Effect of *p*-nitrophenol metabolites on microbial-cell electro-optical characteristics. *FEMS Microbiol. Lett.* **214**, 81–86.
15. Guliy, O.I., Ignatov, O.V., Shchyogolev, S.Yu., Bunin, V.D., Ignatov, V.V. (2002) Quantitative determination of organophosphorus aromatic nitro insecticides by using electric-field cell orientation in microbial suspensions. *Analytica Chimica Acta.* **462**, 165–177.
16. Miroshnikov, A., Fomchenkov, V.M., and Ivanov, A.Yu. (1986) *in* Electrophysical Analysis and Cell Separation, pp. 40–65. Nauka Publishers, Moscow [in Russian].
17. Gimsa, J., Wachner, D. (1998) A unified resistor-capacitor model for impedance, dielectrophoresis, elecrorotation, and induced transmembrane potential. *J. Biophys.* **75,** 1107–1116.
18. Gascoyne, P., Pethig, R., Satayavivad, J., Becker, FF., Ruchirawat, M. (1997) Dielectrophoretic detection of changes in erythrocyte membranes following malarial infection. *Biochim.Biophys.Acta* **1323**, 240–252.
19. Deng, L.W., Malik, P., Perham, R.N. (1999) Interaction of the globular domains of pIII protein of filamentous bacteriophage fd with the F-pilus of *Escherichia coli*. *Virology* **253**, 271.
20. Click, E.M., and Webster, R.E. (1997) Filamentous phage infection: required interactions with the TolA protein. *J. Bacteriol.* **179**, 6464–6471.

RECENT ADVANCES IN ELECTROCHEMICAL AND PHOTOCHEMICAL TRANSDUCTION STRATEGIES FOR IMMUNOSENSORS BASED ON ELECTROPOLYMERIZED FILMS

S. Cosnier
Laboratorie d'Electrochimie Organique et de Photoimie Redox, UMR CNRS 56030 Institut de Chimie Moleculaire de Grenoble FR. CNRS 2607, Universite Joseph Fourier Grenoble, France

Abstract: Electropolymerized films have received considerable attention in the development of biosensors and biochips and are rapidly advancing. This paper reviews the recent advances and the scientific progress in the electrochemical immobilization procedures of biological macromolecules on electrodes and optical fibers via electrogenerated polymer films bearing affinity sites or photoactivable groups. The biomolecule immobilization was based on the auto-formation of avidin bridges between biotinylated films and biotinylated biomolecules. Some advances in the electrochemical transduction of immunoreactions involving, the permeability, the conductivity or the photochemical and redox properties of electropolymerized films are also presented.

1. INTRODUCTION

For three decades, there is a growing interest in the design of biosensors aimed at detection, diagnosis and determination in the fields of food and water quality control, health, safety and environmental monitoring. Owing their adaptability, simple use in relatively complex samples and the possibility to fabricate fast portable analytical devices, biosensors based on an electrochemical transduction constitute the main category; indeed, several amperometric biosensors are being utilized commercially. Deposition of biological macromolecule has been achieved in many different ways such as

D. Morrison et al. (eds.), Defense against Bioterror: Detection Technologies, Implementation Strategies and Commercial Opportunities, 165–173.

physical adsorption, cross-linking, covalent binding and entrapment in gels or membranes. Nevertheless, the stable and reproducible immobilization of biological macromolecules on an electrode surface with complete retention of their biological activity is a crucial problem for the commercial development of biosensors. In particular, the exponential development of biochips and miniaturized biosensors implies the emergence of non-manual methods which provide the reproducible deposition of biomolecules with controlled spatial resolution. Besides photopatterning, screen-printing and spreading methods, the immobilization of biomolecules in or on electrogenerated polymer films is one of the few methods which allows the deposition of biomolecules with controlled spatial resolution [1-4]. In addition, electropolymerization that is compatible with bulk manufacturing procedures, provides polymeric coatings with an electrochemically controlled thickness. This was exploited to modulate the amount of biological molecules immobilized on the transducer surface. The quality of the latter (absence of defects and chemical stability) constitutes an attractive advantage for the regularity at the molecular level of additional functionalization of the electrode via the polymer film. In contrast to self-assembled monolayers (SAM), the electrochemical formation of conducting and non-conducting polymers can easily be performed over a wide range of electrode materials. Moreover, the electrogenerated polymers are stable in organic solvents and hence constitute an attractive host material for transferring enzymatic activities in nonaqueous media. This biomolecule immobilization is classified as covalent linkage, physical entrapment and attachment by affinity interactions. The latter approach is mainly based on the use of the well-known avidin-biotin linkage. The extremely specific and high affinity interactions between the glycoprotein avidin and four biotins, a vitamin (association constant $K_a = 10^{15}\ M^{-1}$), lead to strong associations similar to the formation of covalent binding [5]. These interactions have been extensively used for binding biological species to surfaces in various fields such as immunohistochemistry, enzyme-linked immunoassay (ELISA) and DNA hybridization. The combination of electrochemical addressing with the high-affinity interaction of avidin-biotin provides an affinity-driven immobilization protocol of biomolecules that fully retains their biological activity [6-10]. More interestingly, binding of a protein monolayer displaying excellent accessibility to each immobilized protein, has been recently achieved by using avidin-biotin bridges between electropolymerized biotinylated polypyrrole films and biotinylated bacteria and antibodies [11,12]. In particular, these polymers were exploited for the biomolecule immobilization on electrodes and optical fibers and for the electrochemical or photochemical transduction of antigen-antibody reaction via enzyme conjugated secondary antibody. Another new interesting research direction in reagentless biomolecule deposition lies in the photografting of proteins onto electrogenerated photoreactive polymers. Such an innovative approach combines the advantages of photolithography with those of the electrochemical addressing of polymers films.

The new approaches and concepts for biomolecule immobilization on electrodes and optical fibers based on electrogenerated polymers functionalized by affininity or photoactivable groups will be reviewed.

2. RESULTS AND DISCUSSION

The use of the remarkable strong avidin-biotin interactions was applied to the successful elaboration of assemblies containing multilayers of enzymes via the initial formation of bridges between biotinylated polypyrrole films and biotinylated enzymes. Since avidin molecule can bind biotin with a stoichiometry of four biotin residues per protein, its association with polymerized biotin leaves three unused sites on the avidin. These unused sites have been used to attach biotinylated biomolecules (Fig.1).

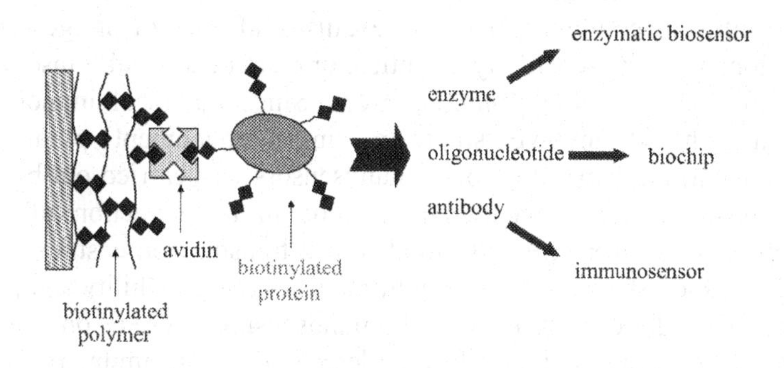

Figure 1. Schematic description of the immobilization procedure of biotinylated proteins on an electrode surface

Although this step-by-step approach has been extensively used for biosensor applications in aqueous solution, this immobilization method has never been tested in organic media. We recently reported the first use of these bioaffinity interactions in conjunction with a biotinylated polypyrrole film for the transfer of polyphenol oxidase (PPO) activity in chloroform [13]. Owing to the high accessibility to the immobilized enzymes, the multilayered PPO assemblies can be protected from the organic solvent by the anchoring of a biotinylated sodium alginate derivative. This approach facilitates, at the molecular level, the intimate association between the immobilized enzymes and the hydrophilic polysaccharide gel. This gentle environment mimics in organic solvent a pseudo-aqueous phase thus inducing a marked improvement of the biosensor performance [14]. With the aim to enhance the biosensor stereoselectivity towards the initial selectivity of the biomolecules, the design of a biotinylated polymer possessing permselectivity towards the diffusion of optically active compounds was performed by electropolymerization of a chiral biotin derivative substituted by two carbazole groups [15]. Furthermore, the efficient coupling of avidin and biotinylated polyphenol oxidase with the underlying biotinylated polycarbazole film has allowed the formation of an enzyme electrode. The detection of *L*- and *D*-norepinephrines leads to an «enantiomeric excess» of the biosensor response better than that recorded for the enzyme, namely 65 instead of 35 %.

Another potential application of this technique is bacterial immobilization, with retention of the biological activity. This immobilization was demonstrated by fluorescence microscopy and QCM measurements. The latter indicated that bacteria were specifically grafted onto the polymer surface [12]. The immobilized bacteria were successfully wired by freely diffusing reduced methyl viologen and applied to the denitrification of contaminating nitrate anions in ground water.

The intimate combination of the exquisite affinity of antigen-antibody interactions with the sensitivity of optical or electrochemical transducers has led to the emergence of immunosensors. Since the ideal immobilisation procedure should involve solely a single attachment point of the immunoagent, the fabrication of immunosensors was carried out by affinity interactions. Among conventional methods in the detection of antigen-antibody binding events, the electrochemical transduction ensures attractive advantages such as its ease of use in turbid samples, portability and low cost. Besides the direct reagentless immunosensing based on frequency impedance measurements [11], the detection of an immunoreaction is commonly performed by amperometric transduction, following the classical enzyme-linked immunosorbent assay strategy. The latter approach consists in the use of antibodies labeled with enzymes catalyzing the production of electroactive species which are amperometrically monitored at the electrode surface (fig. 2).

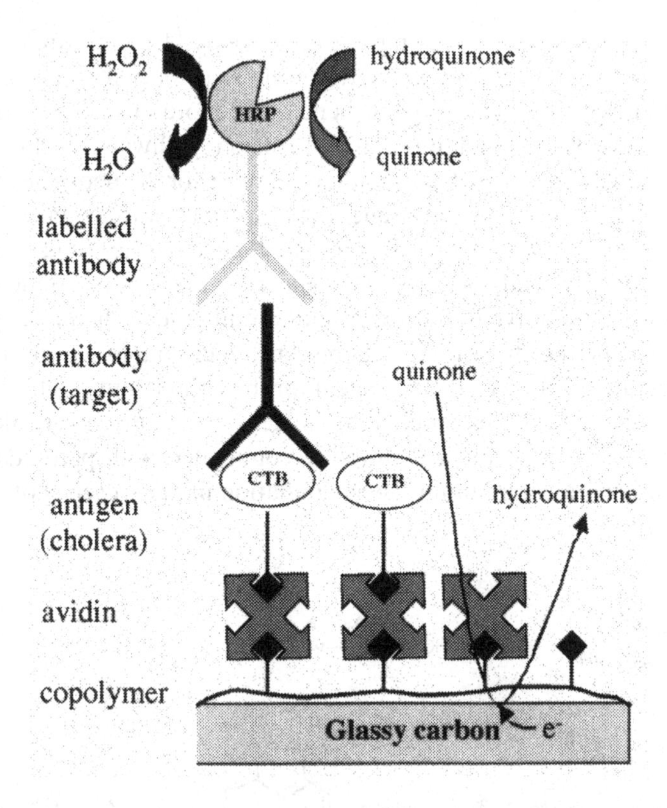

Figure 2 . Schematic representation of the functioning principle of an amperometric immunosensor based on antibody conjugated with peroxidase.

As a consequence, the permeability of the biotinylated polymer film used for the immobilization of antibody or antigen constitutes the key parameter governing the sensitivity of amperometric immunosensors. The electrogeneration of a copolymer film exhibiting a high permeability was thus carried out by electropolymerization of biotin pyrrole for the anchoring of antibody and lactobioamide pyrrole for its hydrophilic character. The model of immunoreaction chosen for the fabrication of an immunosensor was the detection of anti-cholera toxin antibody. The transduction step involved the recognition of the captured target by a secondary antibody marker labeled by peroxidase or biotin. The latter allowed the subsequent immobilization of avidin and biotinylated glucose oxidase or polyphenol oxidase. The amperometric detection of the anti-cholera toxin antibody involved the oxidation of H_2O_2 for glucose oxidase or the reduction of enzymatically generated quinones for peroxidase and polyphenol oxidase. The best detection limit was recorded for the immunosensor configuration based on peroxidase, namely 50 ng/mL of antibody [16]. This may be due to the procedure of enzyme immobilization (one step for peroxidase instead of 3 for the biotinylated enzymes) and hence to a more efficient anchoring of the enzyme marker. In addition, the diffusion pathway for the enzyme products to the electrode surface should be shorter for the peroxidase marker.

This procedure of immunosensor fabrication was extended to the modification of optical fibers. Biotinylated polypyrrole films were thus generated by chemical or electrochemical oxidation on optical fibers modified by a conductive thin layer of indium tin oxide. The presence of the polymer was demonstrated via fluorescent micrographs of immobilized avidin functionalized by a fluorescent probe: rhodamine. The resulting optical immunosensors were applied to the detection of the anti-cholera toxin through the use of a secondary antibody conjugated with peroxidase. This enzyme catalyzes a chemiluminescence radiation in the presence of H_2O_2 and luminol (Fig. 3).

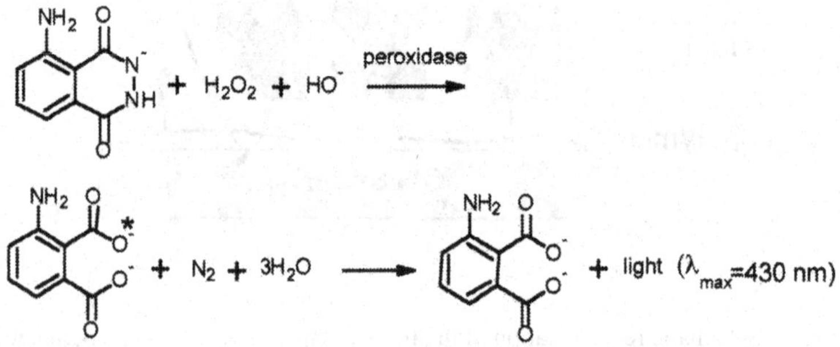

Figure 3. Chemiluminescent reaction catalyzed by peroxidase.

It appears that the fiber-optic immunosensor was up to three orders of magnitude more sensitive than the classical enzyme-linked immunosorbent assay (ELISA) for the determination of anti-cholera toxin [17].

With the aim in view of increasing the sensitivity of immunosensors, one possibility lies in the transformation of the conventional surface process of molecular recognition into a three-dimensional one. The increase of the density of biotinylated biomolecules (antigen or antibody) immobilized on the sensor surface, may be carried out through the design of multilayered avidin structures. Actually, the synthesis of biotin derivatives containing two or three biotin groups was already described [18,19]. These compounds, however, were poorly soluble in water or did not exhibit a rigid structure preventing an intra-molecular binding, namely a connection between two adjacent binding sites of one avidin. In that context, the first example of a biotinylated redox bridge: a tris(bipyridyl)iron(II) complex bearing six preoriented biotin residues was synthesized [20]. Its use in combination with an electrogenerated polymer bearing biotin residues may allow the immobilization of several layers of avidin molecules exhibiting free binding sites. The latter could thus be exploited for a three dimensional anchoring of biotinylated probes (fig.4). In order to develop such an approach, the availability of the biotin groups attached to the bipyridyl ligand for the binding to avidin was therefore examined via the influence of avidin on the electrochemical behavior of the iron complex. After the addition of avidin to an aqueous solution containing the biotinylated complex, the Fe^{II}/Fe^{III} redox system decreased and completely disappeared with time illustrating the formation of avidin-biotin complex in solution.

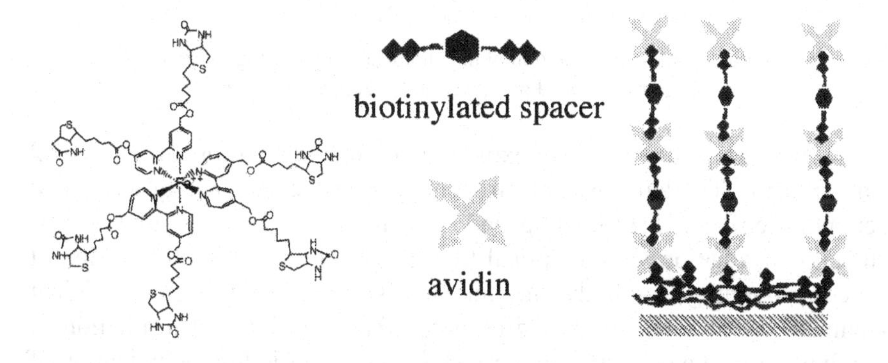

Figure 4. Structure of the biotin-labeled tris(bipyridyl) iron(II) complex and schematic representation of an avidin coating.

In addition, the ability of this complex for the intermolecular binding of two avidin molecules was showed through the successive immobilization of three layers of avidin conjugated to an alkaline phosphatase enzyme. These results demonstrated that the biotin-labeled iron(II) complex constitutes an efficient small building block for the reproducible immobilization of several avidin layers by affinity interactions.

With the aim of developing an alternative transduction approach free of label, a novel biotin-labeled ruthenium (II) tris(bipyridyl) complex functionalized by four pyrrole groups was prepared. The resulting biotinylated redox polypyrrole film allowed both the immobilization of antigens and the detection of immunoreactions via the change of its photoelectrochemical properties.

Finally, a new photo-electrochemical method for the immobilization of biological macromolecules which combines the advantages of photolithography with those of the electrochemical addressing of polymer films was recently described [21]. The synthesis and electropolymerization of a pyrrole-benzophenone derivative provided thus a polypyrrolic film exhibiting photografting abilities for biomolecule immobilization by irradiation at wavelengths ≥ 350nm that are compatible with most biological macromolecules (Fig. 5).

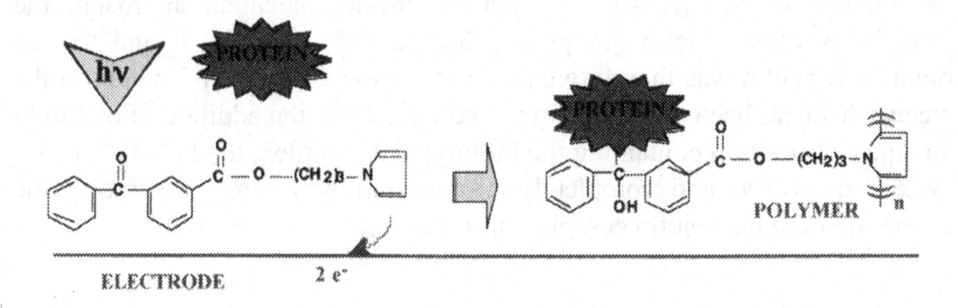

Figure 5. Schematic representation of the photochemical grafting of proteins on an electrogenerated poly(pyrrole-benzophenone) film

This polymeric coating was exploited for the fabrication of an optical immunosensor. The latter was electrogenerated at the end-face of an optical fiber modified by a ITO coating and used for the attachment of hepatitis virus upon irradiation via the optical fiber itself. The successful detection of the corresponding antibody in human blood illustates the efficient attachment of the virus. It should be noted that such internal irradiation of the polymer may be convenient approach for the specific functionalization of fiber bundles acting then as a biochip.

3. CONCLUSIONS

Electropolymerized films bearing affinity binding sites or photoactivable groups present wide potentialities for the biomolecule immobilization with complete retention of their activities and excellent accessibility to the immobilized biomolecule. Moreover, the electrochemical and photoelectrochemical detections combined with the use of electropolymerized films constitute an attractive alternative to fluorescence measurements for the determination of biological recognition processes such as immunoreactions.

REFERENCES

1. Trojanowicz M., Krawczynski vel Krawcyk T., Mikrochim Acta 1995,121, 167-181
2. Palmisano F., Zambonin P.G., Centonze D., Fresenius J Anal Chem 2000, 366, 586-601.
3. Habermuller L., Mosbach M., Schuhmann W. Fresenius J Anal Chem 2000, 366, 560-568
4. Cosnier, S. Anal. Bioanal. Chem. 2003, 377, 507-520.
5. Wilchek, M.; Bayer, E. A. Anal Biochem 1988, 171, 1-32.
6. Cosnier, S.; Galland, B.; Gondran, C.; Le Pellec, A. Electroanalysis 1998, 10, 808-813.
7. Torres-Rodriguez, L. M.; Roget, A.; Billon, M.; Livache, T.; Bidan, G. J. Chem. Soc., Chem. Commun. 1998, 1993.
8. Cosnier, S. ; Le Pellec, A. Electrochimica Acta 1999, 44, 1833-1836.
9. Cosnier, S.; Stoytcheva, M.; Senillou, A.; Perrot, H.; Furriel, R. P. M.; Leone, F. A. Anal. Chem. 1999, 71, 3692-3697.
10. Cosnier, S.; Perrot, H.; Wessel, R.; Bergamasco, J-L.; Mousty, C. Anal. Chem. 2001,73,2890-2897.
11. Ouerghi, O.; Touhami, A.; Jaffrezic-Renault, N.; Martelet, C.; Ben Ouada, H.; Cosnier, S. Bioelectrochemistry 2002, 56, 131-133.
12. Da Silva,S.; Grosjean, L.; Ternan, N.; Mailley, P.; Livache, T.; Cosnier, S. Bioelectrochemistry 2004, 63, 297-301.
13. Cosnier, S.; Mousty, C.;de Melo, J.; Le Pellec, A. ;Novoa, A.; Polyak, B.; Marks, R. S Electrochem. Commun.2001, 3727-732.
14. Cosnier, S.; Mousty, C.;de Melo, J.; Le Pellec, A. ;Novoa, A.; Polyak, B.; Marks, R. S. Electroanalysis in press.
15. Cosnier, S. ; Le Pellec, A. ; Marks, R. S. ; Périé, K. ; Lellouche, J.-P. Electrochem. Commun.2003, 5,973-979.
16. Ionescu, R.; Gondran, C.; Cosnier, Gheber,L.; Marks, R.S. Talanta submitted.
17. Konry, T.; Novoa, A.; Cosnier, S.; Marks, R.S. Anal. Chem. 2003, 75, 2633-2639.
18. Taylor, D., Fukushima, H., Morgan H., Supramolecular Science 1995, 2, 75-87.
19. Wilbur, D., Pathare, P., Hamlin, D., Weerawarma, S. Bioconjugate Chem. 1997, 8, 819-832.
20. Haddour, N.; Gondran,C.; Cosnier, S. J Chem Soc, Chem Commun 2004,324-325.
21. Cosnier, S. J Chem Soc, Chem Commun 2003, 414, 415.

TECHNOLOGICAL PLATFORMS BASED ON MICRO/NANOBIOSENSORS AS EARLY WARNING SYSTEMS FOR BIOLOGICAL WARFARE

L.M. LECHUGA, J.TAMAYO, A.CALLE, M. CALLEJA AND C. DOMINQUEZ
Microelectronics National Center, CSIC, Spain

Abstract: Biological and chemical warfare are fields where new types of analyzers (faster, direct, smaller and cheaper than conventional methods) are demanded. In order to achieve a multibiosensor technological platform that could be used as an early warning system for biological and/or chemical warfare, we are working in the development of two different approaches. A platform based on Optoelectronics biosensors (evanescent wave detection). Two optical biosensors have been already developed: a portable Surface Plasmon Resonance Sensor (actually in commercialization) and an integrated Mach-Zehnder interferometer device. For the second sensor, the use of standard Si microelectronics technology allow the possibility for integration of optical, fluidics and electrical function on one chip in order to obtain a complete lab-on-a-chip. A multibiosensor platform based on Nanomechanical biosensors. Microcantilever biosensors are a new class of high sensitivity biosensors able of performing local, high resolution and label-free molecular recognition measurements. Moreover, nanomechanical biosensors based on microcantilevers have been recently reported as a promisingly alternative to current DNA-chips allowing real-time monitoring of DNA without need of labelling. For that reason, we are working in the development of a portable multibiosensor microsystem based on an array of microcantilevers [2] able to detect analytes with femtomolar sensitivity and ability for discerning single base variations in DNA strands.

1. INTRODUCTION

Biological and chemical warfare (BCW) are fields where new types of analyzers (faster, direct, smaller and cheaper than conventional methods) are

D. Morrison et al. (eds.), Defense against Bioterror: Detection Technologies, Implementation Strategies and Commercial Opportunities, 175–197.

demanded. In order to have reliable diagnostic tools for the rapid detection and identification of biological and chemical warfare agents, new methods allowing label-free and real time measurement of simultaneous interactions (as harmful agent/receptor or DNA recognition) must be developed [1,2]. Biosensing devices, fabricated with micro/nanotechnologies are powerful devices which can fulfill these requirements and have the added draw of being portable to perform "point-of-care" analysis. These devices can also have the multiplex capability in a biodetection platform for performing the identification of pathogens in a faster way than in routine clinical analysis. Nowadays, about 30 microbial pathogens and toxins are considered biological warfare agents [3]. The early detection of these weapons is one of the best defense against bioterror. A multibiosensor system could be designed to perform such 30 detections simultaneously if the receptor molecule is available for each of them. In addition, biosensor devices can allow the direct determination of such pathogens in the presence of complex backgrounds. Biological agents as bacteria (anthrax, tularemia, Q-fever), viruses (smallpox, hemorrhagic fever viruses) or toxins (mycotoxins, SEB, T2) can be determined using biosensor devices which could operate on the battlefield or in civil installations [3-5]. This technology can operate also as an error-proof detection system that can stave off the bioterror threat and could assist emergency personnel in quickly identifying and quantifying BCW toxins during the window of time between the attack and the presence of first symptoms. In this way, the dispensing of antibiotics or vaccines against the specific pathogen to the civil population could be possible in a reduced time scale [1].

The biological receptor for the detection of BCW agents could be or either DNA strands that, for example, can bind to a specific pathogen present in the environment or either antibodies that can recognize, for example, specific sites on bacteria or bind to surface proteins [4]. Biosensors to be developed could be able to work with both types of receptors and to trigger a signal after their specific biomolecular interaction.

In order to achieve a multibiosensor technological platform that could be used as an early warning system for biological and/or chemical warfare, two different approaches [6] will be reviewed:

(A) A platform based on optoelectronics biosensors (evanescent wave detection). Two optical biosensors have been already developed: a portable Surface Plasmon Resonance Sensor (actually in commercialization) and an integrated Mach-Zehnder interferometer device. For the second sensor, the use of standard Si microelectronics technology allows the possibility for integration of optical, fluidics and electrical function on one optical sensing circuit in order to obtain a complete lab-on-a-chip. A limit of detection close to femtomolar is achievable with this sensor in a direct format [7, 8].

(B) A platform based on nanomechanical biosensors. Microcantilever biosensors are a new class of high sensitivity biosensors able of performing local, high resolution and label-free molecular recognition measurements. Moreover, nanomechanical biosensors based on microcantilevers have been recently reported as a promising alternative to current DNA-chips allowing real-time monitoring of DNA without need of labeling. For that reason, we are working in the development of a portable multibiosensor microsystem based on an array of microcantilevers [9] able to detect analytes with femtomolar sensitivity and ability for discerning single base variations in DNA strands.

2. RECEPTOR IMMOBILIZATION AT NANOMETER SCALE

For biosensing purposes, a layer of receptor molecules (proteins, DNA) that are capable of binding the analyte molecules in a selective way has to be previously immobilized on the biosensor surface. The complementary analytes flowing over the surface can be directly recognized by the receptor through a change in the physico-chemical properties of the sensor. In this way, the interacting components do not need to be labeled and complex samples can be analyzed without purification.

The immobilization of the receptor molecule on the sensor surface is a key point for the final performance of the sensor. The immobilization procedure must be stable and reproducible and must retain the stability and activity of receptor. This way can allow high sensitivity levels and miniaturization. Generally, direct adsorption is not adequate, giving significant losses in biological activity and random orientation of the receptors. One of the most promising strategies is the covalent immobilization on gold-coated surfaces using thiol self-assembled monolayers (SAM) [7]. For example, a widespread method is functionalization of ss-DNA with an alkane chain termed in a thiol (-S-H) or disulfide group (-S-S). Sulphurs form a strong bond with gold, thus thiol-derivatized ssDNA spontaneously forms a single self-assembled monolayer upon immersion on clean gold surfaces. This can be applied also for silicon surfaces, using silane monolayers covalently attach to the SiO2 or Si3N4 sensor surfaces [7]. Several aspects must be taken into account in the development of the immobilization procedures as the non-specific interactions, an optimized surface density of the receptor in order to prevent steric hindrance phenomenon or the regeneration of the receptor.

We have developed immobilization procedures at the nanometer-scale which try to fulfill all the requirements described above, based on thiol-chemistry, silanization or esterification depending on the type of sensor surface and the application [7]. Different examples will be shown in the following during the description of each biosensor platform.

3. OPTOELECTRONICS BIOSENSOR PLATFORM

The ideal biosensor for counter-terrorism must have high sensitivity, must be easy to use, and be cheap, reliable and compact for incorporation in a portable system. One very promising type of sensor is the evanescent wave device based on refractive index sensing technology that as, for example, includes the surface plasmon resonance sensor or the interferometer sensor.

The optical sensing approach offers many advantages than its electrical counterpart, as the absence of risk of electrical shocks or explosions, its immunity to electromagnetic interferences, a higher sensitivity and a wider bandwidth [10]. Moreover, by using optical fibers to guide light into and out of the device, remote sensing is also possible. In addition, the optical transducers have a potential for parallel detection making possible array or imaging detection.

The advantages of the optical sensing are significantly improved when this approach is used in an integration schema [11]. The technology of integrated optics allows the integration of many passive and active optical components (including fibers, emitters, detectors, waveguides and related devices, etc...) onto the same substrate, allowing the flexible development of miniaturized compact sensing devices, with the additional possibility of fabrication of multiple sensors on one chip. Then, integration offers additional advantages to the optical sensing systems such as miniaturization, robustness, reliability, potential for mass production with consequent reduction of production costs, low energy consumption and simplicity in the alignment of the individual optical elements. It is also possible the inclusion of Global System for Mobile Communications/ General Packet Radio Service (GSM/GPRS) (in the near future Universal Mobile Telecommunications System (UMTS)) modules for networking and communication capabilities. The last developments in the field of integrated optics have resulted in an innovative class of microoptical sensors exhibiting biosensing sensing performances comparable to that of sophisticated analytical laboratory instrumentation [11].

In an optical waveguide the light travels inside the waveguide, confined within the structure by Total Internal Reflection (TIR). A detailed study of how the light travels inside the waveguide shows that the light is transmitted through a model of the electromagnetic field called "guided modes" (as it is shown in Fig. 1). Although light is confined inside those modes, there is a part of it (evanescent field, EW) that travels through a region that extends outward, around a hundred nanometers, into the medium surrounding the waveguide (see Fig. 1). This EW field can be used for sensing purposes.

When a receptor layer is immobilized onto the waveguide, as it is shown in Figure 1, exposure of such a surface to the partner analyte molecules produces a (bio)chemical interaction, that takes place into the surface of the waveguide and induces a change in its optical properties that is detected by the evanescent wave. The extent of the optical change will depend on the concentration of the analyte and on the affinity constant of the interaction, obtaining, in this way, a quantitative sensor of such interaction. The evanescent wave decays exponentially as it penetrates the outer medium and, therefore, only detects changes that take place on the surface of the waveguide, because the intensity of the evanescent field is much higher in this region [12]. For that reason it is not necessary to carry out a prior separation of non-specific components (which is necessary in conventional analysis) because any change in the bulk solution will hardly affect the sensor response. In this way, evanescent wave sensors are selective and sensitive devices for the detection of very low levels of chemicals and biological substances and for the measurement of molecular interactions in-situ and in real time [10].

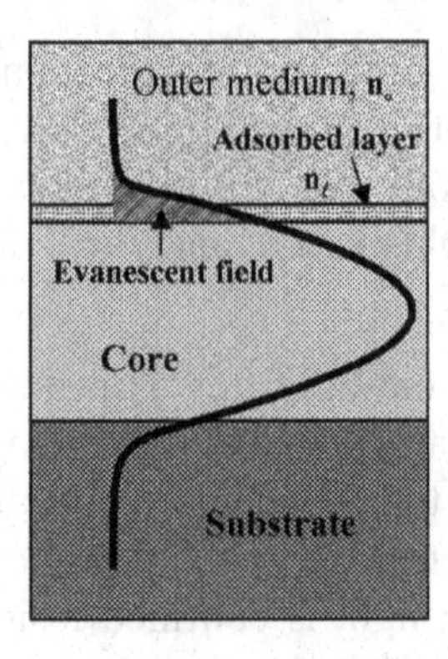

Figure 1. Light propagation inside a monomode optical waveguide. The fraction of the mode that travels through the outer medium (evanescent field) is indicated in the figure.

3.1 Surface Plasmon Resonance Biosensor

One of the best known and more developed EW biosensor is the Surface Plasmon Resonance (SPR) sensor, because of its sensibility and simplicity [7,13,14]. Surface plasmons are elementary excitations, which result from a collective oscillation of the free-electron plasma at a metal-dielectric film interface. In a SPR sensor a thin metal film (usually Au) is evaporated on the dielectric material surface. The sensing mechanism is based on variations of the refractive index of the medium adjacent to the metal sensor surface during the interaction of an analyte to its corresponding receptor, previously immobilized at the sensor surface in the region of the evanescent field. The recognition of the complementary molecule by the receptor causes a change in the refractive index and the SPR sensor monitors that change. After the molecular interaction, the surface can be regenerated using a suitable reagent to remove the bound analyte without denaturing the immobilized receptor.

SPR biosensors based on the principles of solid-phase immunoassays have been used to measure a large variety of small size compounds, as for example environmental pollutants [15,16]. The same principle can be applied to the detection of harmful pathogens. These assays require the use of antibodies (monoclonal or polyclonal), which are the key components of all immunoassays, since they are responsible for the sensitive and specific recognition of the analyte. The application of immunoassays to environmental monitoring also involves the design of hapten derivatives of low molecular weight molecules, such as pesticides or harmful chemical compounds, to determine the antibody recognition properties [17]. Once hapten synthesis and monoclonal antibody production have been accomplished, the use of SPR biosensing technique provides a real-time monitoring of binding interactions without the need of labelling biomolecules.

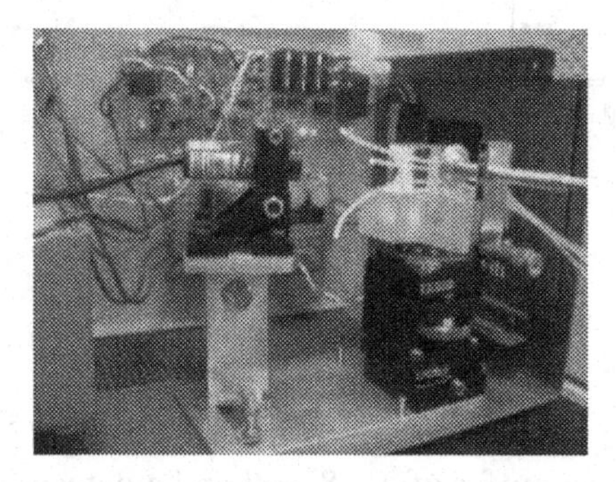

Figure 2. Portable SPR sensor prototype system including sensor, optics, electronics and flow delivery system.

We have developed a home-made and portable SPR sensor prototype (see Fig. 2) as a highly sensitive field analytical method for environmental monitoring [6]. As a proof of its utility towards detection of pathogens, we have determined several pesticides, as the chlorinated compound DDT, and the neurotoxins of carbamate type (carbaryl) and organophosphorus type (chlorpyrifos which resembles SOMAN chemical warfare agent). The chemical structure of these compounds is shown in Figure 3.

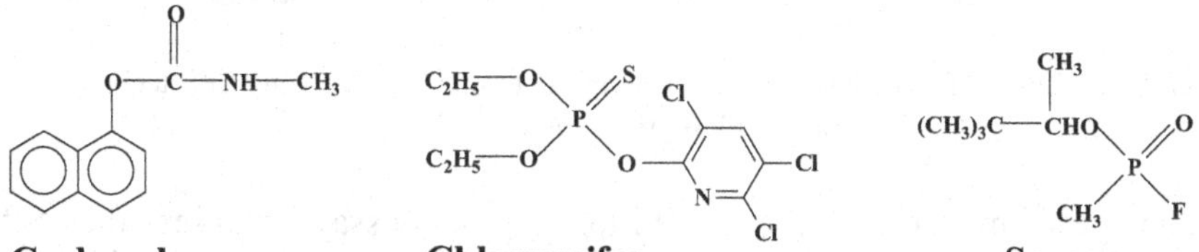

Carbaryl **Chlorpyrifos** **Soman**

Figure 3. Chemical structures of the neurotoxins pesticides used in this study

For the determination of these compounds a binding inhibition immunoassay, consisting of the competitive immunoreaction of the unbound antibody present in an analyte-antibody mixture with the hapten derivative immobilized at the sensor surface, has been applied. With the aim of assuring the regeneration and reusability of the surface without denaturation of the immobilized molecule, the formation of an alkanethiol monolayer was carried out to provide covalent attachment of the ligand to the functionalized carbodiimide surface in a highly controlled way. In Figure 4 a scheme of the binding inhibition immunoassay is shown. For DDT, the assay sensitivity was evaluated in the 0.004-3545µg/l range of pesticide concentration by the determination of the limit of detection (0.3 µg/l) and the I50 value (4.2 µg/l).

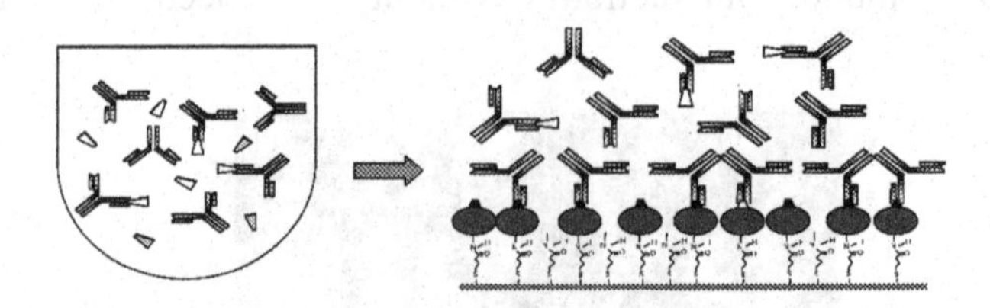

Figure 4. Representation of the binding inhibition immune assay developed to determine DDT.

For carbaryl, the dynamic range of the sensor is 0.12-2 μg/l, with an I50 value for standards in buffer of 0.38 μg/l and a detection limit of 0.06 μg/l. Likewise the immunoassay for chlorpyrifos determination, afforded a high sensitivity (I_{50}= 0.11 μg/l) working in the 0.02-1.3 μg/l range.

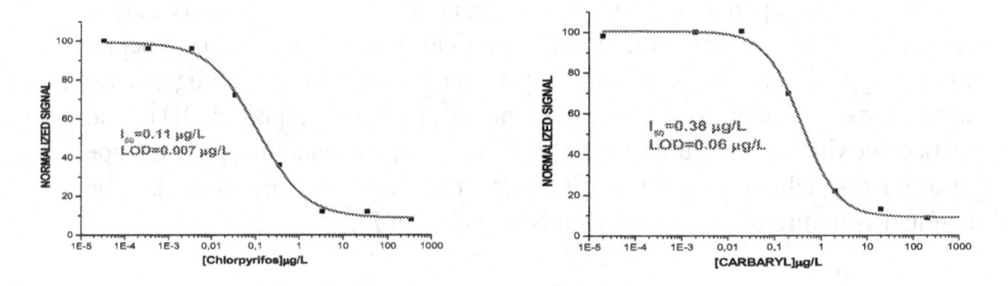

Figure 5. Calibration curve for the immunoassays determination by using SPR technology of (left) Chlorpyrifos (right) Carbaryl

The performance of the inhibition immunoassay enables the SPR biosensor to monitor the immunoreaction between the hapten immobilized on the sensor surface and the monoclonal antibody, from the incubation of a mixed antibody-analyte solution. In addition, the reusability of the sensor was demonstrated after 100 assay cycles, without significant variation of the average maximum signal. The reusability of the sensor combined with the small time of response (approximately 15 min), makes the SPR immunosensing a valuable method for real-time and label-free analysis of environmental samples. This immunnosensing technique together with the portable surface plasmon resonance sensor developed can be applied as a fast and cost-effective field-analytical method for the monitoring of chemical and biological warfare agents if the corresponding receptor is available.

3.2 Mach-Zehnder interferometric biosensor

In a Mach-Zehnder interferometer (MZI) device [8] the light from a laser beam is divided into two identical beams that travel the MZI arms (sensor and reference areas) and are recombined again into a monomode channel waveguide giving a signal which is dependent on the phase difference between the sensing and the reference branches. Any change in the sensor area (in the region of the evanescent field) produces a phase difference (and therein a change of the effective refractive index of the waveguide) between the reference and the sensor beam and then, in the intensity of the outcoupled light. A schematic of this sensor is shown in Figure 6.

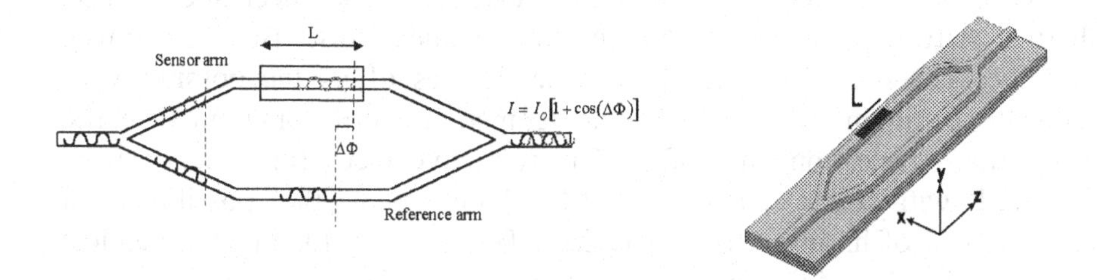

Figure 6. Mach-Zehnder interferometer configuration (left) scheme of the working principle (right) 3-D view of the MZI structure.

When a chemical or biochemical reaction takes place in the sensor area, only the light that travels through this arm will experience a change in its effective refractive index. At the sensor output, the intensity (I) of the light coming from both arms will interfere, showing a sinusoidal variation that depends on the difference of the effective refractive indexes of the sensor (Neff,S) and reference arms (Neff,R) and on the interaction length (L) :

$$I = \frac{1}{2} I_O \left[1 + \cos\left(\frac{2\pi}{\lambda} \left(N_{eff,S} - N_{eff,R} \right) L \right) \right] \tag{1}$$

Where λ is the wavelength. This sinusoidal variation can be directly related to the concentration of the analyte to be measured. For evaluation of specific biosensing interactions, the receptor is covalently attached to the sensor arm surface, while the complementary molecule binds to the receptor from free solution. The recognition of the complementary molecule by the receptor causes a change in the refractive index and the sensor monitors that change. After the molecular interaction, the surface can be regenerated using a suitable reagent in order to remove the bound analyte without denaturing the immobilized receptor as in the case of the surface plasmon resonance sensor.

The interferometric sensor platform is highly sensitive and is the only one that provides with an internal reference for compensation of refractive-index fluctuations and unspecific adsorption. Interferometric sensors have a broader dynamic range than most other types of sensors and show higher sensitivity as compared to other integrated optical biosensors [7, 8]. Due to the high sensitivity of the interferometer sensor the direct detection of small molecules (as for example environmental pollutants where concentrations down to 0.1 ng/ml must be detected) would be possible with this device. Detection limit is generally limited by electronic and mechanical noise, thermal drift, light source instabilities and chemical noise. But the intrinsic reference channel of the interferometric devices offers the possibility of reducing common mode effects like temperature drifts and non-specific adsorptions. Detection limit of 10-7 in refractive index (or better) can be achieved with these devices [8,10] which opens the possibility of development of highly sensitive devices, for example, for in-situ chemical and biological harmful agents detection [18].

For biosensing applications the waveguides of the MZI device must be designed to work in monomode regime and for having a very high surface sensitivity of the sensor arm towards the biochemical interactions. If several modes were propagated through the structure, each of them would detect the variations in the characteristics of the outer medium and the information carried by all the modes would interfere between them. The design of the optical waveguide to be employed, fulfilling the above requirements, and the dimensions of the Mach-Zehnder structure is performed by using home-made modelling programs as Finite differences Methods in non-uniform mesh, Effective Index Method and Beam Propagation Method. Parameters as propagation constants, attenuation and radiation losses, evanescent field profile, modal properties, field evolution,... have been calculated [19]. The final design for the optical waveguides to be employed in the MZI sensor is shown in Figure 7.

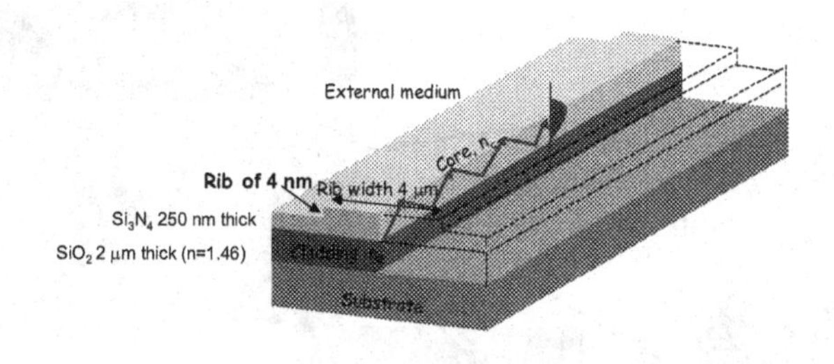

Figure 7. Cross-section of the optical waveguides used in the Mach-Zehnder interferometer. Note that a rib of only 4 nm in needed for monomode and high biomolecular sensitivity characteristics.

The optical waveguides and the MZI structure are fabricated at the Clean Room facilities. The structure is as follows: (i) a conducting Si wafer of 500 µm thickness, (ii) a 2 µm thick thermal Silicon-Oxide layer on top with a refractive index of 1.46, (iii) a LPCVD Silicon Nitride layer of 100 nm thickness and a refractive index of 2.00, which is used as a guiding layer. To achieve monomode behavior a rib structure is defined with a depth of only 4 nm, on the Silicon Nitride layer by a lithographic step. This rib structure is performed by RIE and is the most critical step in the microfabrication of the device. The sensing process (change in the optical properties of the outer medium) will take place during a certain distance, L, in the sensor arm of the interferometer. For that reason, the rest of the MZI is protected from the environment with a covering layer and the sensing window of length L is opened in one of the interferometer branches. The protective layer is a silicon oxide layer with a refractive index of 1.46 deposited by PECVD at 300 ºC. The thickness of this layer is 2 µm, enough to isolate the core from the environment. The opening of the sensor area in the protective layer is done, after a photolithographic process, by wet chemical etching. Several MZI configurations were designed varying the separation between arms and the Y-junction parameters. All the devices are symmetric with a divisor (Y-junction) with circular shape (R=80 mm). The separation between the sensor and reference arms is of 100 µm to avoid coupling between modes travelling through both branches. In one arm, a sensor area of 15 mm length is created. The total length of the device is 35 mm. Finally the sensors are cut in individual pieces containing 14 devices each and polished for light coupling by end-face. A schematic and a photograph of one integrated MZI device is shown in Figure 8.

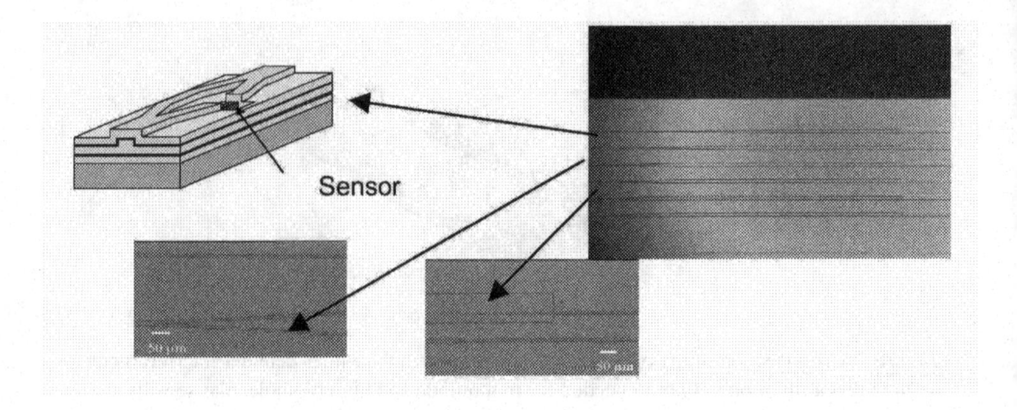

Figure 8. Photographs of the Integrated Mach-Zehnder interferometer: details of the MZI Y-junction and sensor area.

For measuring, the devices are implemented with a microfluidics unit, electronics, data acquisition and software for optical and biochemical testing. Chemical characterisation for evaluating the sensor sensitivity was performed by using sugar solution with refractive indexes varying from 1.3325 to 1.4004 (Δn□0.0002), as determined by an Abbe refractometer operating at 25ºC. The solutions with varying refractive indexes were introduced alternatively, and the change induced in the MZI device can be observed in Figure 9 for one concentration. With these measurements, a sensitivity calibrating curve was evaluated, where the phase response of the sensor is plotted versus the variation in the refractive index, as is depicted in Fig. 10.

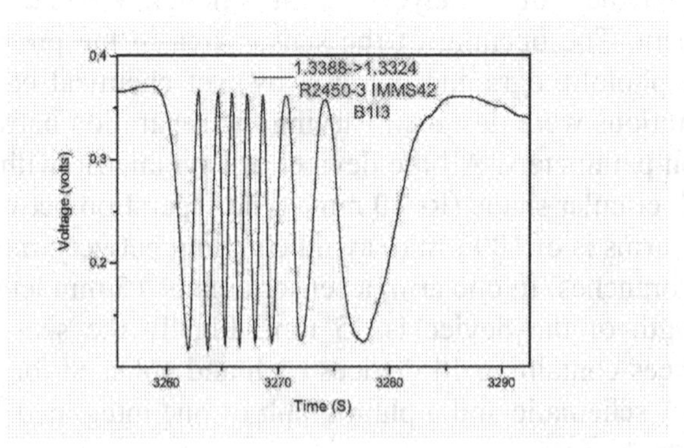

Figure 9. Interferometric signal from a MZI device due to a refractive index change of n = 6,4.10-2

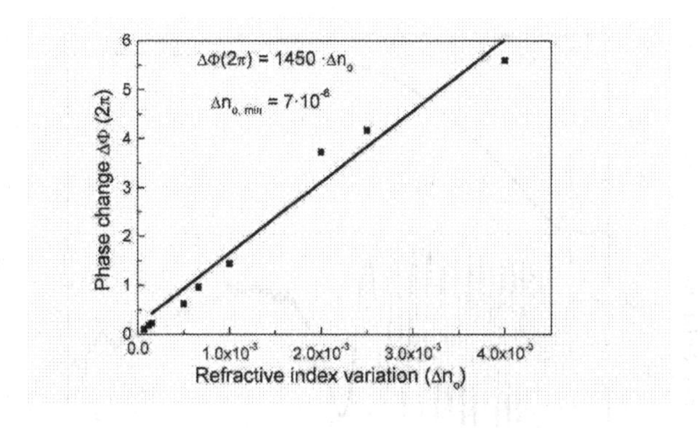

Figure 10. Sensitivity evaluation of a MZI nanodevice by using glucose solutions of varying refractive indices.

For this device, the lower detection limit measured was □no,min = 7.10-6 that means an effective refractive index of □N = 4·10-7. The detection limit has been estimated measuring the signal-to-noise ratio (around 20 dB) and considering the most favorable case in which the interferometer is located at its quadrature point. Under this conditions, we estimate that the lowest phase shift measurable would be around 0.01·2□. The detection limit value corresponds to a surface sensitivity around 2.10-4 nm-1, close to the maximum surface sensitivity reported up to now [20].

We have applied the MZI nanobiosensors for the detection of the insecticide carbaryl which has neutotoxins properties. The immunoassays developed to monitor this pollutant require the use of three different components. The critical part is the monoclonal antibody, since it is responsible for the sensitivity and specificity of the analyte to determine. The second component is the analyte (antigen) to detect, which is unable to produce directly an immune response due to its low molecular weight (~200D). Therefore, it is necessary to design the third compound: the hapten, which can be covalently bound to a large carrier, usually a protein. A hapten is a derivative of the analyte, in this case carbaryl, with similar geometry and structure, which contains an appropriate group (COOH, -NH2) for attachment to the carrier protein (OVA). Once the hapten-carrier conjugate has been formed, antibodies can be produced by animal immunization. To achieve higher specificity and lower cross-reactivity, monoclonal antibodies, obtained unlimitedly by the hybridoma technique are preferred to heterogeneous polyclonal antibodies.

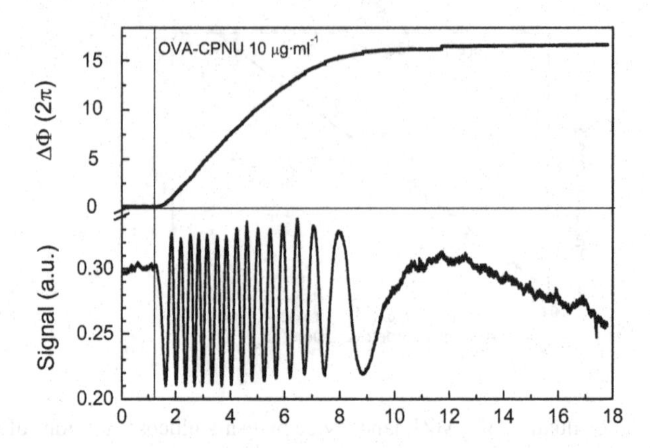

Figure 11. Immobilization of the receptor at nanometer scale by covalent attachment for the detection of the neurotoxin carbamate.

Carbaryl determination requires the immobilization of the hapten-carrier conjugate (carbaryl OVA-CPNU) and the subsequent addition of the monoclonal antibody. The immobilization procedure used is in this case was an esterification process of the silicon nitride sensor MZI surface. A concentration of 10 μg·ml-1 in a buffer solution (PBST, Phosphate Buffered Saline Tween) with pH 7 at a constant flow rate of 20 μl·min-1 was introduced. As it is observed in Figure 12, the phase change, in the first part of the process, is fast and as the surface is being progressively occupied, the phase response $\Delta\Phi$ varies more slowly. The total phase change is 16.2.2π, that corresponds to the adsorption of a homogeneous antigen monolayer of average thickness $d\lambda \approx 3.2$ nm (surface covering of 1.9 ng·mm-2). A concentration of 10 μg·ml-1 of the antibody is flowed through the sensor, giving an additional phase change of $\Delta\Phi$= 5.2π due to the recognition immunoreaction, as it is shown in Figure 13. After rinsing with PBST, the antibodies that have not reacted are washed out, being the net sensor response of $\Delta\Phi$= 0.4.2π (surface covering of 0.33 ng·mm-2).

The main advantage of the development of Mach-Zehnder devices fabricated with standard microelectronics technology comes for the possibility to develop a complete "lab-on-a-chip" by optoelectronics integration in which the light source, photodetectors and sensor waveguides are combined on a single semiconductor package together with the flow system and the CMOS electronics. The reagent receptor deposition can be performed by ink-jet, screen-printed or other technology. A complete system fabricated with integrated optics will offer low complexity, robustness, standardized device and, what is more important, portability. Devices for on-site analysis or point-of-care operations for biological and chemical warfare detection are geared for portability, ease of use and low cost. In this sense, integrated optical devices have compact structure and could allow for fabricating optical sensor arrays on a single substrate for simultaneous detection of multiple analytes. Mass-production of sensors will be also possible with the fabrication of miniaturized devices by using standard microelectronics technology [21].

4. NANOMECHANICAL BIOSENSOR PLATFORM

Nanomechanical biosensors are an excellent example of the application of micro- and nanotechnologies in the development of a new type of biosensors. Microcantilevers, such as those used in Atomic Force Microscopes, have been recently employed as this new class of biosensors [22,23]. The so-called nanomechanical biosensors have demonstrated that they are capable of detecting single-base mismatches in oligonucleotide hybridization without labeling [23] as well as performing protein recognition [24,25] with extreme sensitivity. Among the advantages of nanomechanical biosensors are the potential for performing local, high resolution and label-free molecular recognition measurements on a portable device. Also, the reduced sensor area allows drastic decrease of the reagent consumption.

The working principle for nanomechanical biosensors relies on the induced surface stress that arises when molecules bind to a surface [26]. When a monolayer of receptor molecules is immobilized on one side of the cantilever, a cantilever deflection results from the differential surface stress between opposite sides of the cantilever. Molecular recognition produces also a change of the surface stress of the sensitised cantilever side with respect to the other side, giving a cantilever deflection of few nanometers. Surface stress mainly arises from electrostatic, van der Waals, configurational and steric interactions between the adsorbed molecules. This method, which is often called DC, has allowed the measurement of submonolayer adsorption of different molecular species, changes of pH and salt concentration, and biomolecular interactions.

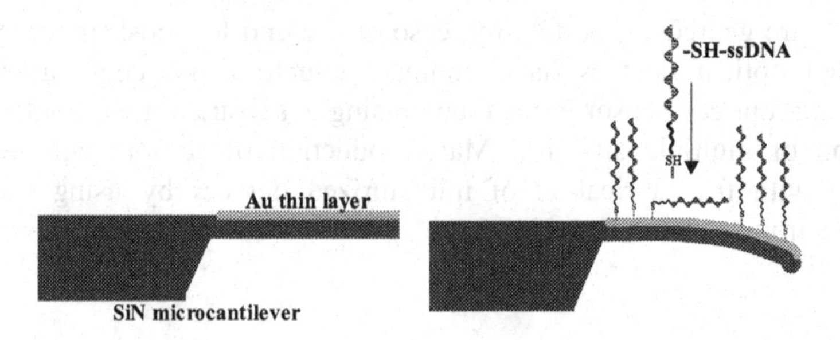

Figure 12. Illustration of the working principle in nanomechanical cantilevers: (left) cantilever with a thin gold layer for covalent attachment of the receptor (right) bending of the cantilever due to adsorption or biomolecular interaction of molecules due to a change on the surface stress.

Cantilever bending (deflection) measurements are carried out by using the well-known optical beam deflection method employed in most of the atomic force microscopes. A laser beam is focused on the free end of the cantilever and the deflection of the reflected beam, which is proportional to that of the cantilever, is measured with a four-segment photodetector. For example, for the DNA hybridization detection, nucleic acids are immobilized on one side of the micromachined lever (active side). Exposure of the cantilever to a sample containing complementary nucleic acid gives rise a cantilever bending (deflection) of a few nanometers. The nanomechanical response is due to the surface stress change of the active side with respect to the other side, in which DNA is not immobilized. The deflection is measured with sub-nanometer resolution by the optical system, in which the laser beam reflects off the back of the cantilever to the position sensitive photo-detector.

Normally, micrometer-sized cantilevers are designed for atomic force microscopy and are fabricated using Silicon technology. But for application as highly sensitive biosensors the cantilevers have to be re-designed carefully according with the dimensions and the mechanical material properties. For the design and fabrication we have follow two approaches: one based on standard Si technology [9] and a novel one, where the cantilevers are fabricated in the polymer SU-8 [27]. In Fig. 13 and 14 photographs of the fabricated cantilevers with both technologies are shown.

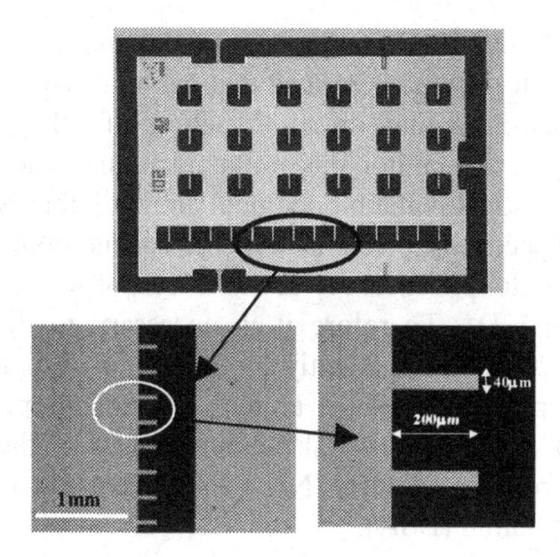

Figure 13. Photographs of an array of 14 silicon cantilevers fabricated by surface micromachining. Arrays of up to 20 have been fabricated with different lengths (from 100 to 500 μm) and widths (from 20 to 40 μm). Cantilever thickness was 300 nm.

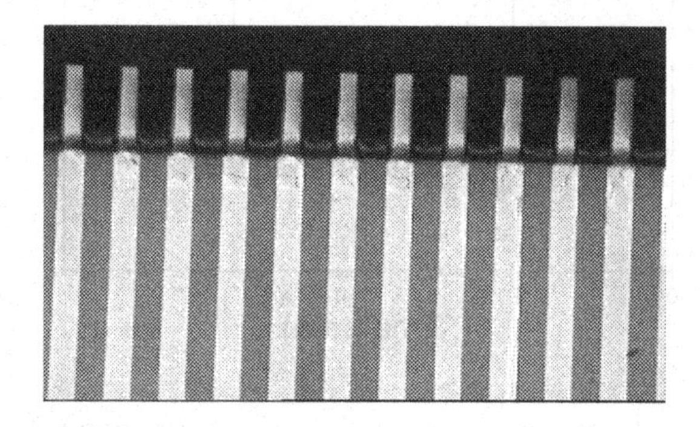

Figure 14. Optical images of a cantilever array fabricated with the polymer SU-8 technology. Arrays of 9, 15, 17 and 33 cantilevers with different lengths (100 μm and 200 μm) and widths (20, 30 and 50 μm) have been fabricated. Cantilever thickness was varied from 1.3 μm to 2 μm.

We have applied the Nanomechanical biosensors for the detection of the pesticide DDT by performing competitive immunoassays using a synthetic hapten conjugated with BSA and its specific monoclonal antibody [25]. DDT is a chlorinated compound with insecticide properties that has been used worldwide for controlling insect pests. Its high hydrophobicity together the great stability to physical, chemical, and biological degradation have resulted in the accumulation of its residues in animal and human tissues, as well as in the environment. We have proved that nanomechanical biosensors allow rapid, highly sensitive and direct detection of DDT. Briefly the immunoassays developed to monitor this pollutant require the use of three components. The most critical is the monoclonal antibody that is responsible of the sensitivity and specificity of the assay. The second component is the DDT, which is unable to produce an immune response due to its low molecular weight (354.5 D). Therefore it is necessary to design a third compound, the hapten, that can covalently bind a carrier, usually a protein, thereby the hapten-carrier can produce animal immunization. The hapten used here has similar geometry and structure to that of the DDT, and contains appropriate groups (-COOH, -NH_2) for attachment to the carrier protein bovine serum albumin (BSA).

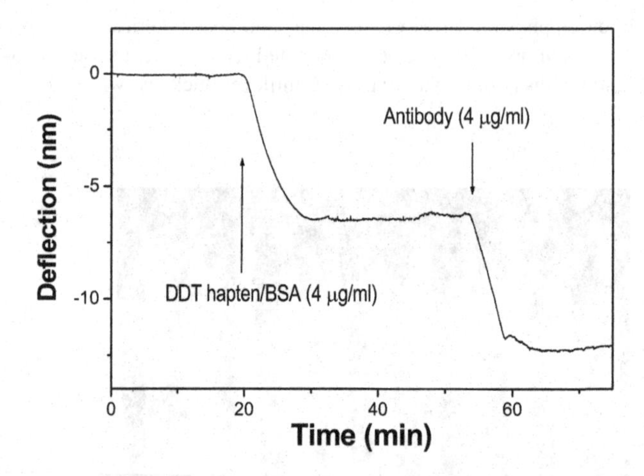

Figure 15. Real-time monitoring of the covalent attachment of the DDT hapten conjugated with BSA on the functionalised cantilever side, and the specific hapten/antibody binding. The experiment was performed in PBS solution.

The gold-coated cantilevers were exposed to 2 mM cystamine dihydroclhoride during 30 min and washed in milliQ water. Cystamine is an amine-terminated thiol that produce a densely packed layer on the gold-coated surface. Then, 2.5% glutaraldehyde was flowed over the cantilevers for 30 min. Rinsing was performed with milliQ water. The glutaraldehyde, with an aldhehyde group in both extremes of the molecule, acts as covalent linker between the amine-functionalised gold surface and primary amine groups of the hapten protein carrier. The deflection induced in the cantilevers by these steps can be observed in Figure 15.

A synthetic hapten of the pesticide DDT conjugated with BSA was flowed over the cantilever (Fig. 16). The injection of the DDT hapten derivative produced a downward bending of the cantilever, indicating a compressive surface stress acting on the gold layer. The DDT hapten is strongly attached to the sensitised cantilever side through the covalent bond between the aldehydes coating the gold surface and the primary amine groups of BSA. Before the pesticide detection assay, the cantilever was exposed to 1M ethanolamine to saturate remaining aldehyde sites and to minimize no specific adsorption on the bottom surface of the cantilever. After rinsing with PBS, the cantilever was exposed to the monoclonal antibody to the synthetic hapten. The antibody/hapten recognition on the cantilever is again translated into a downward bending. Injection of non-specific antibodies did not produce a significant cantilever deflection. The clean detection signal for an antibody concentration of 25 nM indicates that pesticide detection below the nanomolar range is suitable [25].

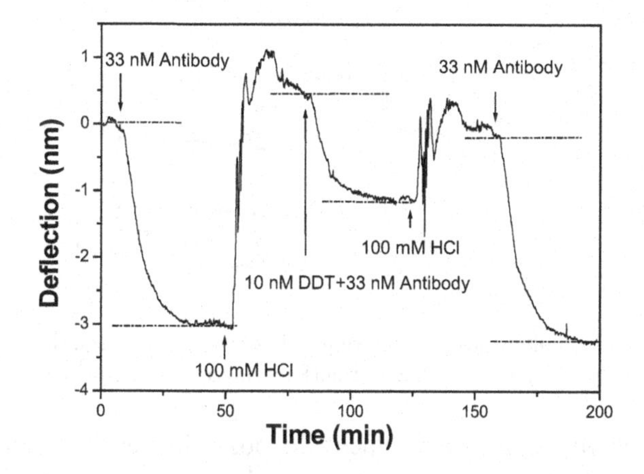

Figure 16. Real-time monitoring of a competitive immunoassay for DDT pesticide detection. The cantilever surface was regenerated with 100 mM HCl (100 μl) to break the hapten/antibody complex.

On the other hand, nanomechanical biosensors based on microcantilevers have been reported as a promisingly alternative to current DNA-chips, allowing real-time monitoring of DNA without need of labeling. Moreover, biosensors based on Nanomechanics can specifically detect single-base mismatches in oligonucleotide hybridization [23]. The nanomechanical biosensing technology readily lends itself to fabrication of microarrays using well-known and standard microfabrication techniques, offering the promising prospect of multiple protein or DNA analysis and allowing the simultaneous detection of biowarfare pathogens. To proof the reliability of such approach, we have used the arrays of Si cantilevers shown above for the real-time detection of (a) immobilization of DNA strands [28] and (b) hybridization with the corresponding complementary DNA strands, as it is shown in Figure 17.

A 12mer ssDNA was derivatized with the alkylthiol SH-(CH2)6 in terminal 5'. Thiols spontaneously form self-assembled ordered monolayers onto the cantilevers array surface giving a pronounced deflection of about 20 nm. After –SH-ssDNA immobilization, the array of cantilevers was exposed to 6-mercapto-1-hexanol (MCH), a 6-carbon chain molecule terminated with thiol (-SH) and hydroxyl (-OH) groups on each of the extremes.

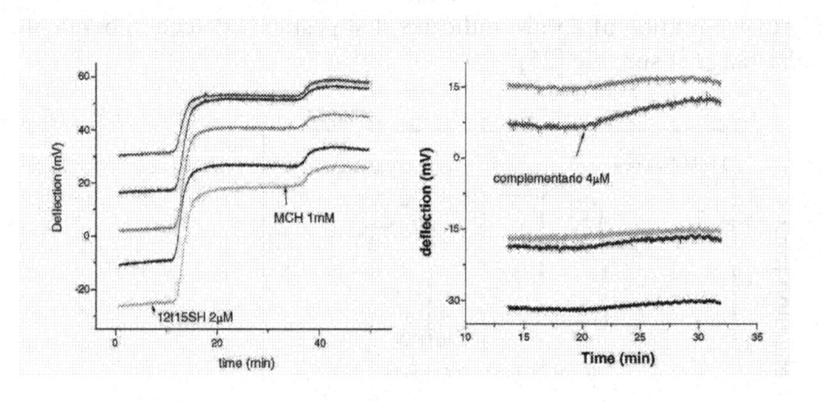

Figure 17. Real time simultaneously monitoring of five microcantilevers. (left) Immobilization and blocking treatment, and (right) hybridisation bonding signal.

The Thiol group of MCH rapidly displaces the possible weaker adsorptive contacts between the nucleotide chain and gold. Since hydroxyl group negligibly interacts with the nucleotide chain, MCH treatment assures that the immobilized ssDNA is only attached to the gold surface through the sulphur atom. This treatment enhances the accessibility of gold-tethered ssDNA molecules for base pairing with complementary nucleic acids, increasing the hybridisation efficiency from less than 10% to 80% approximately [29]. MCH adsorption produces a cantilever deflection of about 10 nm. The array of cantilevers is then exposed to the complementary DNA sequence and the corresponding hybridisation signals were recorded, showing the feasibility of the cantilever technology for the real-time detection of DNA detection.

5. CONCLUSIONS

For the rapid detection and identification of bio and chemical warfare agents, reliable multibiosensor systems allowing label-free and real time measurement of simultaneous interactions must be developed. We have presented the development of two different biosensor platforms: (a) a platform based on optoelectronics biosensors of evanescent wave detection. Two optical biosensors have been already developed: a portable Surface Plasmon Resonance Sensor (actually in commercialization) and an integrated Mach-Zehnder interferometer device made on Si technology (b) A platform based on nanomechanical biosensors made with an array of microcantilevers. The feasibility of the different biosensors platforms have been proved by the immunological recognition of several pesticides, as the chlorinated compound DDT, and the neurotoxins of carbamate type (carbaryl) and organophosphorus type (chlorpyrifos which resembles SOMAN chemical warfare agent). These results open the way for further development of portable and multianalyte platform for the detection of several biological and chemical warfare agents in-situ and in real-time.

6. ACKNOWLEDGMENTS

This work has been supported by the national projects BIO2000-0351-P4, BIO2001-1235-C03-01 and GEN2001-4856-C13-11. M. Calleja acknowledges CSIC-I3P program for financial support. The authors want to thank to Dr. A. Montoya (UPV) for the immunoreagents.

REFERENCES

1. M. D. Wheeler. Photonics on the Battlefield. Photonics Spectra (1999) 124-132.
2. R.A. Lewis. Biophotonics International, Jan/Feb (2002) 40-41
3. E. Croddy, C. Perez-Armendariz and J. Hart. Chemical and Biological Warfare.
4. Ed.Copernicus Books, Springer-Verlag. NY (2002) S.S. Iqbal, M.W. Mayo, J.G. Bruno, B.V. Bronk, C.A. Batt and J. P. Chambers. Biosens & Bioelec 15 (2000) 459-578.
5. C.A. Rowe-Taitt, J.W. Hazzard, K.E. Hoffman, J.L. Cras, J.P. Golden and F.S. Ligler. Biosens. Bioelec. 15 (2000) 579-589.
6. www.imm.cnm.csic.es/biosensores/home.html

7. L. M. Lechuga. Optical Biosensors. In Biosensors and Modern Biospecific Analytical Techniques. Ed. L. Gorton (2004). Comprehensive Analytical Chemistry Series Elsevier Science BV. Amsterdam (The Netherlands).
8. L. M. Lechuga, F. Prieto and B. Sepúlveda. Interferometric Biosensors for environmental pollution detection. In Optical Sensors for Industrial, E (2003). Springer (Springer Series on Chemical Sensors and Biosensors).
9. www.optonanogen.com
10. F.S. Liegler and C.R. Taitt (eds).Optical Biosensor: Present and future. Elsevier, Amsterdam (NL), (2002)
11. C. Domínguez, L.M. Lechuga and J.A. Rodríguez. Integrated optical chemo- and Biosensors. In Integrated Analytical Systems. Ed. S. Alegret. (2003). Elsevier Science BV. Amsterdam (The Netherlands).
12. R.P.H. Kooyman and L.M. Lechuga. In "Handbook of Biosensors: Medicine, Food and the Environment", ed. E. Kress-Rogers, CRC Press, Florida (USA), (1997) 169-196.
13. L.M. Lechuga, A. Calle and F. Prieto. Optical sensors based on evanescent field sensing. Part I. Surface Plasmon Resonance Sensors. Química Analítica 19 (2000).
14. J. Homola, S.S. Yee and G. Gauglitz. Review. Sens. Act. B, 54 (1999) 3-15
15. C. Barzen, A. Brecht and G. Gauglitz. Biosens. Bioelec. 17 (2002) 289-295
16. S.K. Van Bergen, I.B. Bakaltcheva, J.S. Lundgren and L.C. Shriver-Lake. Environ Sci Technol. 34 (2000) 704.
17. Botchkareva, A.E.; Fini, F.; Eremin, S.; Mercader, J.V.; Montoya, A.; Girotti, S., Anal. Chim. Acta 2002, 453, 43-52.
18. D.P. Campbell and J. McCloskey, Interferometric Biosensor, In "Optical Biosensors: present and future". Ed. F. Liegler and C.Rowe, Elsevier, Amsterdam (NL), (2002) 277-304.
19. F. Prieto, A. Llobera, A. Calle, C Domínguez and L.M. Lechuga. J. Lightwave Tech. 18 (7), 966-972 (2000).
20. F. Prieto, B. Sepúlveda, A. Calle, A. Llobera, C. Domínguez, A. Abad, A. Montoya and L.M. Lechuga. Nanotechnology 14, 907 (2003).
21. L.M. Lechuga, B. Sepúlveda, J. Sánchez del Río, F. Blanco, A. Calle and C. Domínguez. Proceedings of Photonics West (SPIE), Integrated Optoelectronics Devices, 5357, 2004.
22. R. Raiteri, M. Grattarola, H-J. Butt, P. Skládal, Sensors and Actuators B, 79, 115-126 (2001)
23. Fritz, M. K. Baller, H. P. Lang, H. Rothuizen, P. Vettiger, E. Meyer, H. J. Güntherodt, Ch. Gerber, J. W. Gimzewski, Science 288, 316 (2000)
24. Wu, R. H. Datar, K. M. Hansen, T. Thundat, R. J. Cote, A. Majumdar, Nature Biotechnology, 19, 856-860 (2001)
25. Álvarez, A. Calle, J. Tamayo, A. Abad, A. Montoya and L.M. Lechuga. Biosensors & Bioelectronics 18, 649 (2003)
26. Wu, H. Ji, K. Hansen, T. Thundat, R. Datar, R. Cote, M. F. Hagan, A. K. Chakraborty, A. Majumdar, Proc. Natl. Acad. Sci. USA 98, 1560 (2001)

27. M. Calleja, J. Tamayo, A. Johansson, P.Rasmussen, L.M.Lechuga, A. Boisen. Sensors Letters, 1 20-24 (2003).
28. M. Álvarez, L.G.Carrascosa, J.Tamayo, A.Calle and L.M. Lechuga. Proceedings of SPIE (The International Society for Optical engineering). Microtechnologies for the New Millennium 2003: Nanotechnologies 5118 (2003) 197-206.
29. T. M. Herne, M. J. Tarlov, J. Am. Chem. Soc., 119, 8916-8920, 1997.

CATALYTIC BEACONS FOR THE DETECTION OF DNA AND TELOMERASE ACTIVITY

Y. XIAO[1], V. PAVLOV[1], T. NIAZOV[1], A. DISHON[2], M. KOTLER[2] and I. WILLNER[1]

The Institute of Chemistry, The Hebrew University of Jerusalem, Jerusalem, Israel[1] and the Department of Pathology, The Hebrew University-Hadassah Medical School, Jerusalem 91120, Israel.[2]

Abstract: Different new methods for the amplified and specific electronic or photonic detection of DNA will be addressed. Specific methods that will be described include:

1. The amplified electrochemical detection of DNA by means of replication of the analyzed DNA on electrode supports, the incorporation of ferrocene redox-active groups into the replica, and the secondary activation of bioelectrocatalytic cascades.

2. The use of rotating magnetic particles as sensing matrices for DNA using electrogenerated chemiluminescence as an optical readout signal.

3. The magneto-mechanical detection of DNA on cantilever sensors exposed to an external magnetic field. In these systems, magnetic particles are attached to the target-replicated DNA associated with the lever, and the lever is mechanically deflected when subjected to an external magnet.

4. The use of semiconductor quantum dots (QDs) for the photonic detection of DNA will be described by using nucleic acid-functionalized QDs that hybridize with the analyte DNA (CdSe-ZnS or CdTe quantum dots). The replication of the analyzed DNA while incorporating an appropriate dye allows, in the next step, to confirm the hybridization of the analyte DNA by fluorescence resonance energy transfer (FRET). The parallel detection of different DNA/RNA will be described by using QDs of different sizes and compositions.

D. Morrison et al. (eds.), Defense against Bioterror: Detection Technologies, Implementation Strategies and Commercial Opportunities, 199–205.

5. Catalytic nucleic acids that bind hemin and mimic peroxidase activity were employed for the amplified detection of DNA (or proteins). The DNAzyme leads to the generation of chemiluminescence in the presence of luminol/H_2O_2. Different DNAzyme-based catalytic schemes for the detection of DNA will be described, including the use of catalytic beacons and Au-nanoparticle/DNAzyme hybrid systems.

1. INTRODUCTION

The discovery of catalytic RNAs (ribozymes) sparked scientific interest directed to the preparation of new biocatalysts.(1,2) Analogous deoxyribozymes (catalytic DNAzymes) are not available in nature, but have been demonstrated synthetically.(3,4) An interesting DNAzyme that revealed peroxidase-like activities is a complex between hemin and a single-stranded guanine-rich nucleic acid (aptamer).(5) This complex catalyzed the oxidation of 2,2'-azino-bis(3-ethylbenzthiazoline)-6-sulfonic acid, ABTS, by H_2O_2. It was suggested6 that the intercalation of hemin into the complex results in the formation of the biocatalyst. We have shown that the hemin/G-quadruplex also mimics peroxidase by the generation of chemiluminescence in the presence of H_2O_2 and luminol.(7) The use of DNAzymes as catalytic labels for biosensing is attractive since non-specific adsorption processes, associated with protein-based labels are eliminated.

Nucleic acid beacons are extensively used as specific DNA sensing matrices. The specific linkage of photoactive chromophores/quenchers to the hairpin termini results in chromophore luminescence quenching. The subsequent lighting-up of the chromophore luminescence by the hybridization of DNA with the hairpins was used as a general motif for the photonic detection of DNA.(8) The quenching of dyes by molecular or nanoparticle quenchers9 or the fluorescence resonance energy transfer (FRET) between dyes was used for the optical detection of the hybridization of the DNA to the beacon.(8) Recently, the labeling of the beacon termini with redox-active units led to electrochemical detection of hybridization to the hairpins.(10) The development of catalytic beacons may provide a major advance in DNA sensing and recently efforts to apply beacon structures for the catalysed sensing of hybridization were reported.(11) Also catalytic DNA coupled to Au-nanoparticles was reported as a colorimetric sensor for lead irons.(12) Here we report on the tailoring of catalytic beacons for the sensing of DNA and telomerase activity originating from HeLa cancer cells. We design hairpin structures that upon opening yield in the presence of hemin a DNAzyme that allows the biocatalytic detection of the hybridization process.

Scheme 1(A) depicts the method for applying the beacon (1) as a catalytic unit for the sensing of DNA (2)(13). The hairpin structure of (1) includes the sequence consisting of segments A and B that in an open configuration form the G-quadruplex with hemin that exhibits peroxidase-like activity. Since segment B is hybridized in the hairpin structure, the formation of the catalytic DNAzyme is prohibited. Hybridization of DNA (2) with the hairpin opens the beacon and the released sequence (components A and B) self-assemble with hemin to form the catalytic DNAzyme that oxidizes ABTS (3) to the colored product (4) by H_2O_2. The hybridization and hairpin opening is detected spectroscopically by following the accumulation of (4) at λ=414

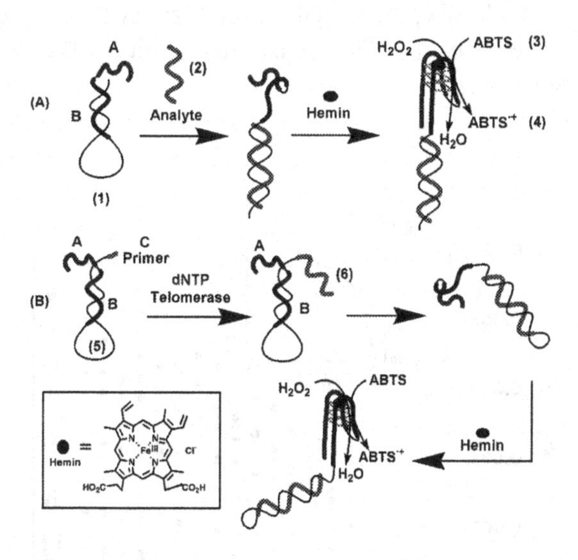

Scheme1: (A) Analysis of DNA by opening of a beacon nucleic acid and the generation of a DNAzyme. (B) Analyzing telomerase activity by a functional DNA beacon that self-generates a DNAzyme.

Scheme1: (nm (ε = 3.6×104 M-1 cm-1). Figure 1, curve (a), shows the time-dependent color evolution upon the analysis of DNA (2) 4.28 μM. Knowing the activity of the pure DNAzyme, we estimate that 85 % of the beacon was opened. The control experiment, curve (c), that follows the spectral changes of the hairpin (1) in the presence of hemin, H_2O_2 and ABTS, and does not lead to any development of a color. Also, the hybridization of (2) with a hairpin structure that lacks the B segment in the "hairpin stem" does not lead to an active DNAzyme. These results indicate that only upon the hybridization of (2) with the beacon (1) and its opening, the DNAzyme that oxidizes ABTS is generated. The extent of opening of the sensing beacons, and thus, the quantity of the generated DNAzyme, is controlled by the concentration of (2). Figure 1, curves (d) to (h) shows the time-dependent evolution of the oxidized product (4), at variable concentrations of the analyzed DNA. As the concentration of (2) increases, the absorbance of (4) is higher. Figure 1, inset, shows the extracted calibration curve obtained upon analyzing variable concentrations of (2) and monitoring the color accumulated after a fixed time-interval of 3 minutes. As expected, the biocatalytic process is enhanced as the concentration of (2) increases. The catalytic beacon reveals specificity and single base mismatches may be discerned. For example, Figure 1, curves (i) and (j) show the time-dependent accumulation of (4) upon analyzing the mutants (2a) and (2b), that include a single-base mismatch relative to the fully complementary analyzed DNA (2). Clearly, the signal for analyzing (2) is 8-fold higher than the signal for the mutants. The sensitivity limit is 0.2 μM (See also supporting information).

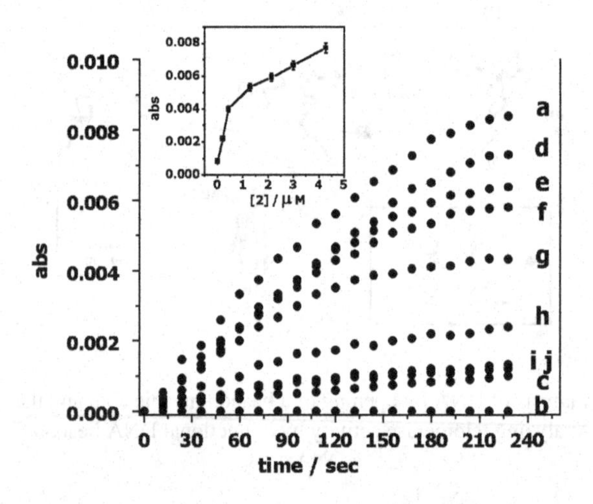

Figure 1. Absorbance changes originating from the formation of (4) upon analysis of: (a) (2), 4.3 μM. (b) Absorbance generated by hemin and (2), 4.3 μM, in the absence of (1). (c) Color formed by hemin and (1) without (2). (d)-(h) Analysis of variable concentrations of (2) corresponding to 3.0 μM, 2.15 μM, 1.30 μM, 0.40 μM and 0.2 μM, respectively. (i) and (j) The analysis of the SNP mutations (2a) or (2b), 4.3 μM. All experiments were performed in the presence of (1), 0.43 μM, hemin, 0.43 μM, ABTS, 3.2 mM and H_2O_2, 3.2 mM in a 0.1M Tris buffer solution, pH=8.1 that included MgC12, 20 mM. Inset: Calibration curve corresponding to absorbance upon analyzing variable concentrations of (2) after a fixed time-interval of 3 minutes.

Telomeres are nucleic acids consisting of constant repeat units at the ends of human chromosomes.14 The telomeres protect the chromosomes, and their erosion during cell proliferation provides a cellular signal for the termination of the cell life cycle. Telomerase is a ribonucleoprotein that replicates the 3'-ends of the linear chromosomes with the telomere repeat units.15 The accumulation of telomerase in cells results in the constant elongation of the telomeres, turning the cells into immortal units. In most cancer and malignant cells increased levels of telomerase were detected, and telomerase is considered as an important biomarker for cancer.16 Several analytical procedures for the determination of telomerase activity were developed, including the telomeric repeat amplification protocol, (TRAP), that involves PCR amplification17 or the functionalization of the telomeres with fluorescent labels.18 We have applied a catalytic beacon as an active component for the analysis of telomerase activity, Scheme1 (B).19 The beacon (5) is designed to include at its two termini two functional nucleic acid components. One end of the hairpin structure ends with a nucleic acid that includes the base sequence that is a part of the DNAzyme in the presence of hemin (part A). The second part of the DNAzyme base sequence (part B) is "hidden" in the hybridized hairpin configuration. At the other end of the hairpin, a nucleic acid segment that is a primer for telomerase, is tethered to the beacon (part C of the beacon). The single stranded loop of the beacon is complementary to the telomere repeat units. Treatment of the beacons with HeLa cancer cell extract in the presence of the dNTP nucleotide mixture, results in the telomerization of the hairpin end. The telomerization was confirmed by gel electrophoresis experiments that showed the extension of the primer attached to the beacon and the formation of telomeres of variable length. The elongated telomere self-generates the sequence for its hybridization with the complementary hairpin loop, and leads to the beacon opening, and to the generation of the DNAzyme. Thus, the telomerase activity is monitored by following ABTS oxidation by H_2O_2 upon opening of the hairpin structure.

Figure 2A, curve (a), shows the time-dependent accumulation of the colored product (4) upon analyzing telomerase originating from 10,000 cells. Figure 2A, curve (b), shows the results of the control experiment where the accumulation of (4) from a system that included a heat-treated deactivated telomerase (10,000 cells). Clearly, only residual formation of (4) is observed presumably, due to non-specific binding of hemin to reaction components. The rate of the telomeres formation is controlled by the amount of telomerase in the sample, and thus the accumulation of (4) is regulated by the number of analysed HeLa cells. Figure 2B, shows the absorbance of (4), obtained upon analysing telomerase activity originating from different numbers of HeLa cells. (The absorbance of (4) is determined after a time-interval of 8 minutes). The detection limit of the HeLa cells is ca. 500 cells.

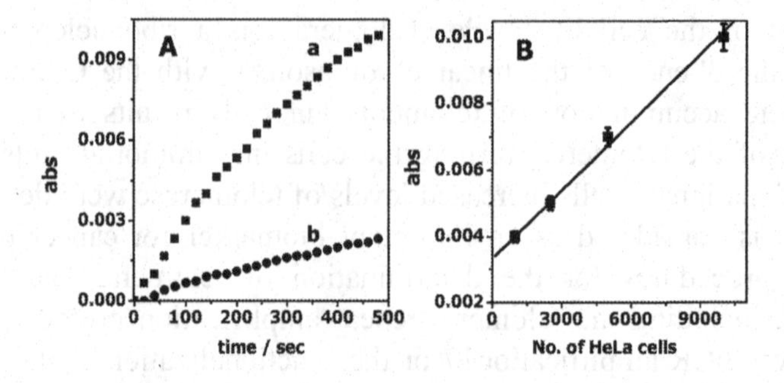

Figure 2. (A): Absorbance changes upon analyzing telomerase activity originating from: (a) 10,000 HeLa cells, (b) 10,000 heat-treated HeLa cells (95°C, 10 minutes). In all experiments the systems consisted of the catalytic beacon (5), 0.04 μM, hemin, 0.04 μM, ABTS, 3.2 mM and H_2O_2, 3.2 mM in 0.1M Tris buffer solution, pH 8.1 that included $MgCl_2$, 20 mM. (B): Calibration curve corresponding to absorbance change of the system upon analyzing variable numbers of HeLa cells.

The present study has introduced a new concept of catalytic beacons for analyzing DNA and telomerase activity. The extension of the concept to include the DNAzyme linked to aptamers20 for the detection of proteins is underway.

2. ACKNOWLEDGEMENT

This research is supported by the Prostate Cancer Charitable Trust (PCCT, U.K.).

3. SUPPORTING INFORMATION AVAILABLE

Sequences and results for analysing the mutants (2c) and (2d) are provided. This material is available free of charge via Internet at http://pubs.acs.org.

REFERENCES

1. Welz, R.; Schmidt, C.; Muller, S. Mol. Biol. 2000, 34, 934-939.
2. Muotri, A.R.; Pereira, L.D.; Vasques, L.D.; Menck, C.F.M. Gene 1999, 237, 303-310.

3. Breaker, R.R. Nat. Biotechnol. 1997, 15, 427-431.
4. Emilsson, G.M.; Breaker, R.R. Cell. Mol. Life Sci. 2002, 59, 596-607.
5. Chinnapen, D.J.F.; Sen, D. Biochemistry 2002, 41, 5202-5212.
6. Travascio, P.; Witting, P.K.; Mauk, A.G.; Sen, D. J. Am. Chem. Soc. 2001, 123, 1337-1348.
7. Xiao, Y.; Pavlov, V.; Gill, R.; Willner, I. ChemBioChem, in press.
8. (a) Tyagi, S.; Kramer, F.R. Nat. Biotechnol. 1996, 14, 303-308. (b) Tyagi, S.; Marras, S.A.E.; Kramer, F.R. Nat. Biotechnol. 2000, 18, 1191-1196.
9. Dubertret, B.; Calame, M.; Libchaber, A. J. Nat. Biotechnol. 2001, 19, 365-370.
10. Fan, C.H.; Plaxco, K.W.; Heeger, A.J. PNAS. USA 2003,100, 9134-9137.
11. Stojanovic, M. N.; de Parda, P.; Landry, D. W. ChemBioChem, 2001, 2, 411-415.
12. Liu J.W.; Lu Y. J. Am. Chem. Soc. 2003, 125, 6642-6643.
13. The following sequences were employed:(1)=5'CCCTACCCAGCCTTAA CTGTAGTACTGGTGAAATTGCTGCCATTTGGGTAGGGCGGGTTGGG3'; (2)=5'AATGGCAGCAATTTCACCAGTACTACAGTTAAGGC 3'; (2a)=5'AATCGCAGCAATTTCACCAGTACTACAGTTAAGGC3'; (2b)=5'AATGGCAGCAATTTCACGAGTACTACAGTTAAGGC3'.
14. Blackburn, E.H. Cell 2001, 106, 661-673.
15. Lingner, J.; Hughes, T.R.; Shevchenko, A.; Mann, M.; Lundblad, V.; Cech, T.R. Science 1997, 276, 561-567.
16. Schalken, J. Eur. Urol. 1998, 34, 3-6.
17. Kim, N.W.; Piatyszek, M.A.; Prowse, K.R.; Harley, C.B.; West, M.D.; Ho, P.L.C.; Coviello, G.M.; Wright, W.E.; Weinrich, S.L.; Shay, J.W. Science 1994, 266, 2011-2015.
18. Patolsky, F.; Gill, R.; Weizmann, Y.; Mokari, I.; Banin, U.; Willner, I. J. Am. Chem. Soc. 2003, 125, 13918-13919.
19. The following sequences were employed: (5)=5'TGGGTAGGGCGGGTT GGGAAA(TAACCC)6AACCCAATCCGTCGAGCAGAGTT3'; (6)=5'AA TCCGTCGAGCAGAGTTAG(GGTTAG)n3'.
20. Nutiu, R.; Li, Y.F. J. Am. Chem. Soc. 2003, 125, 4771-4778.

CRITICAL ELEMENTS OF BIOLOGICAL SENSOR TECHNOLOGY FOR DEPLOYMENT IN AN ENVIRONMENTAL NETWORK SYSTEM

D. IVNITSKI[1], D. MORRISON[1], AND D. J. O'NEIL[2]

[1]*Institute for Engineering Research and Applications, New Mexico Institute of Mining & Technology, Albuquerque, USA and*[2]*CRADA International, Inc., Norcross GA USA*

Abstract: A key factor affected for the ability to protect society against biological and chemical threats is the deployment of an automated sensor monitoring system for use in government and commercial applications. Currently available biosensors have several shortcomings that must be overcome: 1) they are slow to recognize the presence of a pathogen; 2) they are not suitable to discriminate simultaneously a full set of pathogenic vs. non-pathogenic microorganisms in the environment; 3) they cannot be monitored or operated by remote control. The effective testing of biological agents requires new biosensor technology, which should be extremely sensitive, miniaturized, reliable, fast and require less maintenance than current equipment for monitoring of wide spectrum of biological warfare agents in real time.

Recent advances in areas such as microarray technology, microelectromechanical systems, microfluidics, and microseparations present new technological possibilities for producing fast, universal, extremely sensitive, and inexpensive "smart" sensing systems for field application.

The paper is divided into four segments: an overview of the known challenges associated with biological agent detection; discussion of general detection requirements such as ambient environment, selectivity, sensitivity, sampling, shelf life and remote operation; overview of new portable biodetectors for field

D. Morrison et al. (eds.), Defense against Bioterror: Detection Technologies, Implementation Strategies and Commercial Opportunities, 207–220.

applicaton; and lastly, the summary of the future avenues for the electrochemical biosensor systems.

1. INTRODUCTION

The threat of biological weapons has been magnified in recent years due to advances in molecular biology, genetic engineering and related technologies as well as in the development of more efficient delivery and dispersion systems. The advances of biotechnology have facilitated their production and the costs of a funding program have been reduced significantly. According to the Stockholm International Peace Research Institute [1], the comparative costs to produce civilian casualties are "$2,000 per square kilometer with conventional weapons, $800 with nuclear weapons and $1 with biological weapons". Both civilian and military sources predict that, in the next decade, the threat from proliferation of biological weapons will increase. Biological weapons (BW) are possibly the most insidious of the weapons of mass destruction [2-6]. Biological weapons aerosols are usually invisible, odor- and taste-free, and are difficultly detectable due to condensation of liquid droplets on the skin or uniform (low vapor pressure.) Unlike a chemical agent attack a biological attack does not cause an immediate overt physical reaction. An incubation period of 24-48 hours is frequently required before a victim displays serious symptoms that indicate exposure to a biological agent. Consequently both attack on a victim and the spread of the attack is intensive and widely distributed prior to detection. Biological warfare agents, which include bacteria, viruses, fungi, and other living microorganisms, are also many times deadlier than chemical agents. Unlike chemical agents, many biological agents reproduce, multiply inside the host, and are easily passed from person to person, thus spreading throughout a population..

Biological warfare agents may be classified into two broad categories: living microorganisms, and toxins.

The living biological warfare agents include *bacteria*, such as those causing anthrax and the plague or tularaemia, and *viruses*, such as those responsible for diseases such as smallpox, yellow fever or Ebola, and *rickettsiae*, which causes Q fever, and *fung that* act primarily on crops and responsible for potato blights for example.

Toxins are the non-living products of plants or micro-organisms and include ricin and botulinum toxin. They can also be produced by chemical synthesis. Ricin presents a threat not just because of its highly toxic effects in humans, but also because of the wide availability of its source material, the castor plant. Furthermore, the techniques for manufacturing ricin are reasonably well known, and have often been described in the open literature. However, the use of ricin to cause mass casualties would require either its aerosolization by means of a dispersal device or its addition to food and beverages as a contaminant.

The medical effects of biological agents are diverse and are not necessarily related to the type of agent. Some cause pneumonia. Others can cause encephalitis or inflammation of the brain. Each one causes a different complex of symptoms, which can either incapacitate or kill its victim. The effective dose required to induce illness or death may be as great as tens of thousand of organisms as in the case of anthrax, or just a few as with tularemia. With the exception of exposure to a toxin, a period of several days or even weeks may pass before the onset of symptoms and the ultimate effect. This incubation period is the time necessary for the microbe or viral agent to establish itself in the host and replicate. Toxins, on the other hand, are a product of living organisms and behave similar to chemical agents. Botulinum toxin is the most toxic substance known to man. Without supportive care, inhalation of nanograms (10-9 grams) of this agent will cause progressive muscular paralysis leading to asphyxiation and death. Biological agents can be produced quite easily and do not require that much technical capability and expertise to produce. Some potential warfare agents can make their victims very sick without necessarily killing them. Examples include the microorganisms that cause tularemia, Q fever, and yellow fever. After suffering debilitating illness, victims of these diseases often recover, although not always. Other agents are more likely to be lethal. The bacteria that cause bubonic plague and the virus that causes smallpox can kill large number of untreated people. Early antibiotic treatment usually cures plague victims, and smallpox vaccinations before exposure to the viruscan prevent the disease.

2. CHALLENGES ASSOCIATED WITH DETECTION OF BIOLOGICAL WARFARE AGENTS

The rapid detection and accurate identification of biological threat agents in the environment is a key element of biodefense strategy. Considerable effort is being expended on the development of new biosensor technologies and systems to ensure prompt detection and response to a chemical or biological attack. Upon and during attack by biological warfare agents biodetectors preferably should be decentralized and networked to allow the definition of the perimeter of attack and moving fronts, while also having the capability of being moved into the field or area of attack where rapid diagnosis and monitoring can be undertaken. Ideally a response network system should incorporate reliable and portable alarm-type biodetectors and more sophisticated detectors/identifiers. The alarm-type detectors should provide generic discrimination of environmental species, e.g.., biological vs. non-biological species, pathogenic vs. non-pathogenic agents and they should be useful as "triggers" to act both as precursors and to complement a sophisticated detector/identifier system. Effective biodetectors should be capable of operating at anytime and anywhere. Sensor design for deployment in a network should take advantage of remote, wireless communications, such as those used successfully in environmental monitoring and reporting and in telemedicine applications [7-10].

Development of reliable networked biodetectors is a complex undertaking because any of thousands of different biological threat agents could be used in an attack. The complexity is compounded by the fact that there is a huge background of naturally-occurring bio-aerosols, a myriad of non-pathogenic microorganisms constantly present in the environment of a biological threat release. Because many pathogenic organisms differ little from normal flora, a practical detection system has to discriminate accurately and precisely between closely related organisms. "False positives," as well as "false negatives," must be eliminated or drastically minimized to allow reliable biosensing.

There is an enormous demand for new emerging sensor technologies that speed up testing under field conditions. Those novel technologies under development for use in a clinical laboratory environment often are unsuitable for application. However, some may be amenable to modification for field use. A variety of new approaches to biological detection are being pursued. They include (1) replacing the antibody as the identifying molecule with small peptides or combinatorially-derived molecules, (2) replacing the reporter system which notifies the optical reading system that a molecular event has occurred with more robust readouts, (3) using molecular information on the surface of bacterial cells to identify pathogens that are unrelated to antigenic information, and (4) developing a new reagentless approaches for continues monitoring of biological agents in real-time. Nucleic acid approaches for detection and identification of biological warfare and infectious disease agents have been recently evaluated for their potential [16].

Chemical detectors (CDs) are much more developed than biological detectors (BDs) [11-16]. Chemical detectors are able to provide information about chemical agents within seconds or minutes, in near real-time. They generally use transducer technologies including electrochemical, piezoelectric, colorimetric, and optical systems. Very few biological weapons detectors are commercially available. For the most part they ar particle counters. The current biosensor technologies are inherently bulky, complex, and relatively slow. The major technical challenges of the existing biological sensors are in the areas of sample collection, detection and identification, agent discrimination, interferent (i.e., false positive and negative alarms), size/weight and power consumption, including remote/early warning sensing system. Conventional biodetection systems for field application have several shortcomings that must be overcome: 1) they are slow to recognize the presence of a pathogen; 2) they are not suitable to discriminate simultaneously a full set of pathogenic vs. non-pathogenic microorganisms in the environment; 3) except in laboratory settings, they lack adequate sensitivity 4) the systems are often transportable rather than portable, and are cumbersome for field operations; 5) they require highly trained personnel to properly operate them; 6) they cannot be monitored or operated by remote control; and 7) their purchase, maintenance and operation are expensive. Other limitations are the complexity of the instrumentation, multi-step assay processes and the time consuming procedures that are always required.

3. MAIN REQUIREMENTS FOR BIODETECTORS SYSTEM FOR DETECTION AND IDENTIFICATION OF BIOLOGICAL WARFARE AGENTS

The ideal sensor technology would accomplish three aims, namely detection, identification and warning, sufficiently quickly, accurately and with high sensitivity. The effective testing of biological agents requires new sensor technology that will be extremely sensitive, reliable, and fast (Fig. 1). It should be miniaturized, use few consumables (i.e., have a smaller logistics footprint), and require significantly less maintenance than current equipment used to monitor chemical and biological (C&B) warfare agents, and detect in near real-time.

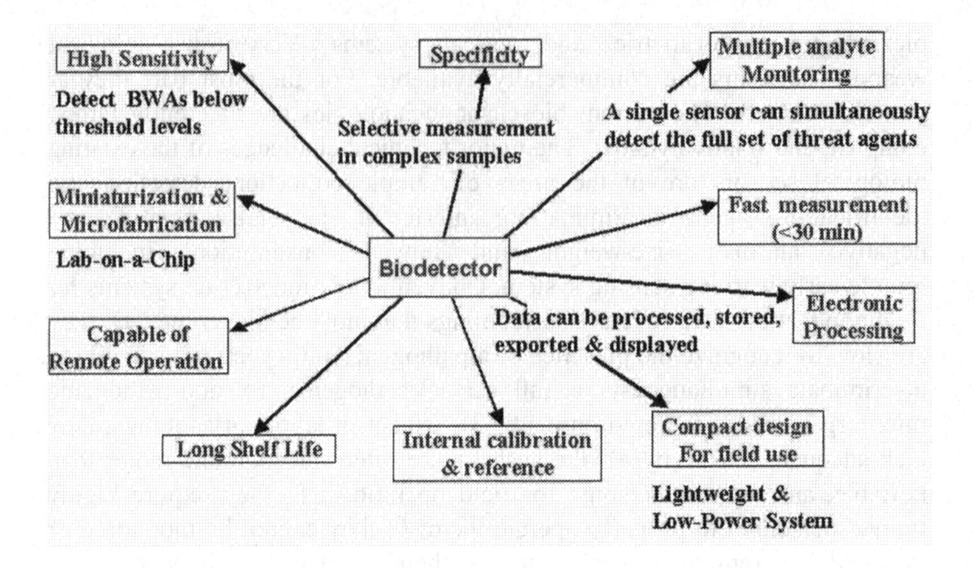

Figure 1. Main requirements of biodetector systems for detection and identification of biological warfare agents

A sensor should be able to detect chemical and biological agents at threshold concentrations in a minimum of 5-10 minutes. It must have the specificity to distinguish target microorganisms in complex samples. That is, an effective sensor must discriminate among real threat agents, interferents and chemical or biological "chaff" (innocuous materials introduced to create false positives). Obviously the incorporation of all the features within a sensor is very complicated and challenging. Solutions to the technological challenge may be realized by the creation of a sensor system and network based on new sensor technologies such as array-based biochips, emerging “lab-on-a-chip” devices, a “Micro Total Analysis System”, and/or PCR-based technologies [12-16]. In recent years, the miniaturization of biochemical and physical processes and their integration onto a single microchip has become a dominant goal of sensor research and development. Recent advances in such areas as microarray technology, microelectromechanical systems, microfluidics, and optoelectronics present new technological possibilities for producing fast, extremely sensitive and inexpensive “smart” sensing systems for field application. Advances in microfabrication methods of silicon chips make it possible to replicate and produce sensor and biosensor arrays coated with specific sensing components with a high degree of reliability, and at a low cost. Miniaturization opens the door for placing complete analytical systems in hostile or remote environments. Sensors based on digital technology have emerged. Their electro-mechanical components are being implemented in silicon via the technology of microelectromechanical systems. Sensors are being developed to incorporate a standard communication interface that enables them to automatically identify themselves and describe their function when they are plugged into a network system.

4. CURRENT BIOSENSOR RESEARCH EFFORTS AT THE NMIM&T INSTITUTE FOR ENGINEERING RESEARCH AND APPLICATIONS

Research at the Institute for Engineering Research and Applications at the New Mexico Institute for Mining & Technology (NMIM&T) has focused on a new generation of highly sensitive and specific devices providing a generic modular sensing platform for development of a wide variety of portable devices for field application.

Figure 2 shows the general configuration of a new hand-held electrochemical biosensor developed in our laboratory [17, 18]. The biosensor consists of a flow-through amperometric detector coupled with a micropipette for injecting a fixed quantity of sample liquid into the flow detector, and an electronic block. The electronic block incorporates an amplifier, a specially designed peak detector, a microprocessor and a liquid crystal display. The peak detector is a peak hold device, designed to track and hold the peak of analog input signals with rise times. The micropipette, amplifier, peak detector, and microprocessor were built in-house.

The amperometric detector includes platinum wire working electrode with an electroactive area of 0.314 ± 0.001 cm^2, platinum wire counter ($A = 2.0$ cm^2) and reference Ag/AgCl disk electrodes. A polarization potential of 0.0 V versus Ag/AgCl is applied between the working and reference electrodes. During electrochemical measurements the selected volume of sample containing an electroactive product of enzymatic reaction is injected into the capillary tube of the working chamber of the biosensor by pressure and immediate release of the micropipette plunger. Technical characteristics of the biosensor include (a) an injected sample volume 30 μl, (b) a response time of 1-3 s, and (c) a sampling frequency of up to 60 samples per hour. The biosensor is compact. Data can be processed, stored, exported, and displayed in multiple ways.

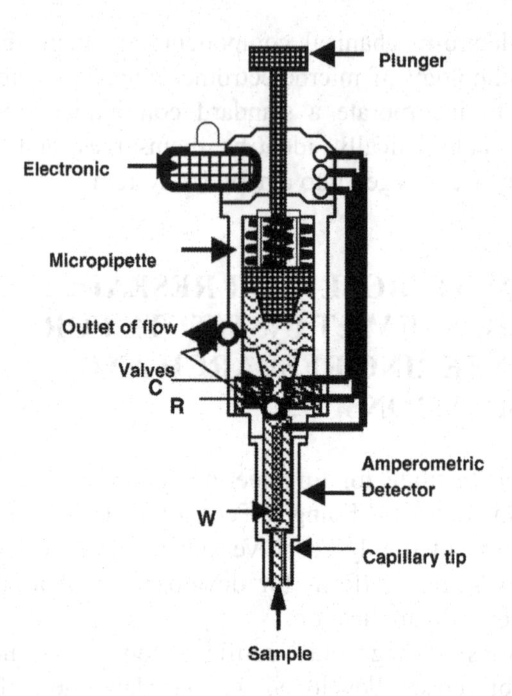

Figure 2. Schematic layout of the hand-held amperometric biosensor: W- working electrode; R- reference electrode; C-counter electrode

This biosensor affords major benefits due to the combination of speed and sensitivity of assay. Current and earlier approaches have usually relied upon the use of a two dimensional format (planar type), in which a suitable set of receptor elements were immobilized on the surface of a planar substrate. A limitation of flat surface geometries is that mass transport of target molecules to the surface is inefficient. This fact negatively impacts the dynamic range, lower detection limit and speed of analysis. The sensitivity and the dynamic range of the NMIM&T hand-held biosensor achieved significant improvement by using a porous working electrode with high surface area. The large electrode surface area provides immobilization of much larger amounts of probe molecules, thus allow fast measurement with high sensitivity simultaneously. The electrode material provides a three-dimensional hydrophilic environment similar to that for bimolecular interactions in a free solution.

The hand-held biosensor has been applied for assay of salivary peroxidase, and for enzyme immunoassay of human luteinizing hormone and human chorionic gonadotropin. A lower detection limit of peroxidase of 0.5 ng/ml (defined as the concentration which gives a signal with $S/N > 2$) was realized. The response time of the sensor was recorded in a range of 1-3 sec. The reproducibility was excellent with a 2.4 % relative standard deviation.

The method for quantitative hormone determination was based on a method of forming a sandwich between the hormone to be detected and two specific monoclonal antibodies, each of which were directed against a different epitope on the hormone molecule. The first antibody was directed against the ß-subunit of the molecule and was attached to a surface of the polystyrene plate, whereas the second antibody was directed against the α-subunit, and was labeled with peroxidase. The immunological principle of the original ELISA remained unchanged, except for the fact that the electroactive product of enzymatic reaction was injected into a capillary tip and detected amperometrically at the working electrode surface of the sensor. It was found that the concentrations of the hormones in physiological situations were within the linear range of the standard curve of the assay. The linear dynamic range ranged from 1.0 to 80 mUI/ml. For comparison, conventional pregnancy tests generally detect pregnancy with lower detection limit of 20 mIU/ml for hCG and 25 mIU/ml for hLH. The proposed hand-held amperometric sensor is ten times more sensitive than the conventional pregnancy tests. The combination of the hand-held device with ELISA allows complete exclusion of sample matrix effects by washing away all components, which might negatively affect the final results of assays. Immunological and enzymatic reactions are performed separately on polystyrene plates and there is passivation of the electrode surface. As a result the irreversible deposition of impurities and formation of insoluble films of oxides on the working electrode surface of the amperometric sensor are excluded. The lifetime of the hand-held amperometric device without need of regeneration of the electrode surface was shown to be at least twelve (12) months. Regeneration of the platinum electrode surface was performed with concentrated HCl for 10 minutes at 100°C and resulted in complete restoration of the initial electrode activity.

In summary the major advantages of the hand-held biosensor are compact design, ease of operation, and the independence of the analytical signals from optical characteristics of solid and liquid phases. In combination with commercially available kits, the biosensor has excellent potential for the enzyme immunoassay of hormones, drugs, viruses, antibodies and bacteria. Furthermore, by conjugation of enzymatic and thiocholine-hexacyanoferrate (III) reactions, the biosensor has an excellent potential for detection and identification of pesticides and other toxic chemicals. Since both pesticides (organophosphates) and nerve gases are chemically similar and work on the same general principal, the biosensor might be utilizable also for detection a broad range of chemical nerve agents.

During the last three years our scientific group at New Mexico Tech has been involved in an comprehensive sensor and protective technology program entitled "Environmental Systems Management, Analysis, and

Reporting neTwork®" (E-SMART®). It is a multidisciplinary program which represents a cooperative effort between the U.S. Air Force Research Laboratory (Tyndall AFB FL,) General Atomics (San Diego CA,) New Mexico Tech, Oklahoma State University and several scientific centers in Russia. One key component of the program has been to locate, evaluate and develop emerging biosensor technologies for detection and identification of biological warfare agents. New Mexico Tech was responsible for technical oversight and collaborative activities for that phase of the E-SMART® program. As a result of the collaboration two engineering models and several prototypes of (1) an Electro-optical Analyzer and (2) a Laser-Based Point Detector for detection of pathogens, were designed, developed and tested [19, 20].

The Electro-optical (EO) Analyzer (Fig. 3) was developed by a team led by Dr. Victor Bunin.of the State Research Center for Applied Microbiology, Obolensk, Russian Federation.

Figure 3. Electro-Optical Analyzer (CELAN)

The principal units of the EO Analyzer are a sample preparation unit, mixer, electrooptical flow cell, AC field generator, processor, operator interface, a microcontroller for liquid stream transfers and a personal computer. Software was designed to allow total control of the complete operation, including video digitalization, object recognition, tracking, calculations, and data storage as well as statistical analysis of the data. The principle of analysis is based upon AC electrokinetic effects, which depend on dielectric properties of bioparticles, their composition, morphology, phenotype, the medium, and the frequency of an applied electrical field. The electro-optical technique includes a number of interrelated processes: 1) action of electric field on the suspended bioparticles, 2) generation of induced charges on the boundaries of cell structures and cell surface, and 3) cell transition into an oriented state, which leads to anisotropic optical properties of the cell suspension. Such effects induce fluctuations in light scattering and relevant variations in optical density result in the production of EO spectra for individual species of bioparticles. In a subsequent step changes of optical density as a result of immunological reaction between cells and specific antibodies produce EO spectra that are used to confirm the presence of individual cell types,. It is a sensitive technique that can be used for the rapid detection of microorganisms at the single organism level.

A second device, portable Laser-based Point detector (LPD) has been developed by a team led by Dr. Irina Moskalenko of the Kurchatov Institute, Moscow). , The LPD is used for detection and identification of pathogens based on laser-induced fluorescence ultraviolet (UV-LIF) spectral measurements (Fig. 4). Preliminary experimental results suggested that LIF UV "spectra" based on "3-D" measurements (spectrum, amplitude, fluorescence decay behavior) held strong potential for discrimination and identification of bacteria. The introduction of the third parameter (kinetics of decay) was used to produce 3-D spectra, overcoming the limitations to "fingerprinting" and discrimination of individual species in mixtures that had been reported in earlier UV_LIF investigations.

Figure 4. . Laser-Based Point Detector

The instrument consists of three major components: (a) an excimer laser operating at 248 nm, (b) a multichannel optical detector (MOD) capable of providing spectral data, and (c) a computer and software to analyze the data from the MOD. The detection process consists of two basic stages: (a) the collection of spectral data and database creation for selected microorganisms at varying concentrations and (b) identification of unknown samples of microorganisms using the database and pattern recognition software program. Detection and identification is based on three ultraviolet laser-induced fluorescence (UV-LIF) parameters, which are processed mathematically to yield characteristic signatures for individual microorganisms. Independent laboratory testing of the Laser-based Point Detector prototype with four microorganisms alone and in mixtures has shown that the point detector at this stage of development has rapid response time (<1 min), an encouraging level of reliability (c. 75%) and good potential for reliable discrimination of different microorganisms. Work is proceeding on improvements for overcoming technical impediments to achieving high levels of reliability and the lowering of detection limits. The LPD technique itself has the advantages of simplicity, low cost, and rapidity and has the potential to be extended for direct (label-free) detection of bacteria, thereby eliminating the need for antibodies.

5. CONCLUSIONS

Analysis of the current status of biosensor technology has shown that despite the great R&D effort spent on developing biosensors in recent years, only a few biosensors for detection of biological warfare agents under field conditions are commercially available or are approaching commercialization. Biosensors simply have not yet been produced with the necessary properties for reliable and effective use in the field. A biosensor system must have the specificity to distinguish the target bacteria in a multi-organism matrix and within a complex bioaersosol background, the adaptability to detect mutliple analytes, to discriminate between benign bioparticles and threat agents, the sensitivity to detect bacteria directly and on-line without pre-enrichment, and the rapidity to give real-time results (<5 min.) Additionally the biosensor must have a relatively simple and inexpensive configuration. Remaining technical problems facing biosensor development include the interaction of matrix compounds, methods of sensor calibration, a requirement for reliable and low maintenance functioning over extended periods of time, sterilization (particularly for clinical applications), reproducible fabrication of numerous sensors, the ability to manufacture the biosensor at a competitive cost, in a disposable format, and for a clearly identified and sustainable market. Networked biosensors must also be easy to regenerate after each measurement or be self-regenerating, and continue to yield reproducible results. The requirement for human participation in all phases of the operation of a biosensor should be minimized or eliminated not only to avoid contamination but to reduce operational costs. Hence, automation should be sought as an inherent attribute of the biosensor.

REFERENCES

1. Graham S. Pearson, The Threat of Deliberate Disease in the 21st Century, Henry L.Stimson Centre Report, No.24, January 1998.
2. Committee on R&D Needs for Improving Civilian Medical Response to Chemical and Biological Terrorism Incidents, Institute of Medicine. Chemical and Biological Terrorism: Research and Development to Improve Civilian Medical Response, National Academy Press, Washington, DC 1999, 279 p
3. Tucker JB. Chemical/Biological Terrorism: Coping with a New Threat. Politics and the Life Sciences (1996) 5:167-184.
4. Wiener S.L. Strategies for the Prevention of a Successful Biological Warfare Aerosol Attack. Military Medicine (1996) 161, 5: 251-256.
5. Atlas R.M. The Medical Threat of Biological Weapons. Critical Reviews in Microbiology, (1998) 24,3:157-168.
6. Malcolm Dando, Biological Warfare in the 21st Century: Biotechnology and the Proliferation of Biological Weapons, Brassey's, U.K.: London, 1994.
7. Greenfield, R.A., B.D. Lutz, M.M. Huycke, and M.S. Gilmore. 2002. Unconventional biological threats and the molecular biological response to biological threats. Am. J. Med. Sciences 323: 350-357.
8. Beeching, N.J., D.A.B. Dance, A.R.O. Miller, and R.C. Spencer. 2002. Biological warfare and bioterrorism. Brit. Med. J. 324: 336-339.
9. Deisingh A.K., and M. Thompson. 2002. Detection of infectious and toxigenic bacteria. Analyst, 127: 567-581.
10. Walt, D., and D.R. Franz. 2000. Biological warfare detection. Anal Chem. 72: 739A-746A.
11. Iqbal, S.S., M.W. Mayo, J.G. Bruno, B.V. Bronk, C.A. Batt, and J.P. Chambers. 2000. A review of molecular recognition technologies for detection of biological threat agents. Biosens. Bioelectron. 15: 549-578.
12. Rowe, C.A., L.M. Tender, M.J. Feldstein, J.P. Golden, S.B. Scruggs, B.D. MacCraith, J.J. Cras, and F.S. Ligler. 1999. Array biosensor for simultaneous Identification of bacterial, viral, and protein analytes. Anal. Chem. 71: 3846-3852.
13. McDonald, R., T. Cao, and R. Borschel. 2001. Multiplexing for the detection of multiple biowarfare agents shows promise in the field. Mil. Med. 166: 237-239.
14. Uhl, J.R., C.A. Bell, L.M. Sloan, M.J. Espy, T.F. Smith, J.E. Rosenblatt, and F.R. Cockerill. 2002. Application of rapid-cycle real-time polymerase chain reaction for the detection of microbial pathogens: the Mayo-Roche rapid anthrax test. Mayo Clin. Proc. 77: 673-680.

15. Stratis-Cullum, D.N., G.D. Griffin, J. Mobley, A.A. Vass, and T. Vo-Dinh. 2003. A miniature biochip system for detection of aerosolized *Bacillus globigii* spores. Anal. Chem. 75: 275-280.
16. D. Ivnitski, D. O'Neil, A. Gattuso, R. Schlicht, M. Moore, M. Calidonna, R. Fisher. (2003). (Review). Nucleic acid approaches for detection and identification of biological warfare and infectious disease agents, BioTechniques, 35 (4), 862-869.
17. D. Ivnitski, R. Sitdikov, N. Ivnitski. (2004). Hand-Held Biosensor for Saliva and Other Fluid-Based Diagnostics. Anal. Chim. Acta, 504, 265-269.
18. D. Ivnitski, R. Sitdikov, N. Ivnitski. (2003). Non-Invasive Electrochemical Hand-Held Biosensor as Diagnostic Indicator of Dental Diseases. Electrochemistry Communications, 5/3, 225 – 229.
19. Bunin V. D., O. V. Ignatov, O. I. Guliy, A. G. Voloshin., L. A. Dykman, D. O'Neil, D. Ivnitski. (2004). Studies of *Listeria Monocytogenes*-Antibody Binding Using Electroorientation. Biosensors & Bioelectronics 19, 1759-1761.
20. Bunin V. D., O. V. Ignatov, O. I. Guliy, I.S. Zaitseva, D. O'Neil, D. Ivnitski. (2004). Electrooptical analysis of the E. coli-Phage M13K07 interaction in real-time. Anal. Biochemistry, 328, 181-186.

ELECTROCHEMICAL IMMUNOSENSOR FOR DETECTION OF FRANCISELLA TULARENSIS

P. SKLADAL[1], Y. SYMERSKA[1], M. POHANKA[1], B.SAFAR[2], and A. MACELA[3]

[1]*Department of Biochemistry, Masaryk University, Kotlářská 2, 61137 Brno;*
[2]*Military Technical Institute of Protection, Brno;*
[3] *Purkyně Military Medical Academy, Hradec Králové (Czech Republic)*

Abstract: An immunosensor for rapid and sensitive detection of *Francisella tularensis* (Ft) was developed using screen-printed sensors with multiple (4) Au working electrodes. The sensing zones were modified with a biolayer of anti-Ft polyclonal antibody attached through the covalently immobilized Protein A. The sandwich assay format was employed with anti-Ft IgG-peroxidase conjugate as tracer. For electrochemical measurements, an automated portable detector was developed; it consisted of a digitally controlled potentiostat and four embedded miniperistaltic pumps. The flow-through measurement involved sequential incubations with sample (10 min), tracer (10 min) and substrate mixture (iodide and hydrogen peroxide). The enzymatically-produced iodine was detected electrochemically at - 50 mV. The limit of detection was 100 Ft cells/ml; the response was significantly higher (3x) when using the living cells compared to the heat inactivated ones.

D. Morrison et al. (eds.), Defense against Bioterror: Detection Technologies, Implementation Strategies and Commercial Opportunities, 221–232.

1. INTRODUCTION

Francisella tularensis (Ft) is the gram-negative microorganism causing the disease tularemia. Ft biovars with different levels of virulence were described [1,2]. This bacterium belongs to the list of potential bioweapons, therefore several bioanalytical methods are focused both on detection of the bacterial cells in various sample matrices and on identification of either microbial cells or antitularemic antibodies in the infected individuals. For the latter case, immunochemical methods as Enzyme-Linked Immunosorbent Assay (ELISA) were reported 20 years ago [3]. For potential production of anti Ft antibodies, different proteins with antigenic determinants were considered [4,5] and monoclonal antibodies were developed against e.g. the 43 kDa outer membrane protein [6] and lipopolysaccharide [7]. The former ones were specific against Ft, while other antibodies against capsular components exhibited a broader specificity to other species [8]. A micropoint enzyme assay carried out on membrane with visual evaluation clamed similar sensitivity as ELISA for anti Ft antibodies [9]. A chemiluminiscence immunoassay declared limit of detection equal to 1 ng/ml of tularemic antigens [10].

In addition to immunochemical methods, specific DNA sequences of Ft are detected. The polymerase chain reaction (PCR) method for Ft cells was introduced quite early and the detection of 1000 cfu/ml in blood was reported [11]. This method allows early identification of infection, as anti Ft antibodies appears two weeks after onset of the disease, and cultivation of the causative agent tends to be avoided due to a risk of infection [12]. For clinical samples, this approach was found more sensitive than traditional culture tests and it was also evaluated on aerosols [13]. Different PCR methods were compared [14,15] including the real-time PCR [16] and the highly sensitive PCR-enzyme immunoassay combination [17]. A promising alternative to bioanalytical methods seems to be the pyrolysis of microorganisms followed by mass spectrometric detection [18].

From the practical point of view, the studies comparing several methods of detection are highly interesting. The capture ELISA was able to detect 10^3 to 10^4 cfu/ml, immunochromatography was 100x less and PCR 10x more sensitive during analysis of tissue samples [19]. On the other hand, the same group of methods was tested in field on water samples, and immunochromatography was found as the most versatile one [20]. At present, a critical need also exists for rapid, sensitive and field-deployable detection systems for infectious agents; such requirements are satisfied by various types of biosensors. For Ft detection, several biosensing approaches were tested. The portable automated fiber optic biosensor RAPTOR (Research International) employs the fluoroimmunoassay format, detection of $5 \cdot 10^5$ Ft cfu/ml was achieved in 10 min [21]. The diffractive grating-based biosensor (BDG) was able to detect $3 \cdot 10^4$ cfu/ml, measuring time was around 30 min [22]. Alternatively, the conventional microplate ELISA formats are coupled to signal amplification systems as time-resolved fluorescence (DELFIA, Perkin Elmer [23]) and electroluminescence (M1R device, BioVeris). An important advantage of such systems is the simultaneous analysis of several target agents [24].

In this contribution, the electrochemical immunosensing system for determination of *Francisella tularensis* will be described. This approach combines the high sensitivity of electrochemical measurements with excellent specificity of antibodies [25]. In addition, rather simple electrochemical instrumentation is favorable for construction of portable sensing systems.

2. MATERIALS AND METHODS

2.1 Chemicals

Anti Ft rabbit polyclonal antibodies (Ab) were developed in the Military Medical Academy, Hradec Králové. The crude serum was purified using the protein L column CBind L (Fluka) [26] and thus obtained IgG fraction was further used. Peroxidase (POD, horse radish), 5-aminosalicylic acid (ASA), cystamine and glutaraldehyde were supplied by Sigma. All other chemicals were from Lachema (Brno). The antibody - peroxidase conjugate (Ab-POD tracer) was obtained by coupling the periodate-oxidized peroxidase with IgG [27]. The *Francisella tularensis* LVS cells were used in the killed form (in 0.5% phenol) suitable for long-term storage. The living Ft cells were obtained from a freshly cultivated culture and used either in the viable state or partially heat-inactivated (reduced virulence).

2.2 Preparation of Immunosensors

The electrochemical system was produced by screen-printing (Fig. 1). The 4 gold-paste based working electrodes (1 mm diameter) were placed on an alumina support (8 x 50 mm). The gold surface was cleaned with acetone and a self-assembled monolayer was formed during a 3-hour incubation with cystamine (20 mg/ml in water). After washing with water, the amino groups were activated for 2 hours with glutaraldehyde (2.5% solution in 50 mM phosphate buffer, pH 7.0).

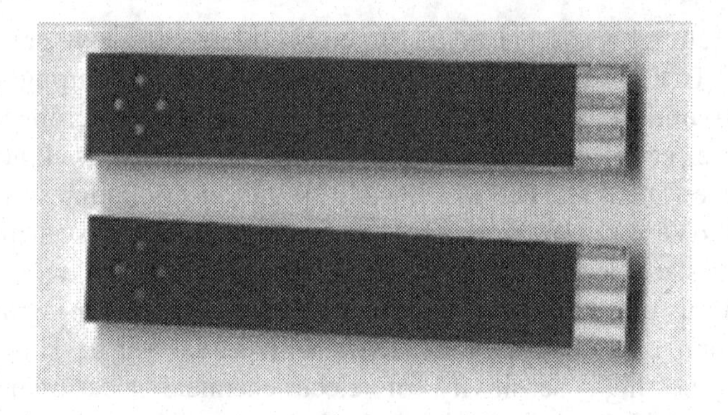

Figure 1.

Figure 1. The four-channel screen-printed sensors. Produced according to our design in BVT Technologies, Brno. The ceramic substrate (alumina) was sequentially coated with contact layer (silver), working electrodes (either Pt or Au based inks) and insulating layer.

Thus obtained reactive aldehyde groups were used to couple protein A through its surface lysine aminogroups; incubation with protein A (0.5 mg/ml in phosphate) proceeded overnight in the refrigerator. After washing, antibodies were attached to protein A (overnight incubation, 0.3 mg/ml Ab). The sensing layer (2x) contained the anti tularemic polyclonal antibody (IgG) and the reference layer (2x) was coated with a non-specific anti albumin IgG. The completed immunosensors were stored in dry conditions.

Prior to use, the measuring area of the sensor was incubated for 1 hour with 0.1 % bovine serum albumin in order to saturate any non-specific binding sites.

2.3 Measuring Procedure

The immunosensor was fixed in a thin-layer flow-through cell made from Plexiglas, and the internal volume was 8 μl. The cell also contained an embedded silver pseudoreference electrode (diameter 3 mm). The initial measurements were carried out using a stand-alone 4-channel electrochemical detector MEB (Ing. J. Kitlička, Brno), a miniature peristaltic pump PP1.3 (BVT Technologies, Brno) and manual switching of solutions was employed. Later, the combined electrochemical system MultiLab 6150 with included minipumps was used. The flow-through rate was 50 μl/min. Immunoreactions (injected zones of sample and tracer) and washing steps were carried out in 50 mM phosphate buffer pH 7.0 containing 1 mg/ml albumin. The response was measured in 50 mM acetate buffer pH 4.5 containing 5 mM H_2O_2 and this was mixed with a zone of 10 mM KI. The iodine produced by the peroxidase label was detected amperometrically at -50 mV versus the silver pseudoreference electrode.

3. RESULTS AND DISCUSSION

3.1 Electrochemical immunosensing system

The previously developed MEB device (multichannel electrochemical biosensor) was equipped with miniature peristaltic pumps in order to obtain the enhanced portable system, MultiLab 6150, suitable for semiautomatic measurements. The device (Figure 2) is completed with a flow-through cell containing the exchangeable immunosensor and a holder of reagent vials and waste container. The following modules compose electronics of the detector:

- potentiostat - 4-channel, digitally controlled (ranges 12 nA, 120 nA, 1.2 μA, 12 μA, common reference, the same working potential for all channels, two-electrode mode)
- analog to digital (16 bit) and digital to analog (12 bit) converters
- 2 to 4 miniperistaltic pumps, variable flow rate (10 to 200 μl/min, 16 steps)
- pump microcontroller board (PIC1)
- central microprocessor (PIC2) controlling measurements and communication with external computer (RS232, options USB and Bluetooth)
- rechargeable battery pack (around 10 hours of stand-alone operation)

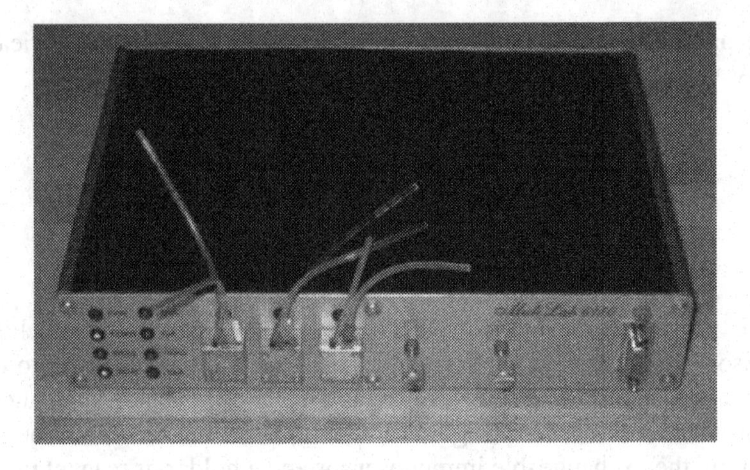

Figure 2. *The prototype electrochemical immunosensor detector MultiLab 6150. Designed and constructed by J. Kitlička, Brno. The front panel contains LED controls. 3 pumps, 2 screws for the flow-through cell and vial holder fixation and a connector for the immunosensor.*

The modular concept allows the system to be extended with additional existing modules, including: single-channel highly sensitive potentiostat, 8-channel potentiostat with variable working potentials, piezoelectric detector, physical sensors detector, and impedimetric detector.

3.2 Optimization of the Immunosensor Performance

The electrochemical immunosensor employed the sandwich assay format schematically shown in Figure 3. The antibody layer was immobilized on the gold electrode surface, and the target microbes captured during incubation with sample were subsequently labeled with the peroxidase-antibody tracer during the second incubation. The assay was heterogeneous, as washing steps were included after each incubation. Finally, the surface bound peroxidase activity was determined electrochemically, and the enzymatically-produced iodine was measured at a negative potential.

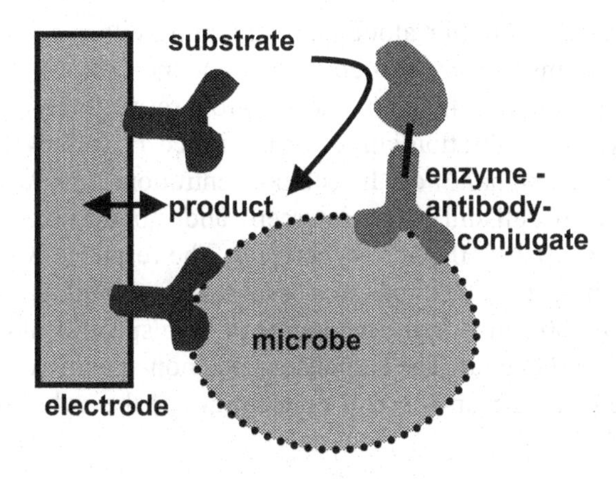

Figure 3. *Sandwich format of the immunosensor for detection of microbes.*

To simplify the development phase, it was decided to use each immunosensor only once; the appropriate regeneration procedure necessary for the potential repeated use was not studied. The experiments were carried out using the killed Ft cells to simplify the necessary safety measures. The initial optimization of the assay procedure started with a composition of the buffer used for immunochemical reactions. In order to suppress the background signal observed in the absence of any Ft cells (control responses), the following mixtures were tested (Table 1):

Table 1. Effect of the immunoassay buffer composition. The responses (Δi, μA) obtained for immunosensors incubated for 10 min in the absence (control) and presence of Ft cells, incubation with tracer was 10 min.

Buffer / Sample	Control	Ft ($5 \cdot 10^5$ cells/ml)
50 mM phosphate pH 7.0	0.63	0.96
+ 150 mM NaCl	0.49	0.42
+ 150 mM NaCl + 0.01 % Tween 20	0.60	0.28
+ 1 mg/ml bovine serum albumin	0.29	0.54

The incubation time with the sample was optimal at 10 min. A longer incubation did not improve the response, however, a shorter period might be adopted in future to shorten the overall assay time. Similarly, the incubation with tracer was fixed at 10 min,. In the case of the tracer, a higher response (1.4x) was observed after 30 min; this would however make the analysis time substantially longer. Furthermore, presaturation of the sensing surface was considered in order to reduce the background signal. For this purpose, solutions of phosphate buffer alone and containing either 5 mg/ml albumin or 5% defatted milk were tested (30 min incubation); the obtained background values were 0.42, 0.35, and 0.54 μA, respectively.

Finally, the calibration curve for the killed Ft cells was constructed as a dependence the signal Δi on the cell concentration: $\Delta i = A + B \cdot \log(cFT)$, ($\Delta i$ in μA, *c*FT in cells/ml). The intercept and slope values were A = -0.43±0.31 and B = 0.495±0.074, respectively. The response was linear up to 10^6 cells/ml, limit of detection was estimated at 500 cells/ml; at the concentration of 100 cells/ml, the responses of both specific and non-specific channels were nearly equal. The response of the non-specific channel did not exhibit any dependence on the cell concentration, the value of slope was B = -0.008±0.017.

3.3 Determination of *Francisella tularensis*

The final part describes evaluation of the developed immunosensor on living Ft cells. These experiments were carried out in the specialized certified category 4 microbiological laboratory. The experimental response curves obtained for several samples containing increasing concentration of Ft cells are shown in Figure 4 for both partially inactivated (A) and fully viable (B) cells.

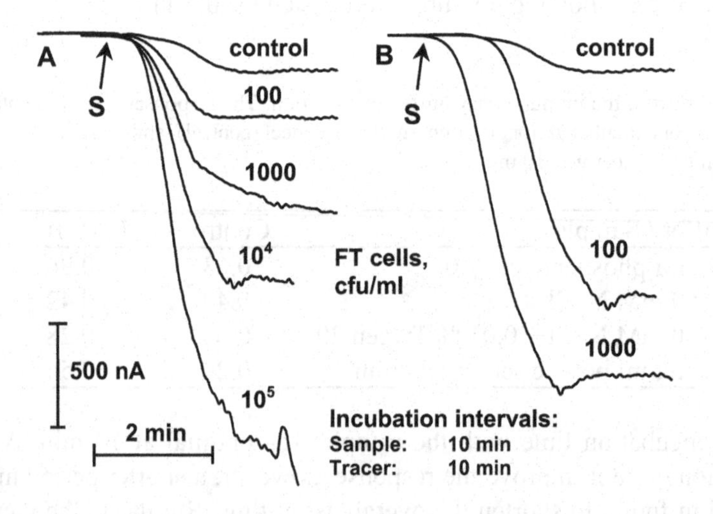

Figure 4 Response traces to the substrate mixture for immunosensors after incubation (10 min in flow) with different samples of inactivated (A) and viable (B) *F. tularensis* cells (concentrations indicated near the curves). Control indicates samples without any cells. The carrier buffer was 50 mM acetate pH 4.5 containing 5 mM H_2O_2, S indicates addition of 10

mM KI into the flowing medium. The traces of current from the specific channel with anti Ft antibodies are shown.

Figure 1. The initial part of the amperometric traces before addition of the second substrate (iodide) was between 0 and -0.05 μA. The control values obtained in the absence of any cells should be considered as the background signal due to the remaining non-specific adsorption of the tracer and partially also due to the non-enzymatic reaction of the substrate mixture. However, the change of signal in the presence of 100 cfu/ml of Ft was clearly visible. The values of current from Figure 1 together with the values from the non-specific reference channel are summarized in Figure 5; the data that were obtained 2 min after introduction of the substrate mixture into the flow-through cell.

Surprisingly, the viable cells provided significantly (2x) higher responses compared to the inactivated cells. This result is most probably due to the partial thermal destruction of the surface antigens of the *F. tularensis* cells. Therefore, the binding of the inactivated cells with the immobilized antibodies is weaker and smaller amount of cells became captured. Similarly, the tracer binding to the captured cells might be reduced, too. This observation will be further investigated in future.

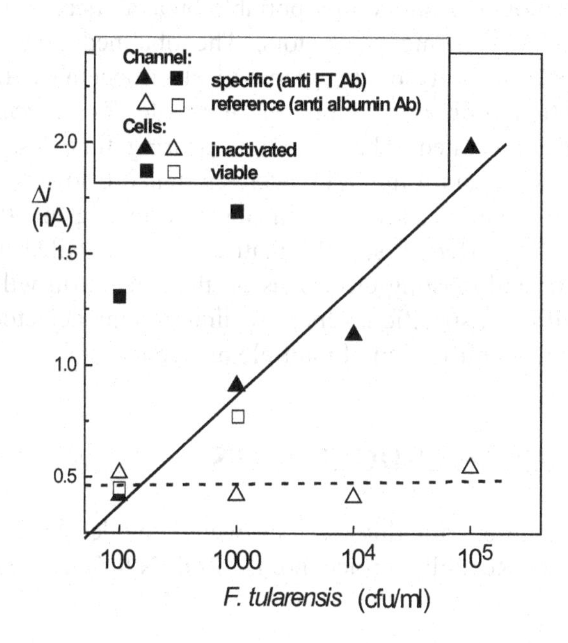

Figure 5. Calibration of the immunosensor for Francisella tularensis. The lines indicate linear regressions $\Delta i = A + B\log(cFT)$, the parameters were: $A = -0.61 \pm 0.30$ and $B = 0.492 \pm 0.083$

(full line, the channel with anti FT Ab) and A = 0.45±0.13 and B = 0.0067±0.037 (dashed line, the non-specific reference channel with anti albumin Ab).

However, as can be seen from Figure 5, the response of the reference channel was increased for the viable cells, too. When comparing the specific/reference channels ratio, this was almost the same at 1000 cfu/ml, i.e. 2.17 and 2.08 for inactivated and viable cells. A dramatic difference was observed at 100 cfu/ml, when the ratio was 0.80 and 3.1 for inactivated and viable cells, respectively. Therefore, the presence of 100 cfu/ml of viable Ft cells was clearly demonstrated. On the other hand, for the inactivated cells, the reliably detectable concentrations of Ft cells start at 1000 cfu/ml.

The achieved results are fully comparable to other biosensor methods applied for *F. tularensis* detection [21,22] and achieve even the sensitivity of ELISA [19] and PCR [11] assay formats mentioned above.

4. CONCLUSIONS

It was demonstrated that the screen-printed electrodes represent a convenient measuring platform suitable for construction of disposable immunosensors for detection of bacteria. This technology allows development of low-cost and easily portable bioanalytical systems based on multichannel electrochemical detectors. The obtained preliminary results confirmed high sensitivity of the proposed electrochemical immunosensor and the detection limit at 100 cfu/ml of viable *F. tularensis* cells was successfully demonstrated. The overall measuring time less than 30 min seems to be feasible and it might be even shortened if lower sensitivity will be satisfactory. Future research will include testing of the developed immunosensor in different sample matrices and detailed validation. In addition, the several existing electrodes on the same strip will be modified with several different specific layers providing immunodetectors suitable for the simultaneous identification of multiple bioagents.

5. ACKNOWLEDGEMENTS

Financial resources for this research were provided by the Ministry of Defense of Czech Republic (project no. ISPROFIN 907910-6620).

REFERENCES

1. Morner, T., Mattsson, R., Forsman, M., Johansson, K.E. and Sandstrom, G. (1993) Identification and classification of different isolates of Francisella-tularensis. *J. Vet. Med. B* **40**, 613-620.
2. Kormilitsyna, M.I. and Meshcheryakova, I.S. (1996) The new vaccine strains (or variants) of Francisella tularensis. *FEMS Immunol. Med. Microbiol.* **13**, 215-219.
3. Umnova, N.S., Shakhanina, K.L., Meshcheryakova, I.S. and Pavlova, I.P. (1984) Determination of tularemia antibodies by means of the solid-phase ELISA. *Zh. Microbiol. Epidemiol. Immunobiol.* (4), 79-83.
4. Štulík, J., Černá, J., Kovářová, H. and Macela, A. (1989) Protein heterogeneity of Francisella-tularensis - detection of proteins with antigenic determinants. *Folia Microbiol.* **34**, 316-320.
5. Havlasová, J., Hernychová, L., Halada, P., Pellantová, W., Krejsek, J., Štulík, J., Macela, A., Jungblut, P.R., Larsson, P. and Forsman, M. (2002) Mapping of immunoreactive antigens of Francisella tularensis live vaccine strain. *Proteomics* **2**, 857-867.
6. Bevanger, L., Maeland, J.A. and Naess, A.I. (1989) Competitive enzyme-immunoassay for antibodies to a 43,000-molecular-weight Francisella-tularensis outer-membrane protein for the diagnosis of tularemia. *J. Clin. Microbiol.* **27**, 922-926.
7. Fulop, M.J., Webber, T., Manchee, R.J. and Kelly, D.C. (1991) Production and characterization of monoclonal-antibodies directed against the lipopolysaccharide of Francisella-tularensis. *J. Clin. Microbiol.* **29**, 1407-1412.
8. Bhatti, A.R., Wong, J.P. and Woods, D.E. (1993) Production and partial characterization of hybridoma clones secreting monoclonal-antibodies against Francisella-tularensis. *Hybridoma 12*, 197-202.
9. Yurov, S.V., Pchelintsev, S.Y., Afanasyev, S.S., Vorobyev, A.A., Urakov, N.N., Cherenkova, G.V., Fedortsov, K.K., Krasnoproshina, L.I., Vlasov, G.S. and Denisova, N.B. (1991) Use of the micropoint enzyme-immunoassay with visual indication for the detection of tularemia antibodies. *Zh. Microbiol. Epidemiol. Immunobiol.* (3), 61-64.
10. Vidziunaite, R., Mikulskis, P. and Kulys, J. (1995). Chemiluminescent immunoassay (CLIA) for the detection of brucellosis and tularemia antigens. *J. Biolumin. Chemilumin.* **10**, 199-203.
11. Long, G.W., Oprandy, J.J., Narayanan, R.., Fortier, A.H., Porter, K.R. and Nacy, C.A. (1993) Detection of Francisella-tularensis in blood by polymerase chain-reaction. *J. Clin. Microbiol.* **31**, 152-154.
12. Sjostedt, A., Eriksson, U., Berglund, L. and Tarnvik, A. (1997) Detection of Francisella tularensis in ulcers of patients with tularemia by PCR. *J. Clin. Microbiol.* **35**, 1045-1048.
13. Zhai, J.H., Yang, R.F., Lu, J.C., Zhang, G.L., Chen, M.L., Che, F.X. and Hong, C. (1996) Detection of Francisella tularensis by the polymerase chain reaction. *J. Med. Microbiol.* **45**, 477-482.
14. Johansson, A., Ibrahim, A., Goransson, I., Eriksson, U., Gurycova, D., Clarridge, J.E. and Sjostedt, A. (2000) Evaluation of PCR-based methods for discrimination of Francisella species and subspecies and development of a specific PCR that distinguishes the two major subspecies of Francisella tularensis. *J. Clin. Microbiol.* **38**, 4180-4185.

15. De la Puente-Redondo, V.A., Del Blanco, N.G., Gutierrez-Martin, C.B., Garcia-Pena, F.J., Ferri, E.F.R. (2000) Comparison of different PCR approaches for typing of Francisella tularensis strains. *J. Clin. Microbiol.* **38**, 1016-1022.
16. Shaw, J.J., McCleskey, F.K., Beninga, K.K., Candler, W., Redkar, R.J., DelVecchio, V.G. and Goode, M.T. (1998) Rapid detection of tularemia with real-time PCR. *Infect. Med.* **15**, 326-330.
17. Higgins, J.A., Hubálek, Z., Halouzka, J., Elkins, K.L., Sjostedt, A., Shipley, M. and Ibrahim, M.S. (2000) Detection of Francisella tularensis in infected mammals and vectors using a probe-based polymerase chain reaction. *Am. J. Tropical Med. Hygiene* **62**, 310-318.
18. Basile, F., Beverly, M.B., Voorhees, K.J. and Hadfield, T.L. (1998) Pathogenic bacteria: their detection and differentiation by rapid lipid profiling with pyrolysis mass spectrometry. *Trends Anal. Chem.* **17**, 95-109.
19. Grunow, R., Splettstoesser, W., McDonald, S., Otterbein, C., O'Brien, T., Morgan, C., Aldrich, J., Hofer, E., Finke, E.J. and Meyer, H. (2000) Detection of Francisella tularensis in biological specimens using a capture enzyme-linked immunosorbent assay, an immunochromatographic handheld assay, and a PCR. *Clin. Diagn. Lab. Immunol.* **7**, 86-90.
20. Berdal, B.P., Mehl, R., Haaheim, H., Loksa, M., Grunow, R., Burans, J., Morgan, C. and Meyer, H. (2000) Field detection of Francisella tularensis. *Scand. J. Infect. Dis.* **32**, 287-291.
21. Anderson, G.P., King, K.D., Gaffney, K.L. and Johnson, L.H. (2000) Multi-analyte interrogation using the fiber optic biosensor. *Biosens. Bioelectron.* **14**, 771-777.
22. O'Brien, T., Johnson, L.H., Aldrich, J.L., Allen, S.G., Liang, L.T., Plummer, A.L., Krak, S.J. and Boiarski, A.A. (2000) The development of immunoassays to four biological threat agents in a bidiffractive grating biosensor. *Biosens. Bioelectron.* **14**, 815-828.
23. Peruski, A.H., Johnson, L.H. and Peruski, L.F. (2002) Rapid and sensitive detection of biological warfare agents using time-resolved fluorescence assays. *J. Immunol. Meth.* **263**, 35-41.
24. Taitt, C.R., Anderson, G.P., Lingerfelt, B.M., Feldstein, M.J. and Ligler, F.S. (2002). Nine-analyte detection using an array-based biosensor. *Anal. Chem.* **74**, 6114-6120.
25. Skládal, P. (1997). Advances in electrochemical immunosensors [review]. *Electroanal.* **9**, 737-745.
26. Thomas, T.M., Shave, E.E., Bate, I.M., Gee, S.C., Franklin, S. and Rylatt, D.B. (2002) Preparative electrophoresis: a general method for the purification of polyclonal antibodies. *J. Chromatogr. A* **944**, 161-168.
27. Catty, D. and Raykundalia, C. In: *Antibodies Vol. 1* (Ed.: Catty, D.), IRL Press, Oxford, 1991, p. 97.

BIOSENSORS AND NANOTECHNOLOGICAL IMMUNOCHIPS FOR THE DETECTION AND MONITORING OF CHEMICAL AND BIOLOGICAL AGENTS

S. VARFOLOMEYEV, I. KUROCHKIN, A. EREMENKO, E. RAININA AND I. GACHOK

The Lomonosov Moscow State University, Faculty of Chemistry, Chemical Enzymology Department, Lenin's Hills, 1/11, Moscow, 119992 Russia Institute of Biochemical Physics, Russian Academy of Science

Abstract: The elaboration of highly sensitive and express methods for quantitative and qualitative detection and monitoring of chemical warfare agents (CWA), organophosphate and carbamate pesticides, compounds with delayed neurotoxicity, and pathogenic microorganisms and viruses is discussed. The application of potentiometric and amperometric biosensors, automatic biosensors discriminating the neurotoxins of different classes, is performed. The information about biosensors detecting the compounds with delayed neurotoxicity through the evaluation of "neurotoxic esterase" activity in the blood is presented. The use of immunochip technology for the detection of pathogenic microorganisms and viruses is demonstrated. The enzymatic methods of destruction of organophosphorus neurotoxins are considered as the base of new defense technology

D. Morrison et al. (eds.), Defense against Bioterror: Detection Technologies, Implementation Strategies and Commercial Opportunities, 233–243.

1. INTRODUCTION

The contemporary level of investigations in chemistry, biology and molecular biology notably extends the possibilities for creation of new compounds and systems potentially applicable as agents of chemical and biological lesion. The factors promoting this situation are the following:

- growing volume of available information concerning the structure of compounds of various classes and their physiological activity,
- broad application of everyday and agricultural neurotoxins and pesticides comparable on toxicity with traditional toxicants,
- elaboration and development of new methods of synthesis including the enzymatic ones,
- elaboration and broad application of genetic engineering methods for sufficiently simple transfer of genes of biological supertoxicants into the nonpathogenic human microflora.

In view of this, a cardinal objective seems to be the creation of new reliable highly sensitive and fairly speedy methods for control, qualitative and quantitative assays and monitoring of supertoxicants of various classes. Likewise, the important aim appears to be the elaboration of new methods for protection against chemical and biological lesions. In terms of the overall task of chemical and biological safety, a pressing purpose seems to be a notable improvement of analytical methods and minutes affording the researchers:

- to increase $10\text{-}10^4$-fold the efficiency of analytical instruments and assay methods and, hence, to speed up the analytical procedure,
- to sensitize the methods 100-1000-fold,
- to notably reduce the cost of one assay,
- to arrange the system of non-stop control and monitoring of water, air and foods,
- to elaborate the adequate methods for individual control.

At present, there are evident approaches affording us to solve this problem. These approaches include the following analytical technologies:

- biosensors [1-14];
- bioluminescent analysis;
- immuno-analytical protocols;
- enzymatic, DNA- and immunochip technologies,
- nanotechnological methods based on application of the scanning probe microscopy [15-19].

Recently, Chemical Enzymology Department (Chemistry Faculty, The Lomonosov Moscow State University) jointly with colleagues from other institutions conducted a battery of researches on creation of the analytical assay for the detection of toxins of various classes and pathogenic microorganisms.

2. NEUROTOXINS AS CHOLINE ESTERASE INHIBITORS. BIOSENSORS FOR DETECTION AND DISCRIMINATION OF NEUROTOXIC COMPOUNDS

Pesticides widely applied presently are largely the derivatives of two classes i.e. organophosphorus pesticides, phosphoric, phosphonic and thiophosphorilic acids as well as carbamates, carbonic acid derivatives. Extremely toxic compounds of these classes are chemical warfare agents (CWA): Sarin (LD_{50}=0.075 ml/L*min), Soman (LD_{50}=0.03 ml/L*min), and Vx (LD_{50}=0.01 mg/L*min). Both organophosphorus neurotoxins and carbamates are highly effective inhibitors of choline esterase. In some cases, the toxicity of carbamates is close to that of CWA [20] .

Below are examples of some most known neurotoxins, choline esterase inhibitors:

1. Chemical Warfare Agents:

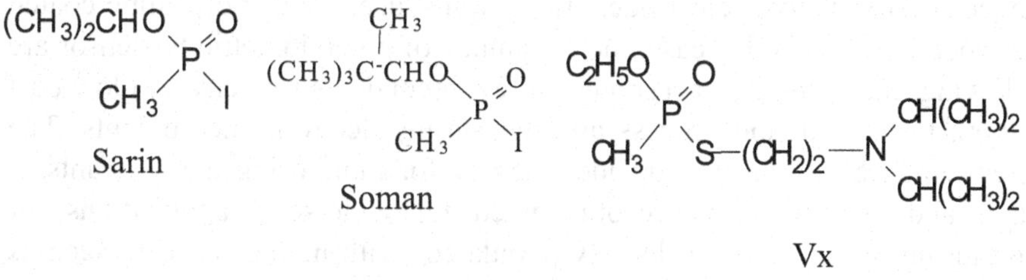

Sarin

Soman

Vx

2. Organophosphorus pesticides:

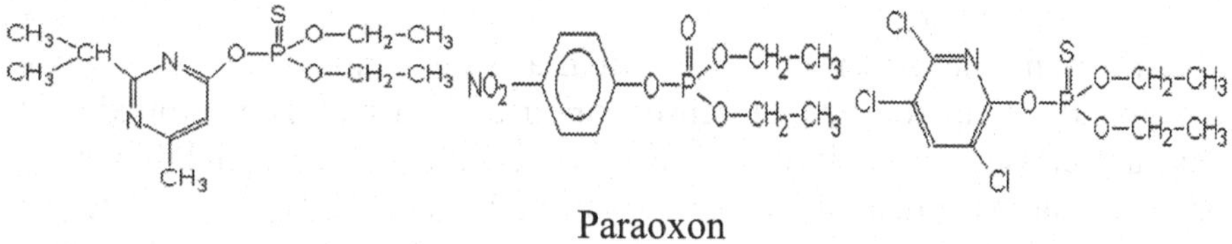

Paraoxon

Diazinon

Chlorpyrivos

3. Caramates: Cararyl, Carbofuran

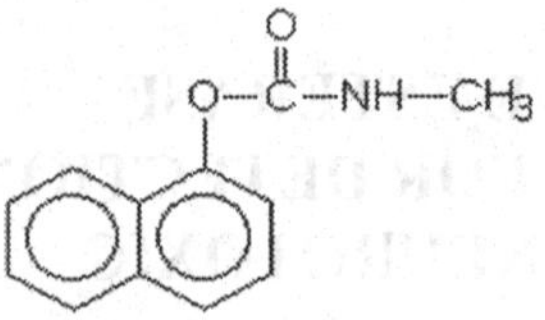

Carbaryl

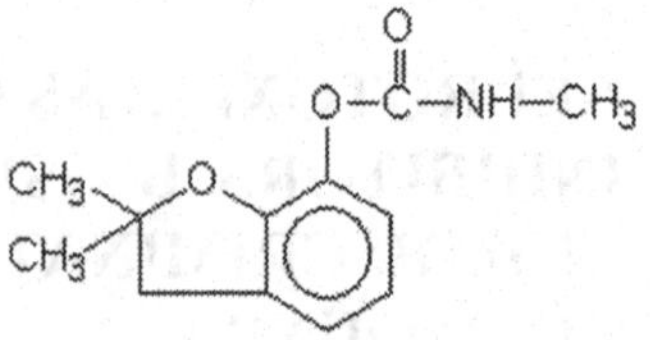

Carbofuran

To quantify and monitor the neurotoxins of various classes, the potentiometric and amperometric biosensors were elaborated.

2.1 Potentiometric Biosensor and Amperometic Biosensors, Using Organophosphate Hydrolases

Organophosphate hydrolase (EC 3.1.8.1) is capable of hydrolyzing the P-I, P-S and P-O bonds in all presently known organophosphorus neurotoxins [21-24]. Since the hydrolysis products are strong acids, the starting compound can be detoxicated by the shift of potential of a pH-sensitive device (pH-sensitive electrode, field transducer and polyamine-coated electrode [25,26]. In this case, the advantage of potentiometric biosensor are following: the high rate of response (a few seconds), simplicity and low cost of both the production process and measuring electronic instruments. The potentiometric biosensor is an ideal sensor for monitoring the toxicants in water and air; it can be recommended for wide-scale applications for monitoring of air flows in densely populated (with humans) areas (such as metro stations).

Drawbacks of potentiometric biosensor include a rather low sensitivity (up to 10^{-7} M) and a possibility to detect only organophosphorus neurotoxins since the biosensor is insensitive to carbamate neurotoxins.

Amperometric biosensor for detection and discrimination of neurotoxic compounds. Amperometric biosensors are used to detect the neurotoxins of various classes. The level of butyryl choline esterase inhibition is proportional to the quantity of neurotoxins in the sample. Activity is detected in accordance with the rate of appearance of hydrogen peroxide as a result of multienzymatic transformation of the substrate (butyryl choline). The first approach supposes the application of peroxidase at the last reaction step with direct electron transfer from the electrode to the enzyme active site [27-29]. The second approach deals with amperometric registration of hydrogen peroxide with platinum electrode. The biosensor quantifies neurotoxins (pesticides and toxins) in the range of extreme permissible concentrations and lower. The sensitivity limit is up to 10^{-12} M. The neurotoxins of various chemical classes (ions of heavy metals, organophosphates and carbamates) are discriminated by application of specific enzymatic and chemical kits. An automatic robot performing all analytical operations with no operator participation was elaborated. The concept of the analysis in based on the application of two biosensor elements:

- The first reporter element provides registration of the total neurotoxic effect by measurement of a degree of inhibition of choline esterase.
- The second, selector or disriminating element, provides selection of organophosphates connected inhibition of choline esterase by application of organophosphate hydrolase, which deletes the organophos.

The chemistry of analysis includes the following reactions:

Reporter reactions

1. $\text{Butyryl Choline} \xrightarrow{\text{Choline esterase}} \text{Choline} + \text{Butyryc acid}$
2. $\text{Choline} + 2O_2 + H_2O \xrightarrow{\text{Choline oxidase}} \text{Betain} + H_2O_2$
3. H_2O_2 detection by pt electrode (600 mV *vs* AG(AgCl)

Diserimninatig reactions:

$$(R_1O)(R_2)P(=O)X + H_2O \xrightarrow[\text{hydrolase}]{\text{organophosphate}} (R_1O)(R_2)P(=O)OH + HX$$

Organophosphate hydrolase eliminates the active phosphates from analytes and does not react with carbamates.

On the difference between inhibition activity before and after action of organophosphate hydrolase can be a quantitative estimate of the content in the sample of organophosphorus compounds (chemical warfare agents, pesticides) and carbamates.

Table 1. Technical characteristics of electrochemical biosensor analyzer

Measuring range of choline	2.5-100 μM
Sample volume	120 μl
Calibration	one point (50 μM choline solution)
Operation stability	12 days or 1000 measurements 3 months at +4°C
Storage stability	45 samples per hour
Sample throughput	90 sec
Time of sample measurement	22 min
Total time for inhibitor assay	

The analyzer was applied for the detection of pesticides and chemical warfare agents in the water, sails, and foodstuffs. The characteristics of analysis for widely spread neurotoxins and pesticides is represented in table 2.

Table 2. Detection of widely spread neurotoxins

Inhibition	Linear range	Detection limit	Inhibition rate constant $M^{-1}min^{-1}$
DFF	1-50 nM	1 nM	$(3,0\pm0,2)10^6$
Paraoxon	20-250 nM	20 nM	$(7,0\pm1,0)10^5$
Carbaryl	2.5-25 μM	2,5 μM	$(6,2\pm0,7)10^3$
Carbofuran	0.07-1.6 μM	0.07 μM	$(7,2\pm0,5)10^4$
Diazinon	0.3-20 nM	0.3 nM	$(1,0\pm0,2)10^6$
Chlorpyrifos	1-8 nM	1 nM	$(4,2\pm1,3)10^7$

The data of analysis of chemical warfare agents is given in Table 3

Table 3. Biosensor detection limits for Chemical Warfare Agents (10 minutes analytical mode)

Agent	Detection Limit, nM	Permited concentration in drinking water, nM
Sarine	0,2	70
VX (Russian)	0.9	-
VX (USA)	1,5	19

All measurements were verified by the gas chromatography method. The method provides the unique possibility for detection of CWA and pesticides at concentrations below the ecological permitted limit. The biosensor was elaborated for chemical safety, environmental monitoring (water, soil and air), qualitative control of agricultural production and food safety.

3. DELAYED NEUROTOXICITY

Certain organophosphates were found to induce a delayed neuropathy in humans and susceptible species sevral weeks after an initial toxic insult, this effect being unrelated to acetyl cholinesterase inhibition. To illustrate the substances of this class, mipafox and some other compounds are listed below:

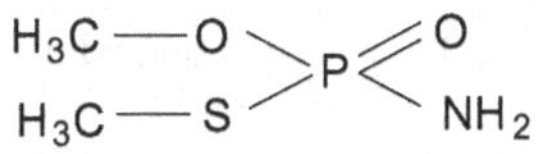

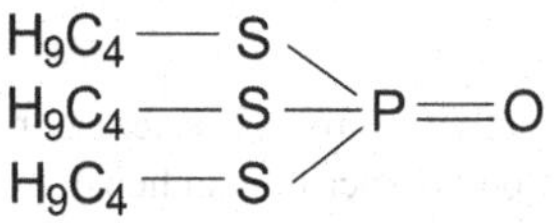

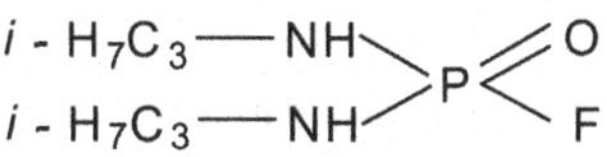

$$(H_3C{-}O)(H_3C{-}S)P(=O)NH_2 \qquad (H_9C_4{-}S)_3P{=}O \qquad (i\text{-}H_7C_3{-}NH)_2P(=O)F$$

metamidophos butyphos mipafox

Many organophosphates commonly used as insecticides reveal a delayed neurotoxicity. The organophosphates induced Delayed Neuropathy is characterized by following:

- Not related with acetylcholine esterase inhibition.
- Can be initiated by chronic low-level organophosphates exposure.
- Irreversible paralysis of limbs as a result of organophosphates – induced long axon degradation.
- Symptoms are observed after latent period from 1 to 3 weeks after initial toxic insult.

It is initiated by the organophosphorylation and specific modification of the neuronal protein known as neuropathy target esterase (neurotoxic esterase, NTE). A new biosensor method for NTE activity measurements based on the combination of NTE enzymatic hydrolysis of phenyl valerate with electrochemical phenol assay was developed [30,31]. Good correlation was obtained between brain and lymphocyte NTE inhibition, as well as between brain and blood NTE inhibition, and lymphocyte and blood NTE inhibition suggests blood NTE activity as a biochemical marker of neuropathic organophosphate exposure. The inhibitory NTE activities in brain, lymphocytes and whole blood were shown to correlate. The inhibition of NTE was caused by the action of O,O-dipropyl-dichlorvinyl phosphate for 24 h. (Fig. 1.)

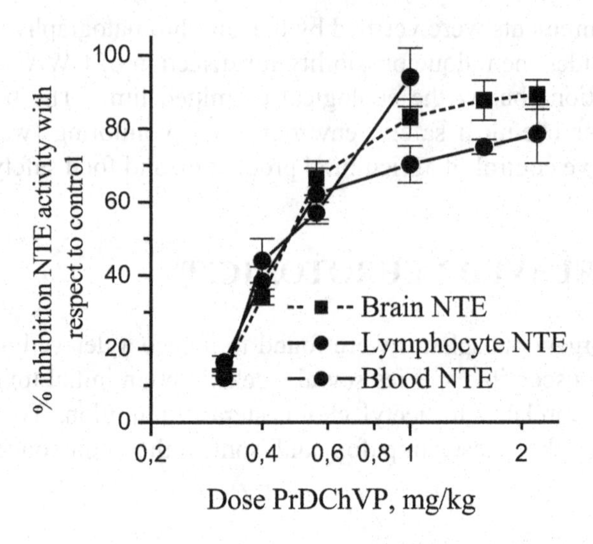

Figure 1. Correlation between the active neurotoxic esterase in brain, lymphocytes and whole blood (experimental hens).

This biosensor gives the unique method for the medico-biological monitoring of personnel. Method based on combination of NTE catalyzed hydrolysis of phenyl valerate with phenol detection by tyrosinase biosensor.

The analysis includes the following chemical reactions:

Phenylvalerate $\xrightarrow{\text{NTE}}$ phenol

phenol + 1/2 O_2 $\xrightarrow{\text{Tyrosinase}}$ Catechol

graphite electrode $\xrightarrow{e^-}$

1/2 O_2

Tyrosine

o-Quinone

H_2O

The technical characteristics of these methods are listed in Table 4.

Table 4. Biosensor for detection of Delayed Neuropathy

Parameter	Characteristics
Measuring range of phenol in flow	100 nM – 20 μM
Operational stability	7 days

Storage stability	At least 6 months at 4°C
Impresision	5%
Sample throughput	20 analysis per hour
Calibration	Automatic one point calibration
Standard solution	0.5 μM phenol solution
Sample injection	Autosampling

The NTE biosensor can be applied to the control of a broad group of compounds on their potential delayed neurotoxicity, for monitoring the occupational exposure of humans to neurotoxic organophosphates, and particularly for field use and for developing and improving the methods of early diagnostics of delayed neuropathy. At present, the biosensor assay for measuring NTE activity in blood is applied to control the health condition of the personnel working at the enterprises for chemical weapon destruction.

4. NANOTECHNOLOGICAL IMMUNOCHIPS FOR THE DETECTION OF SUPERPATHOGENIC MICROORGANISMS AND VIRUSES

The principles of express detection of microorganisms and viruses by the use of an atomic force microscope were elaborated. The method is based on the creation of an immunochip with high-specific antibodies and the detection of adsorbed cells by the scanning probe microscope. The typical immunochip surface obtained by Langmuir-Blodgett (LB) method using ampliphilic polyelectrolytes is shown in Fig.2.

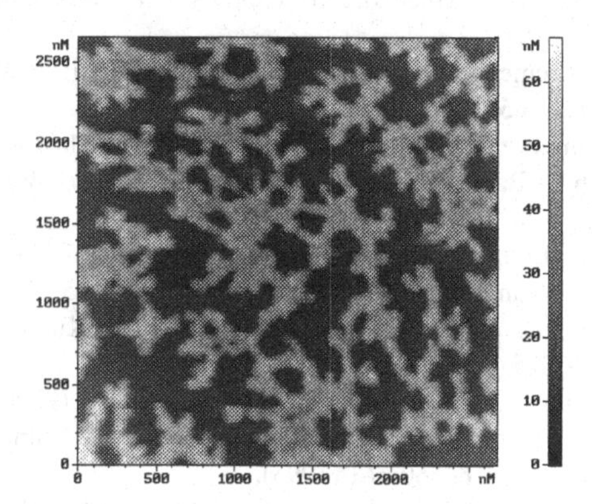

Figure 2. Atomic force image of immunochip coated with LB films from antibodies and amphiphilic polyelectrolytes

The assay protocol includes: (1) obtaining an affinity surface (in the general case, the antibodies, receptors and DNA(RNA)-probe can be employed); (2) performance of specific adsorption of the searched object in the analyte; (3) scanning the surface; and (4) image analysis and identification of the biological agent using pattern-recognition technique.

The possibilities of the method were demonstrated for the analysis of *Coxiella burnetti*, *Smallpox vaccine virus*, and *Yersinia pestis*. The sensitivity of the method is down to the single cell level (Fig.3).

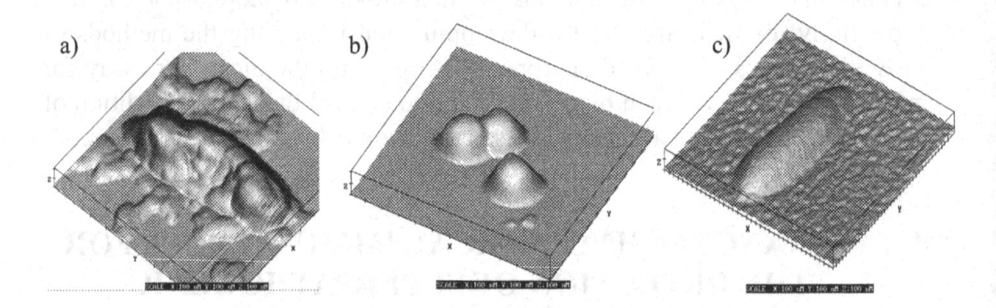

Figure 3. Three-dimensional atomic force microscope image of Coxiella burnetti (a), Smallpox vaccine virus (b), and Yersinia pestis (c).

REFERENCES

1. J. Wang, Y.Lin, A.V. Eremenko, I.N. Kurochkin, M.F. Mineyeva *Anal.Chem.*, **65**, 513-516 (1993).
2. J. Wang, E. Dempsey, A. Eremenko, M. Smyth. *Anal. Chim. Acta,* **279**, 203-208 (1993).
3. J. Wang J., Y. Lin, A.V. Eremenko, A.L. Gindilis, I.N. Kurochkin. *Anal. Lett.*, **26**, 197-207 (1993).
4. Makower, A.V. Eremenko, K. Streffer, U. Wollenberger, F.W. Scheller. *J. Chem. Technol. Biotechnol.*, **65**, 39-44 (1996).
5. F.F. Bier, E. Ehrentreich-Foerster, F.W. Scheller, A. Makower, A. Eremenko, U. Wollenberger, Ch.G. Bauer, D. Pfeiffer, N. Michael. *Sensors & Actuators.B.*, **33**, 5-12 (1996).
6. Ch.G.Bauer, A.V. Eremenko, E. Ehrentreich-Foerster, F.F. Bier, A. Makower, H.B. Halsall, W.R. Heineman, F.W. Scheller. *Anal. Chem.*, **68**, 2453-2458 (1996).
7. A.V. Eremenko, A. Makower, C.G. Bauer, I.N. Kurochkin, F.W. Scheller. *Electroanalysis,* **9**, 1-5 (1997).
8. Makower, A. Barmin, F. Scheller. *Biospektrum*, **6** , 119-121 (2000).
9. F.W. Scheller, A. Makower, F. Bier, U. Wollenberger, A. Ghindilis, A. Eremenko, D. Pfeiffer. *GIT Fachz. Lab.*, **6**, 560-561 (1995).
10. Makover, A. Barmin, T. Morzunova, A. Eremenko, I. Kurochkin, F. Scheller. *Anal. Chim. Acta*, **357**, 13-20 (1997).

11. F.F. Bier, E. Ehrentreich-Foerster, R. Doelling, A.V. Eremenko, F.W. Scheller. *Anal. Chim.Acta*, **344**, 119-124 (1997).
12. A.V. Eremenko, C.G. Bauer, A. Makower, B. Kanne, H. Baumgarten, F.W. Scheller. *Anal. Chim.Acta,* **358**, 5-13 (1998).
13. C.G. Bauer, A.V. Eremenko, A. Kuehn, K. Kuerzinger, A. Makower, F.W. Scheller. *Anal.Chem.*, **70**, 4624-4630 (1998).
14. I.N. Kurochkin. *Adv. in Biosensors*, **3**, 77-109 (1995).
15. Budashov, I. Kurochkin, A. Denisov, V. Skripnyuk, G. Shabanov. *J.Microbiol.& Epidemiology* (Rus.), **6**, 11-15 (1997).
16. Kurochkin, I. Budashov, A. Pavelev, A. Denisov, V. Skripnyuk, G. Shabanov. *Sensornie sistemi* (Rus.), **12**, 122-134 (1998).
17. Pavelev, I. Kurochkin, S. Chernov. *Biol.Membrane* (Rus.), **15**, 342-348 (1998).
18. I.A. Budashov I.A., I.N. Kurochkin, V.V. Tsibezov, S.L. Kalnov, A.K. Denisov, O.I. Kiselyova, I.V. Yaminsky. *Biol.Membranes* (Rus.), **16**, 325-335 (1999).
19. I.A. Budashov, I.N. Kurochkin, V.V. Tsibezov, S.L. Kalnov, A.K. Denisov, O.I. Kiselyova, I.V. Yaminsky. *Membr.Cell Biol.*, **13**, 397-409 (2000).
20. Yu.G. Zhukovskii, S.A. Kutsenko, L.P. Kuznetsova, E.E. Sochilina, E.N. Dmitrieva, M.L. Fartseiger, V.D. Tonkopii, N.N. Korkishko, V.I. Kozlovskaya, V.E. Fel'd. *J.Evol.Biochem.Physiol.* (Russ.) **30**, 177-184 (1994).
21. E.N. Efremenko, V.S. Sergeeva. *Russ.Chem.Bull, Int.Ed.* **50**, 1826-1832 (2001).
22. R.S.Gold, M.E.Wales, J.K.Grimsley, J.R.Wild. In "Enzymes in Action: Green Solution for Chemical Problems", ed. B.Zwanenburg, M.Mikolajczyk, P.Kielbasinski, NATO Science Series, Vol.33, p.263-288 (2000).
23. V.S. Sergeeva, E.N. Efremenko, G.M. Kazankov, S.D. Varfolomeyev. *J.Molec.Catalysis B: Enzymatic*, **10**, 571-576 (2000).
24. E.N. Efremenko, V.I. Lozinsky, V.S. Sergeeva, F.M. Plieva, T.A. Makhlis, G.M. Kazankov, A.K.Gladilin, S.D. Varfolomeyev. *J.Biochem.Biophys.Methods*, **51**, 195-201 (2002).
25. E.I. Rainina, E.N. Efremenko, S.D. Varfolomeyev, A.L. Simonian, J.R. Wild. *Biosens.Bioelectron.* **11**, 991-1000 (1996).
26. A.L. Simonian, E.N. Efremenko, J.R. Wild. *Anal.Chem. Acta* **444**, 179-186 (2001).
27. A.L. Ghindilis and I.N. Kurochkin. *Biosens.&Bioelectronics,* **9**, 353-357 (1994).
28. A.L.Ghindilis, T.G. Morzunova, A.V. Barmin, I.N. Kurochkin. *Biosens. &Bioelectronics*, **11**, 873-880 (1996).
29. S.D. Varfolomeyev, A.Y. Yaropolov, I.N. Kurochkin. *Biosens. & Bioelectronics*, **11**, 863-871 (1996).
30. L.V. Sigolaeva, A.V. Eremenko, A. Makower, G.F. Makhaeva, V.V. Malygin, I.N. Kurochkin. *Chem.-Biol.Interactions*, **119-120**, 559-565 (1999).
31. L.V.Sigolaeva, A.Makower, A.V.Eremenko, G.F.Makhaeva, V.V.Malygin, I.N.Kurochkin, F.W.Scheller. *Anal.Biochem.,* **290**, 1-9 (2001).

BIOSENSOR FOR DEFENCE AGAINST TERRORISM

M. Mascini and I. Palchetti

Dipartimento di Chimica, Università degli Studi di Firenze, Via della Lastruccia 3, 50019 Sesto Fiorentino, Firenze

Abstract: : The apparatus we would like to report is an integrated instrument for rapid (seconds or minutes) detection of neurotoxic compounds in aqueous solution (or gaseous samples like confined air) and genotoxic compounds. The portable instrument can measure in the range of ppm to ppb in seconds or minutes compounds potentially genotoxic or neurotoxic. Neurotoxic devices are based mainly on the use of enzyme Acetylcholinesterase.

Genotoxic devices are DNA electrochemical biosensors that are realized by immobilizing oligonucleotide sequence or the calf thymus DNA on an electrode surface.

1. INTRODUCTION

Chemical warfare agents (CWAs) can be divided into several groups, of which the most important are nerve agents, blistering agents or vesicants, blood agents and incapacitating agents [1, 2].

Nerve agents derive their name from their adverse effects on the nervous system. These are all organophosphates that contain: a phosphonate group that mimics the acetate moiety of the acetylcholine (ACh) molecule; an ester or thioester linkage; and often a positively charged group mimicking the choline moiety. Many commonly used agricultural pesticides are organophosphate compounds (e.g. chlorpyrifos, parathion) [1, 3-4]. The principal effect of nerve agents is the inhibition of the enzyme acetylcholinesterase (AChE), which is essential for terminating the action of the neurotransmitter ACh. The phosphonate moiety of nerve agents mimics the acetate group of acetylcholine; unlike ACh, however, the nerve agent is

D. Morrison et al. (eds.), Defense against Bioterror: Detection Technologies, Implementation Strategies and Commercial Opportunities, 245–259.

capable of modifying the AChE protein in either a slowly reversible or irreversible fashion. ACh is a neurotransmitter at the neuromuscular junction, in the parasympathetic nervous system and in a number of brain regions. Actually, nerve agent intoxication results in an accumulation of endogenous acetylcholine and continual stimulation of the nervous system [1-4]. One nerve agent in particular, sarin, was in the news after its use against the population of the Kurdish village of Birjinni in 1993 and after a terrorist attack in the Tokyo underground system on 20 March 1995. Another well-known nerve agent is VX, which is among the most toxic substances ever produced by man [1]. The second group of CWAs, the vesicants, such as sulphur mustard, affect the eyes and lungs and blister the skin. Sulfur mustard is a bifunctional alkylating agent that reacts rapidly with nucleophiles under physiological conditions, via the intermediate episulfonium ion. Nucleophilic species in the human body include DNA, the tripeptide glutathione, various amino acid residues present in proteins, and water [1, 2]. The third group, blood agents, like hydrogen cyanide, interfere with the oxygen transport capability of blood and may cause death by suffocation. The fourth group, incapacitating agents, have non-lethal physiological effects such as vomiting and /or mental effects [1, 2].

The most frequently used methods for the unambiguous identification of CWAs, their precursor and breakdown products are based on gas chromatography (GC) in combination with mass spectrometry (GC-MS) and/or tandem mass spectrometry (GC-MS/MS), liquid chromatography (LC) coupled with mass spectrometry (MS), and nuclear magnetic resonance (NMR) spectrometry [1,3]. The apparatus we would like to propose is an integrated instrument for rapid (seconds or minutes) determination of neurotoxic compounds in aqueous solution (or gaseous samples) and genotoxic compounds with disposable electrochemical sensors. The portable instrument able to measure neurotoxic or genotoxic agents in the range of parts per million (ppm) to parts per billion (ppb) seconds or minutes. To date the technology has produced a separate neurotoxic device and genotoxic device which have been reported on scientific journals. The prototypes developed in the author's lab in recent years are summarized [5-13].

2. EXPERIMENTAL

Electrodes were printed with a high performance multi-purpose precision screen printer DEK 249 (DEK, UK). Inks were from Acheson (Argon, Italy). The plastic substrate for printing was a polyester film (Autostat HT5, Autotype). Cholinesterase acetyl (AChE) from Electric Eel (EC 3.1.1.7, 530 U/mg protein), the Acetylthiocholine (ATCh chloride salt), and the mediator

7,7,8,8, Tetracyanoquinodimethane (TCNQ) were purchased from Sigma (Milan, Italy). Carbofuran was obtained from Polyscience Corporation (USA). Pesticide standard solutions were prepared daily by dissolving the pesticide in the acetonitrile purchased from Sigma. Calf Thymus DNA, Sodium acetate, acetic acid, KCl, methanol were from Sigma. Water was from a reagent grade ion-exchange MillQ system (Millipore Inc).

2.1 Sensor Development and Biosensor Assembly Bare Screen-Printed Sensor (SPE)

Acreen-printed electrodes are planar devices, based on different layers of inks printed on a plastic or ceramic substrate. The inks were deposited onto a polyester substrate (350 μm as thickness) in a film of controlled pattern and thickness to obtain overlapping layers. At first the silver tracks were printed, and then the graphite pad was positioned over part of the silver track, to obtain the working and the counter electrodes. Finally, the insulating layer was deposited with openings that allow the electrical contact with the circuit at one end and the analyte solution at the other end. The electrochemical cell consists of a round shaped (3mm diameter) graphite working electrode, a graphite counter electrode and a silver reference electrode. After each printing step the ink film was dried at 110°C for 10 min.

2.2 Acetylthiocholine Sensor

The surface of the carbon working electrode was covered with 2 ml of a suspension obtained by mixing 5 ml of 1 10 -3 mol/l TCNQ solution in acetonitrile and 50 ml Nafion. Sensors were stored overnight at room temperature, in the presence of desiccant and used as disposable. The sensor can be used up to 7 days by storing it in a refrigerated dessicant. Longer storage time yielded an unreproducible behaviour (this point requires more careful attention and further experimental results).

2.3 DNA Biosensor

The bare graphite electrode surface was pre-treated by applying a potential of +1.6 V for 3 min. The biosensor was assembled by immobilizing double-stranded calf thymus DNA at fixed potential (+0.5 V vs. Ag screen-printed pseudo-reference electrode, for 120 s) onto the screen-printed electrode surface. During the immobilization step, the strip was immersed in acetate buffer solution containing 20 ppm of double stranded calf thymus

DNA. Then a washing step was performed by immersion of the biosensor in a clean acetate buffer solution for 30 s, at open circuit condition.

2.4 Electrochemical Measurements

Electrochemical measurements were performed with an Autolab PGSTAT 10 electrochemical analysis system, with a GPES 4.5 software package Ecochemie, Utrecht, Holland), in connection with a VA-Stand 663 (Metrohm, Milan, Italy). Using the GPES4 software the signal was smoothed and a baseline correction was generally performed.

2.5 Acetylthiocholine Sensor for the Determination of Neurotoxic Agent

5-50 microliters of a stock solution of AChE enzyme (80 U/μl) and 20 μl of a 5x10 -2 mol/l acetylthiocholine were mixed in 1 ml of buffer phosphate 0.1mol/l at pH 7.5, and the reaction was allowed to proceed for 4 min at room temperature. Then 100 μl of this solution were deposited onto the planar surface of the TCNQ-Nafion SPE and after 1 min. DPV measurement was performed from + 0.35 to + 0.7V vs. Ag pseudo-reference electrode; with a pulse amplitude of 50 mV, a scan rate 25 mV s -1 and a pulse width of 60 ms. The DPV current peak obtained at 0.5 V which was exploited as analytical signal. For the determination of neurotoxic agent 50 μl of pesticide standard solutions or river water samples were added to 1 ml of 0.1 mol/l phosphate buffer solution pH 7.5 with acetylcholinesterase (1U/ml) and incubated for 10 min. Then 2ml of the 5.10 -2 mol/l acetylthiocholine solution was added. After 5 min, 100 μl of this solution was directly dropped on the surface of the sensor. The oxidation current peak obtained by DPV after 5 min was measured (I_2) and compared with the oxidation current value obtained without pesticide (I_1). The percent of inhibition (I%) was obtained according to the following formula: I % Q =100 $[(I_1 - I_2)/I_1]$. The total anticholinesterase activity of a sample is expressed as the amount of a known compound (carbofuran, a carbamate pesticide) producing the same enzymatic inhibition to acetylcholinesterase.

2.6 DNA Biosensor

The interaction step was performed just by placing 20 μl of the sample onto the surface of the DNA biosensor. After 2 min the biosensor was washed, immersed in acetate buffer and a square wave voltammetric scan was carried out to evaluate the oxidation of guanine residues on the electrode surface. The area of the guanine peak (around + 1.0 V vs. Ag screen-printed pseudo-reference electrode) was measured. Genotoxic compounds present in samples were evaluated by changes of the electrochemical signal of guanine. We estimated the DNA modification with the value of the percentage of response decrease (R%) which is the ratio the guanine peak area after the interaction with the analyte (GPAs), and the guanine peak area after the interaction with the buffer solution (GPAb): $R\%=[(GPAs/GPAb)-1]\times 100$. The result of the test for one sample can be obtained within 8 min. The supporting electrolyte for the voltammetric experiments and for any step in the biosensor was acetate buffer 0.25 M pH 4.7, KCl 100 mM. Square-wave voltammetry parameters were: frequency 200 Hz, step potential 15 mV, amplitude 40 mV, potential range 0.2–1.2 V versus Ag-pseudo-reference electrode.

3. RESULTS

3.1 Detection of Neurotoxic Agent

The use of enzymatic biosensors, and especially electrochemical biosensors, for organophosphorus and carbamate pesticides detection has been reported by many authors [5-8]. These are based mainly on the use of enzyme Acetylcholinesterase (AChE); AChE hydrolyses the acetylcholine to choline and acetic acid which are not electroactive; therefore another enzyme can be added in the reaction scheme such as choline oxidase which oxidizes the choline to produce hydrogen peroxide easily detected by amperometry; some other authors report the use of acetylthiocholine as substrate, and the enzymatic product (thiocholine) can be oxidized directly at the electrode surface. The major problem associated with the use the acetylthiocholine as electroactive substrate is the high potential required to oxidize the thiocoline. This can be overcome using an electron transfer mediator, and tetracyanoquidimethane (TCNQ) appeared to exhibit the most suitable characteristics for this application. Due to the presence of four cyano groups and the relative large π conjugation, TCNQ serves as an excellent electron donor and acceptor. Even if many theories are reported in literature, the mechanism of electron transfer is not completely clear. Among the different electrochemical techniques, amperometry has been the most used technique in acetylcholinesterase activity measurement using TCNQ as electron transfer mediator or even when choline oxidase is exploited. However, the disadvantage in the use of amperometry for in-situ environmental analysis is the long working electrode polarisation step, (sometimes there is the need to wait 20- 30 min. to obtain a stable baseline).

This problem can be overcome in some case using a fast voltammetric technique.

The method proposed in the present work is based on the electrochemical determination of cholinesterase enzymatic activity with differential pulse voltammetry (DPV) using a disposable TCNQ-Nafion modified screen-printed electrode. The combination of voltammetric techniques with screen-printed disposable electrodes is very attractive for the development of in-situ analytical tools. Among these techniques DPV is one of the most sensitive and rapid. With this technique we avoided the long period necessary for attaining a stable baseline, and therefore the procedure is really very fast and therefore is very attractive.

3.2 Optimisation of Analytical Parameters

The measurements of enzymatic inhibitors have to follow a zero-order kinetic $[S]>>K_m$, where [S] is the concentration of the substrate and K_m is the Michaelis-Menten constant. For this reason the enzyme activity has to be optimized. This parameter affects the detection limit; higher enzymatic activity results in higher detection limits, lower enzymatic activity results in higher coefficient of variation due to the low reproducibility of the analytical signal. The enzyme concentration was varied in the range 0.2 - 4 U/ml and best results in term of sensitivity and reproducibility were obtained with 1U/ml AchE. These concentrations were used for the rest part of the work. Such values ensure that the enzymatic reaction is complete. It is important to consider that the amount of thiocholine obtained is not dependent from the acetylcholine concentration but upon the enzyme activity. We must assume that in our conditions a steady state is reached and the concentration value of thiocholine depends by the activity of enzyme. The influence of the solution pH was analyzed in the range 4-10. As it is well known, the pH of the solution greatly affects the enzyme activity. When the pH range is 4-6 the current responses change slightly, when the pH is more than 8 the voltammetric response is very small. The maximum currents appear at pH 7-8; thus the pH 7.5 was chosen as optimal value. Then, a calibration curve of acetylthiocholine was obtained with 1 U/ml AchE. The variation coefficient evaluated for four replicates of 10^{-3} mol/l was 10% using 5 min of enzyme-substrate incubation time. The variation coefficient was calculated using different electrodes.

3.3 Pesticide Detection in a Real Sample

In the presence of organophosphorus and carbammic pesticides, that bind the enzyme acetylcholinesterase, inhibiting it, the rate of thiocholine production is reduced with the amount of the pesticides. Using the proposed biosensors, different compounds were tested. In figure 1 A are the reported voltammograms obtained in absence and in presence of two different concentrations of Carbofuran, a carbammic pesticide. For this compound a limit of detection of $1x10^{-9}$ M was found. In figure 2 curves of some organophosphorus pesticides are reported. In table 1 are the reported results obtained with acetylthiocholine biosensor, analysing river water samples. Since the anticholinesterase pesticide concentration in river water is, generally, very low, and in many cases <0.1 µg/L, the samples were preconcentrated by the use of solid-phase extraction (SPE) technique. The samples were preconcentrated 100 times and thus the detection limit of the procedure becomes 9 10^{-11} M. Table 1 illustrates the results obtained with a standard method (gas chromatography) and by the biosensor. There is a good correlation between the two sets of data. However, the results obtained with the sensor were slightly greater than those obtained by the reference method, since the sensor determines a group of cholinesterase inhibitor compounds. In fact, acetylcholinesterase activity is inhibited by several compounds other than pesticides such as heavy metals, toxins, etc. TAA is a general index

Table 1. Comparison of the results obtained by Gas Chromatography (GC-ITDMS) and the acetylthiocholinesterase biosensor. The river water samples were 100 times concentrated by Solid-Phase Extraction; Carbofuran was the reference pesticide.

Gas Chromatography			*Biosensor*	
Samples	Carbofuran ng/L	Carbofuran M	% Inhibition	Carbofuran Equivalent Conc. *M*
1	0.2	2 10^{-13}	----	< 9 10^{-11}
2	0.2	2 10^{-13}	----	< 9 10^{-11}
3	0.2	2 10^{-13}	----	< 9 10^{-11}
4	26	1.2 10^{-10}	40 ± 9	1 10^{-9}
5	24	1.1 10^{-10}	30 ± 7	4 10^{-10}
6	21	1.0 10^{-10}	30 ± 7	4 10^{-10}
7	13	6.0 10^{-11}	15 ± 6	1 10^{-10}
8	33	1.5 10^{-10}	40 ± 9	1 10^{-9}
9	24	1.1 10^{-10}	40 ± 9	1 10^{-9}
10	0.3	4 10^{-13}	----	< 9 10^{-11}
11	0.5	5 10^{-13}	----	< 9 10^{-11}
12	0.2	2 10^{-13}	----	< 9 10^{-11}
13	0.6	8 10^{-13}	----	< 9 10^{-11}

combining these different contributions, although the effect of heavy metal on acetylcholinesterase activity is negligible compared to the effect of pesticides.

In Table 2 are the reported results obtained analysing fruits and vegetables samples; 5g of sample were ground and treated with 15 ml of ethylacetate; then, 5 ml were dried and dissolved in 2 ml of buffer. The sample volume used was 1 ml for the evaluation of pesticides

Table 2. Inhibition values obtained analyzing fruit and vegetables samples.

Samples	Biosensor I%
1 (Grape)	93
2 (Grape)	50
3 (Pear)	14± 1
4 (Fennel)	5±1
5 (Onion)	5±1
6 (Celery)	29±2
7 (Melon)	10±1
8 (Grape)	0
9 (Lettuce)	31±2
10 (Orange)	39±3

3.4 Detection of genotoxic agent

Preliminary studies were performed to identify general assay conditions which affected the electrochemical signal of the guanine oxidation peak, including ionic strength, pH, buffer composition, DNA concentration and form (single stranded and double stranded) [9,10]. With the optimized conditions, several experiments for evaluating the variation of the area of the guanine peak were performed (Table 3). The results are expressed as response percent (R%). Experiments carried out in buffer solution show that the ssDNA guanine peak area is higher than the dsDNA guanine peak area (more than 2 times); this can be explained considering that the guanine base in single-stranded DNA is reported to be more available for oxidation than in double-stranded DNA [9].

Table 3. Compounds tested with single stranded or double stranded calf thymus DNA immobilised on screen printed electrodes.

Compounds tested	Calf thymus dsDNA R%	Calf thymus ssDNA R%
1,2-Diaminoanthraquinone (0.05 mg L^{-1})	-50 ± 7	-10 ± 4
2-Anthramine (0.05 mg L^{-1})	-38 ± 6	-26 ± 9
2-Naphthylamine (1.5 mg L^{-1})	-48 ± 13	-31 ± 10
Acridine orange (0.05 mg L^{-1})	-40 ± 4	-19 ± 5
Aflatoxin B1 (10 mg L^{-1})	–3 ± 1	-15 ± 3
Cisplatin (300 mg L^{-1})	-50 ± 8	-10 ± 2
3-OH-Benzo(a)pyrene (2 mgL^{-1})	-40 ± 5	n.d.

Compounds such as aromatic amines decrease the peak area (even when low concentrations were present). This can be explained with lower availability of guanine for oxidation at the electrode surface; with aromatic amines could be deducted a binding of the guanine with the structure of the DNA which prevents its oxidation. From Table 3 it is noteworthy to point out that 2-anthramine and 2-naphthylamine gave similar decrease with ds and ssDNA while with the other two amines (1,2 di-aminoantroquinase and acridine orange) the decrease of ssDNA is higher than with dsDNA. Therefore, the structure of the compounds should be also considered. Aflatoxins are metabolites produced by some strains of the mould *Aspergillus flavus*. They are among the most potent environment mutagens and are implicated as liver carcinogens. The binding of the aflatoxin B1 to both native and denatured DNA has been demonstrated [11]. We obtained a gradual decrease of the guanine peak in the presence of increasing levels of the aflatoxin B1 with a ss-DNA in the range 10–30 mg L^{-1} . Any effect in this concentration range was observed when dsDNA was used. However, we have to point out that for such compounds the ppm range is not relevant for environmental interest being often present at ppb or lower level. Moreover, since some compounds, such as aromatic amines (that constitute a very important class of environmental pollutants), are electroactive we compared the oxidation signal of these compounds with the effect of them on the oxidation peak of the guanine. In Figure 3 the effect of 2-anthramine on dsDNA coated electrodes is reported. Increasing the concentration we obtained an increase in the amount of aromatic amines collected and oxidized to its cation radical, and a decrease in the dsDNA guanine oxidation peak.

In Table 4, the analysis of some well-known carcinogen and mutagen compounds are reported; in this table the results were compared with a reference genotoxicity test, the Comet Test. Analysing the contents of Table 4, the effect of the 4-nitroquinoline-N-oxide on the DNA biosensor was to decrease the guanine oxidation signal, in a concentration dependent way, to induce DNA damage (detected by the Comet test). These results correlate with data reported in the literature [15] where this compound is classified as carcinogen and mutagen since it produces adducts with DNA bases via oxidative damages. N-methyl-N-nitro-nitrosoguanidine is an alkylating compound which produces O-6-methylguanine and O-4-methylthymine. This molecule showed a slight and highly irreproducible effect on the DNA biosensor (at least at the concentration investigated); the Comet test showed that high concentrations of this molecule induced severe DNA damage but not low concentrations. Moreover, DNA damage induced by 2-anthramine is demonstrated by the Comet test, confirming the DNA biosensor response. As reported in Table 4 all the tests have shown very similar results on real samples: the canal water sample was found to be 'clean' by the tests, while the textile industry effluent contained substances with affinity to the DNA and inducing the DNA damage (as demonstrated by the DNA biosensor response and the Comet test results, respectively).

Table 4. Results of the analysis of genotoxic compound solutions [4-nitroquinoline-N-oxide (4-NQO), 2-aminoanthracene (2-AA) and N-methyl-N'-nitro-nitrosoguanidina (MNNG)] and real samples as reported in [15].

Test Name	4-NQO		2-AA		*MNNG*		Canal water sample	Textile industry effluent water sample
	0.1 µg/ml	5 µg/ml	4 µg/ml	200 µg/ml	4 µg/ml	400 µg/ml		
DNA biosensor	DI	DI	DI	DI	NDI	NDI	NDI	DI
Comet Test	G(-S9)	G(-/+S9)	G(+/-S9)	G(+/-S9)	NG	G(+/-S9)	NG	G(-/+S9)

DI : DNA-interaction (decrease of guanine peak area), NDI: not DNA interaction, G: genotoxic, NG : not genotoxic, -S9 : without metabolic activation, +S9 : with metabolic activation. The Comet test was performed by researchers of VITO Group.

4. CONCLUSION

The work reported indicates that electrochemistry can play an interesting role in the detection of "classes" of chemicals of environmental interest. Such compounds gave a value of neurotoxocity and of genotoxocity which are useful in the defense against terrorism. The detection is very fast (seconds or minutes), and can be automated; these sensors are disposable and simple to use. All features indicate that such sensors can be easily integrated in a single device, obtaining simultaneous information of different parameters.

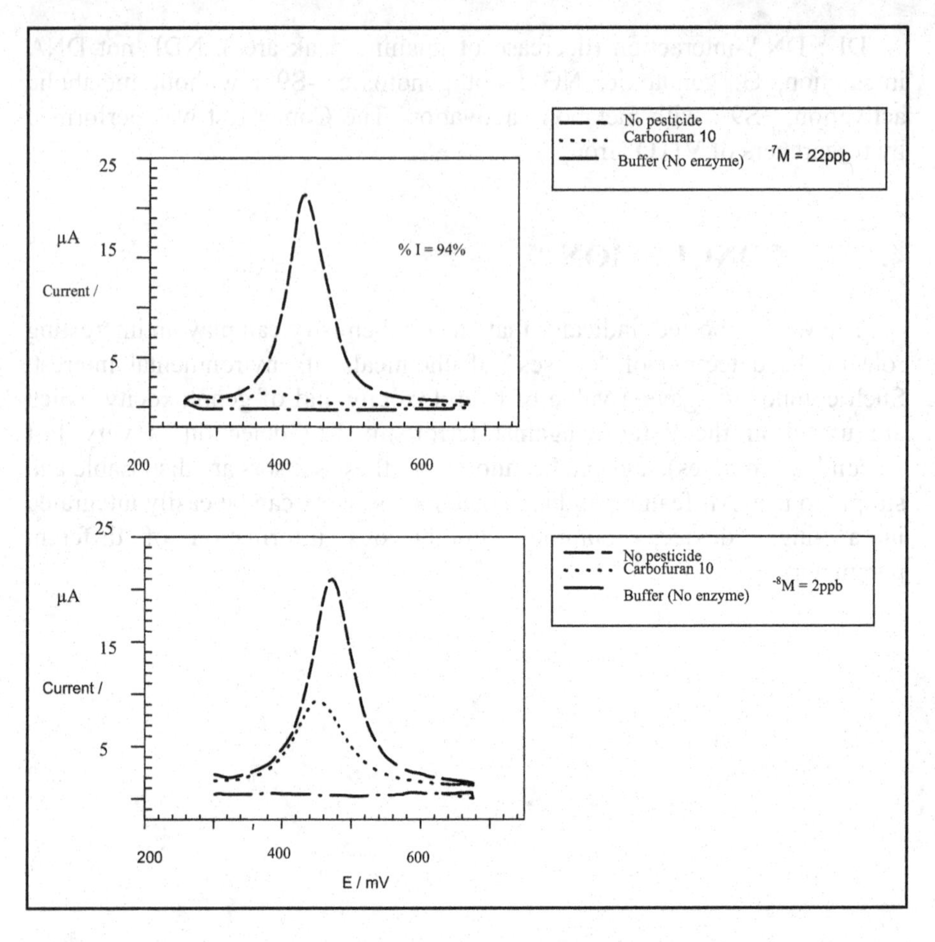

Figure 1. Calibration curve of Carbofuran A) DP voltammograms obtained in absence and in presence of two different concentration of carbofuran, a scan obtained in buffer only without enzyme is also reported.

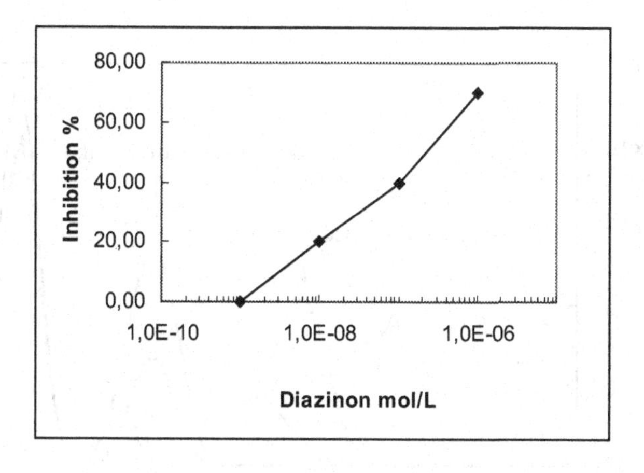

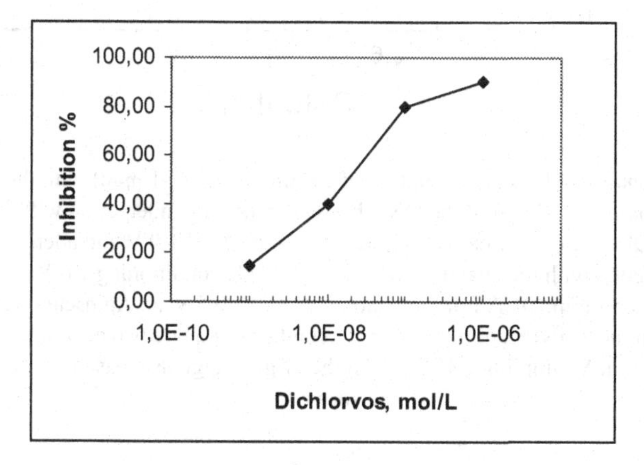

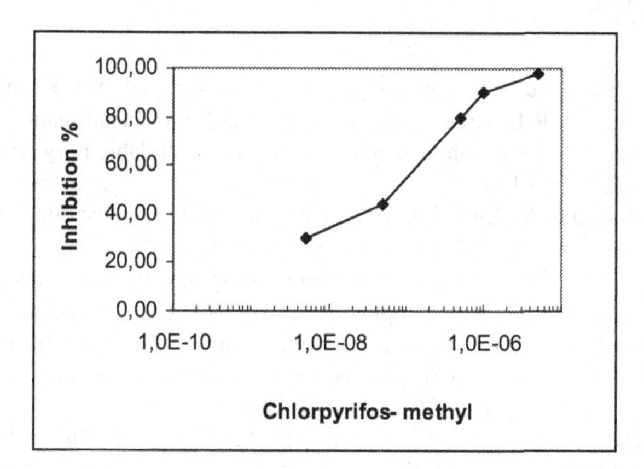

Figure 2. : Calibration curves of different pesticides: Diazinon (organophosphorus with a phosphorothioate group); Dichlorvos (organophosphorus with a phosphate group); Chlorpyrifos-methyl (organophosphorus with a phosphorothioate group)

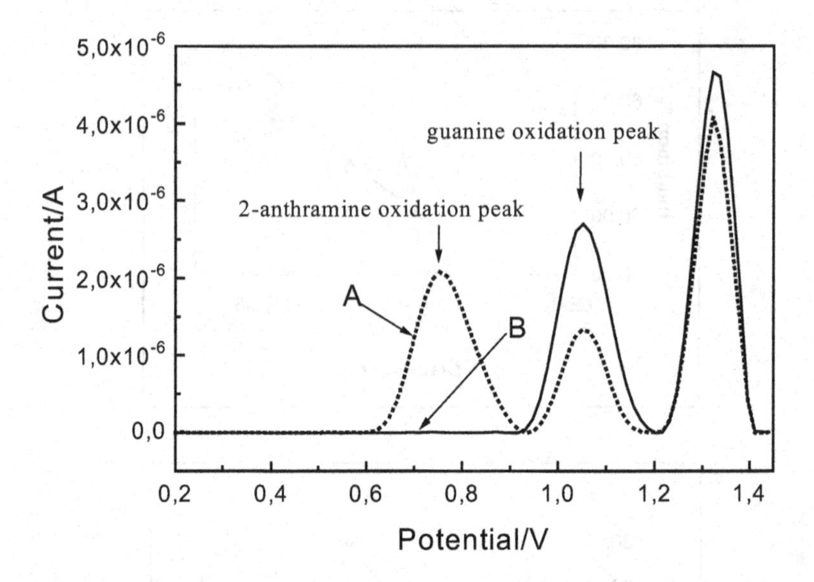

Figure 3. Analysis of standard solution of 2-Anthramine (0.1 ppm) with the ds DNA biosensor. (A) oxidation signal of the DNA biosensor after the interaction with 2-Anthramine (0.1 ppm); (B) DNA biosensor oxidation peaks (guanine ca. + 1.0 V and adenine ca. + 1.3 V), without interaction with the analyte solutions. Electrode conditioning 1.6 V *vs.* Ag pseudo reference SPE, for 3 min: DNA immobilization: 2 min at 0.5 V *vs.* Ag pseudo reference SPE. Interaction with the analyte 2 min open circuit condition. Square wave parameters: freq. 200 Hz, step pot. 15 mV, amplitude 40 mV. On the original signals a baseline correction was performed.

REFERENCES

1. Hooijschuur E W.J., Hulst A.G. de Jong A. L., de Reuver L. P., van Krimpen S.H., van Baar B.L.M., Wils E.R.J., Kientz C.E., Brinkman U.A.Th., Identification of chemicals related to the chemical weapons convention durino an interlaboratory proficiency test, TRAC, 21, 2, 2002, 116-130.
2. Bismuth C., Borron S.W., Baud F J., Barriot P., Chemical weapons: documented use and compound on the horizon, Toxicology Letters 149, 2004, 11-18
3. Noort D., Benschop H.P., Black R.M., Biomonitoring of exposure to chemical warfare agents: A review, Toxicology and Applied Pharmacology, 184, 116-126 (2002)
4. Sanchez-Santed F., Canada F., Flores P., Lopez-Grancha M., Cardona D., Long-term neurotoxicity of paraoxon and clorpyrifos: behavioural and pharmacological evidence, Neurotoxicology and teratology, 26, 2004, 305-317.
5. Cagnini A., Palchetti I., I. Lionti, Mascini M., A.P.F. Turner, Disposable ruthenized screen-printed biosensors for pesticides monitoring, Sensors and actuators B 24-25, 1995, 85-89.

6. Cagnini A., Palchetti I., Mascini M., A.P.F. Turner, Ruthenized Screen-printed Choline Oxidase-Based Biosensors for measurement of anticholinesterase Activity, Mikrochim. Acta 121, 1995, 155-166.
7. Palchetti I., Cagnini A,, M. Del Carlo, Coppi C., Mascini M., A.P.F. Turner, Determination of anticholinesterase pesticides in real samples using a disposable biosensor, Analytica Chimica Acta, 337 (1997) 315-321.
8. Hernandez, .,Palchetti I., Mascini M., Determination on Anticholinesterase activity for pesticides monitoring using Acetylthiocholine Sensor, Int. Journal of Environmental Analytical Chemistry, 78, 3-4, (2000), 263-278
9. Marrazza G., I., Chianella, Mascini M., Disposable DNA electrochemical biosensors for environmental monitoring. Anal. Chim. Acta, 387, 1999, 297-307.
10. G. Chiti, Marrazza G.,, Mascini M., Electrochemical DNA biosensor for environmental monitoring. Anal. Chim. Acta, 427, 2001,155-164.
11. Mascini M., Palchetti I,, Marrazza G.,, DNA electrochemical biosensor. Fresenius J. Anal. Chem., 369/1, 2001, 15-22.
12. Lucarelli F., Palchetti I., Marrazza G., and Mascini M., Electrochemical DNA Biosensor as a Screening Tool for the Detection of Toxicants in Water and Wastewater Samples, Talanta, 56, 949-957, 2002.
13. Lucarelli F., Kicela A., Palchetti I., Marrazza G., and Mascini M., Electrochemical DNA Biosensor for Analysis of Wastewater Samples, Bioelectrochemistry, 58, 113-118, 2002
14. Wang, Rivas G, Luo D, Cai X, Valera FS, Dontha N. Anal. Chem. 68, 1996, 4365.
15. Technical Workshop on genotoxicity biosensing, TECHNOTOX, May 8-12 2000, Mol, Belgium, http://www.vito.be/English/environment/environmentaltox5.htm

BIOSENSORS AND BIOMIMETIC SENSORS FOR THE DETECTION OF DRUGS, TOXINS AND BIOLOGICAL AGENTS

A. P. F. TURNER and S. PILETSKY
Cranfield University, Silsoe, Bedfordshire MK45 4DT, UK; www.silsoe.cranfield.ac.uk; a.p.turner@cranfield.ac.uk

1. INTRODUCTION

Over the past decade, increasing attention has been paid to the development of biosensors to enhance testing for safety and security and a growing market potential has been identified. In biodefense, attention has focussed on technologies for the rapid identification of biological threat agents in the environment and in the human population, while forensic science has been revolutionised by DNA technology. The UK's Royal Society (*Making the UK Safer, April 2004)* stated that *"The most urgent need is for mobile or hand-held instruments for point detection at the scene of an incident for use by responders."* Biosensors are well suited to delivering on-site detection, identification and monitoring in a variety of security and defence situations. Biosensors comprise a biological or biologically derived sensing element intimately associated with or integrated within a physicochemical transducer [1-3]. The net result is a digital electronic signal reflecting the concentration of one or more chosen analytes. This description is reflected in the definition used by the international journal *Biosensors and Bioelectronics* and has been broadly endorsed by the International Union of Pure and Applied Chemistry (IUPAC) following an extensive debate. A substantially similar definition also appears in the

D. Morrison et al. (eds.), Defense against Bioterror: Detection Technologies, Implementation Strategies and Commercial Opportunities, 261–272.

Oxford English Dictionary. Table 1 lists the biological and transducer elements usually used to construct biosensors.

Over 1,500 papers are published each year describing the various permutations of sensing element and transducer [4]. These are applied mainly in medical diagnostics [5], environmental diagnostics [6] and the food industry [7,8]. Of most relevant to defence and security are enzyme electrodes, whole-cell based systems and affinity sensors including both immunosensors and DNA chips.

Table 1. Biological elements and transducers commonly used in biosensors.

Biological Elements	Transducers
Whole (micro)organisms	Electrochemical
Tissues and organelles	Optical
Intact cells	Piezoelectric
Receptors	Thermometric
Enzymes	Magnetic
Antibodies	Micromechanical
Nucleic acids	
Synthetic receptors	

A number of sensors developed in our laboratories, in collaboration with various partners in industry, can find application in a variety of testing areas related to crime prevention and security. These include biosensors, immunosensors, electronic nose technology and molecularly imprinted polymer sensors for on-site testing for:

- Integrated sampling and sensing for minimally invasive blood monitoring
- Food & Water Safety: GMOs, chlorophenols, chloranisoles, nitrosamines, aflatoxins, endocrine disrupters, algal toxins
- Environmental Protection: pesticides, VOCs, formaldehyde, hydrocarbons
- Biological warfare agents
- Security: drugs of abuse, adulteration and poisoning

These advances will support new commercial detection opportunities in civil protection including biometrics, transport security, detection of adulteration of water and food contamination.

Cranfield has a long strived to alleviate the suffering associated with diabetes and to promote innovative new products to serve this market. In the early 80's [9], we invented, designed and developed, together with Oxford University, a new class of home blood glucose monitor based on mediated electrochemistry, which was introduced onto the market by Medisense in 1987. This mediated electrochemical technology went on to become the industry norm, comprehensively replacing the previous reflectance photometry devices and being adopted by all the four major companies that now account for over 90% of the market. Diabetes is the fastest growing chronic disease today, and afflicts approximately seven percent of the adult population of the western world. In order to avoid both the short- and long-term effects of the disease, patients are required to frequently observe their levels of blood glucose and take action to normalise them. A major impediment to good patient care is that today's glucose self-monitoring process is very cumbersome. In a typical test situation, four components are required: a glucose meter, a lancing device, glucose strips (mediated enzyme electrodes) and lancets and the whole test procedure takes at least a minute. The pain and inconvenience involved hinder good compliance and work against improving the long-term care of individuals with diabetes. Therefore, the trend has been to eliminate steps in the testing process. Companies like Bayer and Roche have introduced products designed towards simplification. Both have launched meters where strips are proffered automatically, thus eliminating the need for manual insertion.

Until now; however, no device has come close to delivering a truly one-step procedure. Engineering a one-step blood sensor is complicated, as a chain of events has to occur without fault. First, blood has to be acquired, second, a micro-litre blood sample must be sampled from the surface of the skin and fluidically guided to the sensor and third, an accurate measurement must take place to quantify the analyte in the blood. A failure in any one of these steps would compromise the success of the test. New technology has been developed to overcome these hurdles, notably, an electronic lancing engine to generate a blood drop on the skin surface without 'milking', microfluidics to guide the blood sample to the sensing region, and a new nanolitre sensor to fit the fully integrated format (Figure 1) [10].

The Pelikan [www.pelikantechnologies.com] electronic lancing engine is central to the performance of the integrated device. In the past 20 years, the skin-lancing step has received little attention, and only modest progress was made. The Pelikan device represents the first electronically controlled lancing system, which allows for unprecedented control of the lancing process. The key component of the engine is a series of solenoids, which produce a strong, uniform and very controllable magnetic field inside a tube that houses a metal piston (Figure 2).

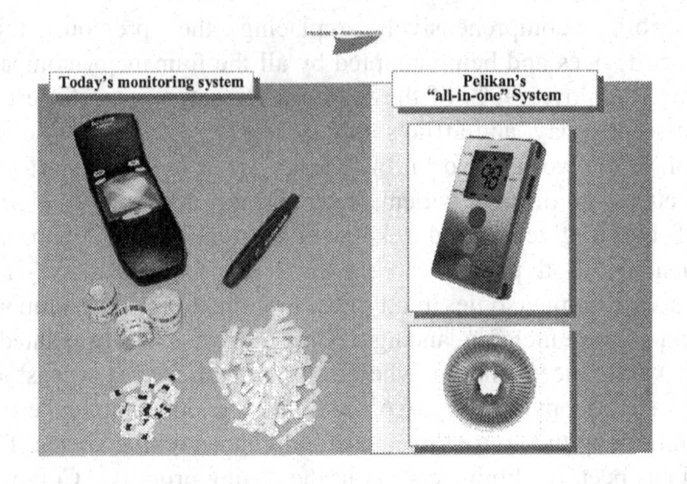

Figure 1. A current home blood glucose measuring biosensor kit (left) compared to the Pelikan system (right), which offers a single disposable disc containing 50 sensors and 50 lancets.

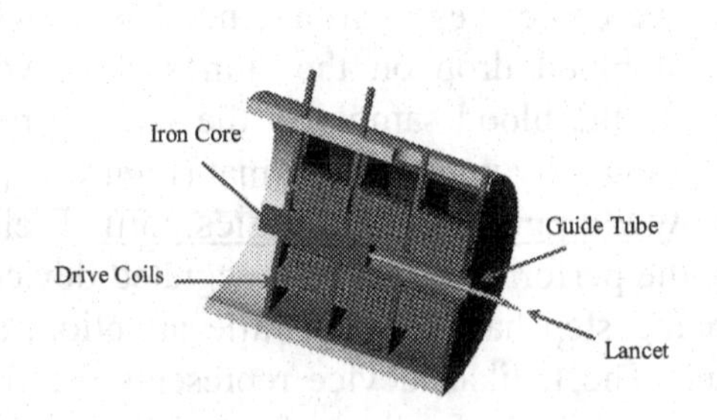

Figure 2. Pelikan's solenoid controlled lancing engine.

The direction and strength of the magnetic field controls the movement of the piston, which is connected to a lancet and can induce this to pierce the skin. Knowledge of the force exerted and the precise position of the lancet allow real-time calculation of the position of the lancet in the finger. While the lancet is advancing into the skin, dynamic properties of the tissue are determined and used to select the best motion profile. This instant adaptation has great impact on the pain / blood yield balance. The ideal end position of the lancet in its forward movement is just below the stratum corneum. PTI developed lancet motion profiles, which allow the wound channel to remain open after lancet retrieval. Thus sufficient blood will flow to the surface of the skin to provide a sample for glucose testing. When the lancet has returned to the disposable it is securely stored and cannot be used again.

The blood on the skin surface is collected via an opening of the integrated disposable, which consists of a disk-shaped array of integrated lancet/sensor chambers. Each lancet/sensor chamber is sterile and hermetically sealed from other lancet/sensor chambers. Each chamber is individually opened just prior to lancing and measurement. The opening of one chamber exposes microfluidic structures that induce the blood to enter the disposable where it reacts with the sensor chemistry.

The integrated glucose monitor quantifies glucose amperometrically by measuring the current produced when glucose is oxidized enzymatically. The sensor requires approximately 400 nL of blood, senses glucose in four seconds and completes the entire testing cycle in less than ten seconds. This is a significant improvement from the 60 – 90 seconds test time that is typical today. A novel sensor design makes the measurement robust against endogenous and exogenous compounds such as haemoglobin, triglyceride, uric acid, food additives and drugs that interfere with some other designs. The instrument's accuracy and precision characteristics make it very well suited for individual management of diabetes. When compared to a hexokinase reference assay fewer than 2% of measurements were in the B regions of the Clarke diagram with the remainder being in the A region (Figure 3).

This technology will offer the user a very different experience from that which they experience with current technology. The portable device allows fifty integrated lancets and sensors to perform glucose tests in a one-step, one-button operation that includes lancing and measurement and lasts less than ten seconds. The product will handle all tasks related to a test, including containing the used lancet and sensor. Automation allows significant reduction in blood requirements, which, along with the smooth electronic lancing, means a significant reduction in pain experienced during lancing.

All of these benefits are likely to promote more frequent testing and disease management. Outside of the diabetes arena, this technology could provide the basis for a multianalyte pocket blood monitor that could be used in a variety of situations ranging from testing for alcohol and illegal substances to verifying infection or poisoning. The integrated sampling and microfluidics could be coupled to a range of biosensors or biomimetic sensors.

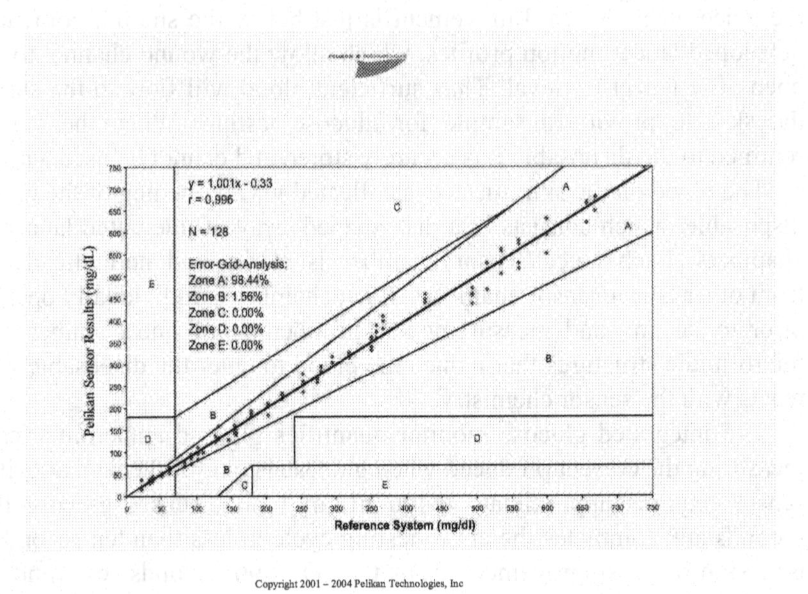

Figure 3. Clarke error grid showing the Pelikan glucose biosensor performance.

The development of suitably robust biosensors for many situations outside of glucose monitoring, has been hindered by several problems associated with the properties of biological material: poor stability, poor performance in organic solvents, at low and high pHs and at high temperature; absence of enzymes or receptors that are able to recognise certain target analytes; problems with immobilisation of biomolecules; and poor compatibility with micromachining technology.

The search for possible solutions to these problems has led to the development of biomimetic systems such as the electronic nose, which shows excellent practical potential for the detection of disease and infections [11]. An alternative approach has been to seek synthetic analogues of natural receptors [12] and antibodies using supramolecular systems such as cryptands, dendrimers, fullerenes, polypeptides, self-assembled monolayers of lipid and lipid-like molecules and molecularly imprinted polymers. Molecular recognition phenomena, which originate from biological systems lie at the core of the most exciting features of the design and functioning of nano- and smart materials. Characteristically these phenomena imply two steps, a recognition process and a signal induced reaction of the system. If nature can produce nanomaterials with recognition and functional properties by evolution, molecular engineers should be able to accomplish comparable, but broader capabilities by design, guided by examples from living systems. One of the most promising areas of biomimetics is Molecular Imprinting [13], which can be defined as the process of template-induced formation of specific recognition sites (binding or catalytic) in a material where the template directs the positioning and orientation of the material's structural components by a self-assembling mechanism. The material itself could be oligomeric (the typical example is the DNA replication process), polymeric (organic Molecularly Imprinted Polymers (MIPs) and inorganic imprinted silica gels) or 2-dimensional surface assemblies (grafted monolayers).

One of the most challenging tasks in developing biomimetic transduction systems based on MIPs is transforming chemical binding into electrical signals [14,15]. The associated technical issues include the development of new monomer molecules with responsive functionalities (e.g. environment-sensitive fluorescent probes), conjugation of binding sites with transducers (molecular wires), utilisation of induced fit common for natural and synthetic receptor molecules in the generation of sensor responses etc. Another key element is the need for rational design of MIPs and a thorough understanding of the effect of polymerisation conditions on their recognition properties. We have introduced a computational approach to the selection of appropriate monomers using a virtual library of functional monomers and screening them for their ability to form a molecular complex with the template [16]. The monomers giving the highest binding score are used for polymer preparation. In addition, we have studied the appropriate polymerisation conditions in a series of fundamental papers, which attempt to better define the molecular imprinting process and hence facilitate the future design of separation and sensing materials [17,18,19]. Key factors in achieving the desired properties include the type and concentration of the monomers used and the polymerisation temperature, pressure, solvent, time and initiation conditions. This level of understanding has permitted the design of a range of MIP-based separation and sensing systems of relevance to defense and security. In one example, the challenge of detecting algal toxins in drinking water was taken up under sponsorship from the European

Union [20]. In this work an artificial receptor for microcystin-LR was synthesised using the computational approach to MIP design. The synthesised polymer was used both as a material for solid-phase extraction (SPE) and as a sensing element in a piezoelectric sensor (Figure 4). Using the combination of SPE followed by detection with a piezoelectric sensor the minimum detectable amount of toxin was 0.35 nM. The use of MIP–SPE provided up to 1000 fold pre-concentration, which was more than sufficient to achieve the required detection limit for microcystin-LR in drinking water (1 nM). This work is the first example where the same MIP receptor has been used successfully for both SPE and the corresponding sensor.

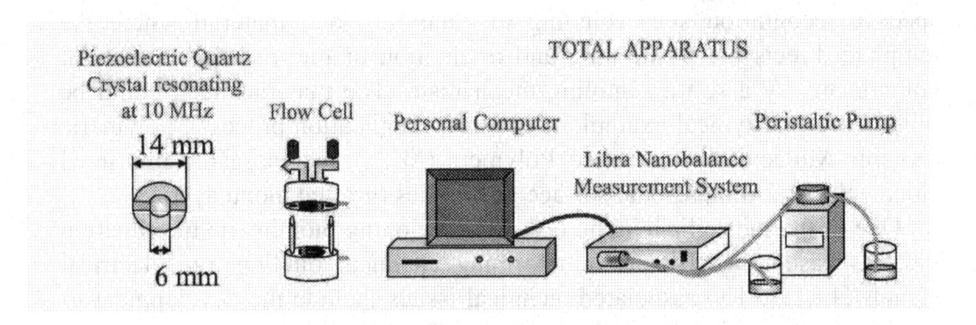

Figure 4. – The combination of a solid phase extraction cartridge and a MIP-based sensor for detection of microcystin-LR. [20]]

Table 2 shows some of the toxins for which MIPs have been computationally designed at Cranfield for use in combination with optical, electrochemical and piezoelectric transducers.

In another programme MIPs have been developed for a variety of explosives including PETN, DRDX, Tetryl and TNT. Table 3 shows some preliminary results for the extraction of Tetryl using acetone and acetonitrile.

In a final example, a programme is underway at Cranfield in collaboration with Siemens and other European partners to design an array

Table 2. Toxin for which MIPs have been developed at Cranfield.

Fungal Toxins	**Algal Toxins**
Nivalenol	Domoic acid
DeoxyNivalenol	Microcystin
Aflatoxin B1	
Ochratoxin A	
Fumonisin B1	
Patulin	

of piezoelectric sensors for drugs of abuse. So far, MIPs have been designed at Cranfield for Cocaine, Methamphetamine, Methadon and Morphine. Figure 5 shows preliminary work illustrating microgravimetric detection of Methamphetamine binding to a computationally-designed MIP.

Table 3. Results of Tetryl extraction using a computationally designed MIP with acetone and acetonitrile

Solvent	**Polymer**	**1st Extraction**	**Efficiency %**	
		Conc (g l-1)		
MeCN	10	0.03057	153	132
	2	0.02536	127	
	3	0.02646	132	
	4	0.02325	116	
Acetone	5	0.02159	108	108
	6	0.02068	103	
	7	0.02430	122	
	8	0.01937	97	

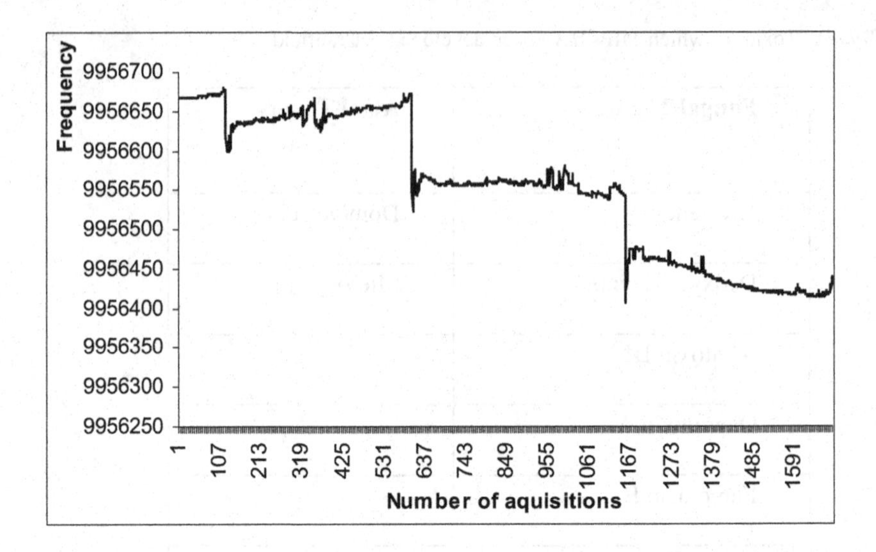

Figure 5. The response of PQC with immobilised MIP on injection of methampehtamine.

MIPs are increasingly considered as potentially viable alternatives to the relatively unstable antibodies and enzymes currently used in biosensor technology. We envisage that in near future molecular imprinting will become a generic technology, interfacing sensor transduction platforms to their chemical and biological targets. In order to challenge the technical barriers to realising this vision, key obstructions to the development of high-performance MIPs need to be identified and overcome and optimal protocols for integrating polymers with optical and electrochemical sensors need to be developed.

2. ACKNOWLEDGEMENTS

The authors would like to thank their teams at Cranfield and at Pelikan in Palo Alto for their extensive contributions to the work briefly reviewed above and in particular to Drs Iva Chianella, Kal Karim, Elena Piletska and Dave Cullen at Cranfield and Dirk Boecker, Dominique Freeman and Klaus Becker at Pelikan Technologies Inc.

REFERENCES

1. Turner, A.P.F., Karube, I. and Wilson, G.S. (1989). *Biosensors: Fundamentals and Applications.* Oxford University Press, Oxford. 770p. Revised, paperback. (Also published in Russian). ISBN: 0198547455.
2. Turner, A.P.F. (2000). Biosensors – Sense and Sensitivity. *Science* **290** (5495), 1315-1317.
3. Laschi, S., Mascini, M & Turner, A.P.F (2002). Biosensors. *Kirk-Othmer Encyclopedia of Chemical Technology* (on-line edition) www.wiley.com
4. Newman, J.D., Tigwell, L.T., Turner, A.P.F. and Warner, P.J. (2004). *Biosensors – A Clearer View*, Cranfield University.
5. Newman, J and Turner, APF (2004). Biosensors for monitoring glucose. *Sensors in Medicine and Health Care (Sensors Applications)* **3, Eds:** Öberg, P. Å.,Togawa, T. and Spelman, F. A. Wiley-VCH. ISBN: 3527295569.
6. Bilitewski, U. and Turner, A.P.F. (2000). *Biosensors for Environmental Monitoring.* Harwood Academic publishers. 1-409pp. ISBN 90 5702 449 7.
7. Tothill, I and Turner, A.P.F (2003). Biosensors. *Encyclopaedia of Food Sciences and Nutrition* (2nd Edition), (Editor in Chief: Benjamin Caballero; Eds. Luiz Trugo and Paul Finglas), Academic Press. pp 489-499. ISBN: 0-12-227055-X.
8. Palleschi, G., Volpe, G and Turner, A.P.F. (2002). New Biosensors for Micobiological Analysis. *Detecting pathogens in Food.* (Ed. Thomas A McMeekin), Woodhead, pp 294-331. ISBN 1 85573 6705
9. Cass, A.E.G., Davis, G., Francis, G.D., Hill, H.A.O., Aston, W.J., Higgins, I.J., Plotkin, E.V., Scott, L.D.L. and Turner, A.P.F. (1984). Ferrocene-mediated enzyme electrode for amperometric determination of glucose. *Analytical Chemistry* **56**, 667-671.
10. Turner, A.P.F. and Boecker, D. (2004). Fully integrated home blood-glucose monitor. *Eighth World Congress on Biosensors*, 24-26 May, 2004, Granada, Spain. Elsevier.
11. Turner, A.P.F and Magan, N (2004). Electronic noses and disease diagnostics. *Nature Microbiology Reviews* **2**, 161-166.
12. Subrahmanyam, S., Piletsky, S.A and Turner, A.P.F (2002). Application of Natural Receptors in Sensors & Assays – A Review. *Analytical Chemistry* **74**, 3942-3951
13. Piletsky, S. and Turner, A.P.F. (2004). *Molecular Imprinting*, Landes Bioscience, Georgetown, TX, USA (in press). www.landesbioscience.com/iu/output.php?id=358
14. Piletsky, S.A & Turner, A.P.F. (2002). New Materials Based on Imprinted Polymers and their application in Optical Sensors, *Optical Biosensors: Present and Future.* Eds: Ligler, F.S. & Rowe Taitt, C.A. Elsevier Science B.V. pp397-425. ISBN: 0444509747.
15. Piletsky, S.A and Turner, A.P.F (2002). Electrochemical Sensors Based on Molecularly Imprinted Polymers. *Electroanalysis.* **14**, 317-323
16. Chianella, I., Lotierzo M, Piletsky, S.A., Tothill, I., Chen, B., Karim, K., Turner, A.P..F (2002). Rational design of the polymer specific for microcystin-LR using a computational approach. *Analytical Chemistry* **74**, 1288-1293.
17. Piletsky, S.A., Piletska, E.V., Karim, K., Freebairn, K.W., Legge, C.H and Turner A.P.F. (2002). Polymer Cookery; Influence of Polymerization Conditions on Performance of Molecularly Imprinted Polymers. *Macromolecules* **35,** 7499-7504.
18. Piletsky, S.A., Guerreiro, A., Piletska, E.V., Chianella, I., Karim, K. and Turner, A.P.F. (2004). Polymer Cookery II: Influence of Polymerization Pressure and Polymer Swelling on the Performance of Molecularly Imprinted Polymers. *Macromolecules* (in press).
19. Piletsky, S.A,, Mijangos, I., Guerreiro, A., Piletska, E.V., Chianella, I., Karim, K. and Turner, A.P.F. (2005). Polymer Cookery III: Influence of Polymerization Time and Different Initiation Conditions on Performance of Molecularly Imprinted Polymers. *Macromolecules (submitted).*

20. Chianella I., Piletsky S. A., Tothill I. E., Chen B., Turner A. P. F. (2003). Combination of solid phase extraction cartridges and MIP-based sensor for detection of microcystin-LR. *Biosensors & Bioelectronics*, **18**, 119-127.

DISCLAIMER

CHEMICAL MULTI-SENSOR ARRAYS FOR LIQUIDS MONOLITHIC INTEGRATION USING MICROELECTRONIC TECHNOLOGY

A.BRATOV AND C. DOMINGUEZ
Centro Nacional de Microelctronica, 08193 Bellaterra, Barcelona, Spain

Abstract: The need for chemical sensor systems is expanding rapidly. By employing an array format the sensor systems with different sensor materials might be more generic and also more universal in their applications. The peculiarity of such array-based systems is that utilizing already available signal processing schemes and pattern recognition methods, it is possible to characterize an analytical sample as a whole. This means that a sensor array development should be application driven and sensitive materials, as well as types of transducers comprising the array, should be considered for each application. Microelectronic technology has large experience in fabrication of chemical sensors based on ion-selective field effect transistors (ISFET) and other types of transducers, like amperometric or conductimetric, utilising thin film technology. Though large ISFETs arrays have been reported (T. Yeow et al. Sensors and Actuators B 44 (1997) 434-440) they are produced on the same silicon substrate that prevents simultaneous measurements of all the devices. To resolve this problem complementary metal-oxide semiconductor (CMOS) technology was proposed to isolate individual ISFETs by a p-n junction (S. Martinoia et al. Biosensors and Bioelectronics 16 (2001) 1043-1050); but this results in high leakage currents and cross-talking of sensors that shows that they must be isolated. The latest achievements in microelectronic micromachining gave rise to production of new starting materials like BESOI (silicon on insulator) that can be used as a base for the development of a liquid multi-sensor array containing different types of chemical and physical sensors and that permit to integrate NMOS and thin film technologies.

D. Morrison et al. (eds.), Defense against Bioterror: Detection Technologies, Implementation Strategies and Commercial Opportunities, 273–289.

1. INTRODUCTION

The need for chemical sensor systems that are required in a vast amount of different practical applications is expanding rapidly. By now there exists a wide diversity of designed chemically selective materials, but only some of these have adequate parameters like selectivity, sensitivity, and durability to be used in practical applications as individual sensors. By employing an array format the chemical sensor systems with different sensor materials might be more generic and also more universal in their applications. The peculiarity of such array-based systems is that utilizing already available signal processing schemes and pattern recognition methods [1] it is possible to characterize an analytical sample as a whole. This means that a sensor array development should be application driven and sensitive materials, as well as types of transducers comprising the array, should be considered for each particular application.

While gas sensor arrays (electronic nose) are well developed and commercialised, little still is done in the field of liquid sensor arrays (electronic tongue). Usually it is not an array on a chip, but several individual sensors measured together. The other drawbacks of existing approach to the electronic tongue development is that reported arrays are normally composed by the same type of sensors, either potentiometric (ion-selective electrodes) [2, 3] or voltammetric sensors [4, 5]. This reduces the amount of useful data that may be obtained from a sample. Moreover, in complex systems that require the use of biosensors the diversity in types of the electrochemical transducers might be essential.

Among different electrochemical transducers that may be used in an array format we may distinguish potentiometric sensors (ion-selective field effect transistors [6] and all solid state ion selective electrodes [7]) that can be used directly to measure ion concentration (activity) or together with enzyme membranes, thus forming subsequent biosensors [8].

Another widely applied electrochemical transducer is an amperometric sensor [9] used in a three-electrode configuration with electrodes typically formed by platinum, glassy carbon or carbon paste. And the last class of electrochemical transducers is impedimetric sensors [10] that are used to measure changes in conductivity or dielectric properties of various sensing matrixes [11]. Those can be metal electrodes (e.g. gold with self assembled monolayers [12]), silicon/silicon dioxide capacitive structures [13], interdigitated electrode arrays [14], etc. All these types of transducers together with a biochemical recognition system form a wide class of biosensors.

2. DESIGN AND FABRICATION OF A MULTI-SENSOR CHIP

Microelectronic technology has large experience in fabrication of electrochemical sensors [15-17] based on ISFET and other types of transducers, like amperometric or impedimetric, combining standard CMOS or n-channel metal oxide semiconductor (NMOS) processes with a thin film technology. As a first prototype we have chosen an array composed by six ISFET devices, interdigitated electrode array and a temperature sensor. Though large ISFETs arrays have been reported [18] they are made on the same silicon substrate that prevents simultaneous measurements of all the devices and requires signal multiplexing. When designing an ISFET sensor array it must be taken into account that switching between different sensors may result in longer response time and drift of the output signal especially in ISFETs with polymer membranes with low electric conductivity. To be able to work with several individual ISFET sensors at the same time, without switching them on and off, these devices should be galvanically isolated one from another. When several devices are fabricated on a conventional silicon wafer they stay interconnected through the common silicon substrate, which presents another problem, if a leakage current appears through some defect in dielectric layer, the whole array will loose its functionality. To resolve this problem and to isolate individual ISFETs of the array Martinoia et al [19] used an approach proposed earlier [20] by embedding each ISFET of the array into a p-well formed on an n-type silicon substrate. The resulting p-n junction should guarantee the electrical isolation of devices. We tried this option forming ISFET and MOSFET in implanted p-wells, as presented in Figure 1. Unfortunately, due to the large surface occupied by the ISFET the resulting leakage currents (Fig. 2) of the p-n junction were too high to guarantee good devise isolation.

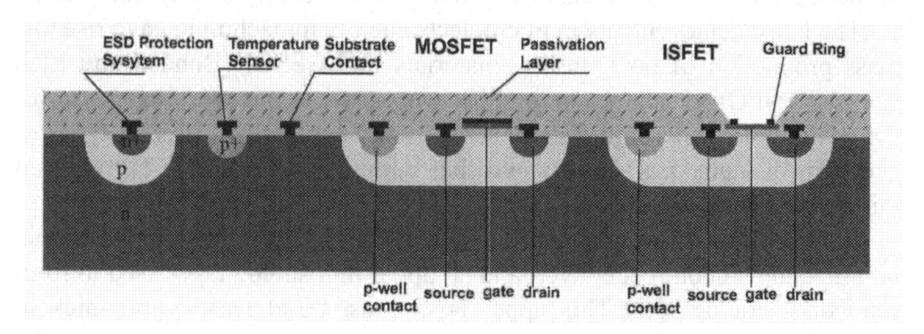

Figure 1. Isolation of an ISFET by a p-wellgure.

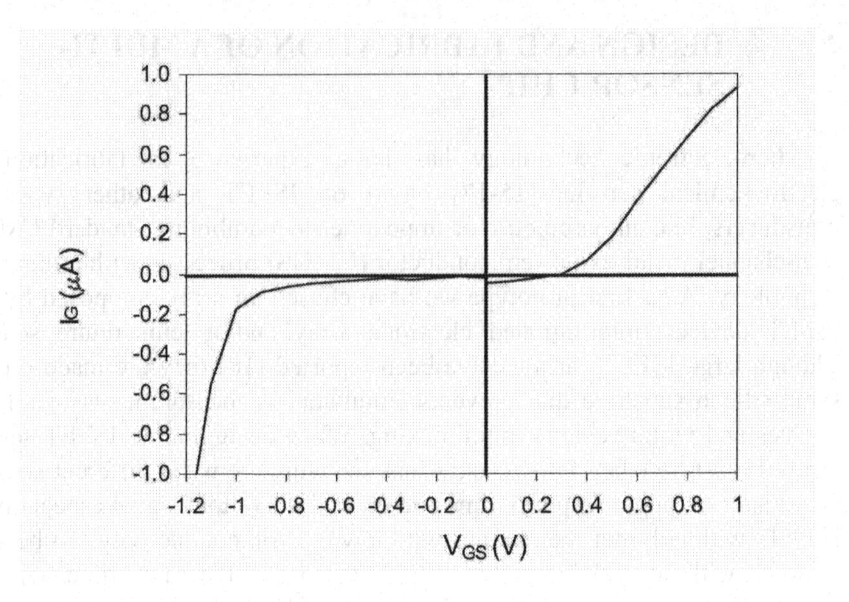

Figure 2. Leakage currents of the ISFET on fig.1 measured in a water solution (VGS polarisation applied between reference electrode and the p-well contact).

Another method to isolate ISFETs was proposed at the beginning of ISFET history [21] in 1981 and used to develop enzyme ISFET sensors [22] [23] is based on utilising Silicon-On-Sapphire (SOS) wafers as a substrate for ISFET formation. Using this approach, firstly, isles of silicon was formed on a wafer and then ISFETs processing was performed in a common way. This resulted in ISFETs isolated one from another and after sawing the wafer dies had to be encapsulated only at the contact pads. Unfortunately SOS wafers were rather expensive and the epitaxial silicon layer was quite thin to guarantee the stable functioning of a FET device. An alternative to this was proposed to use silicon on insulator (SOI) [24] wafers formed by Si/SiO2/Si.

The latest achievements in microelectronic micromachining gave rise to a mass production of new starting materials like BESOI (Bonded and Etch back Silicon On Insulator) wafers that were used in this work as a base for the development of a liquid multi-sensor array containing different types of chemical and physical sensors and that permitted to integrate NMOS and thin film technologies.

BESOI wafers are formed by three layers, thick silicon substrate, intermediate silicon oxide layer, and an upper thin silicon layer used to form semiconductor devices. This upper layer must be at least 5 μm thick to prevent the so called back gate effect, which consists in formation of a parasitic conduction channel between source and drain in silicon at the interface with the imbedded oxide layer.

BSOI wafers used to fabricate ISFET devices had the following parameters

- Upper working layer-Thickness 5 ± 1 μm; Resistance: 30-40 Ω·cm (boron doped)
- Oxide imbedded layer. Thickness 1 ± 0.05 μm; Type- thermal oxideSubstrate silicon layer- Thickness 450 ± 10 μm; Resistance: 30-40 Ω·cm (phosphorus doped)

The first experiments were performed to test the possibility of isolating ISFET on a BESOI wafer by forming a groove that removes silicon around the device, which is schematically presented in Figure 3. Taking into consideration that the groove is 5 μm deep and 10 μm wide in order to prevent some difficulties in performing photolithography steps due to an uneven distribution of a photoresist over the wafer, which can leave uncovered parts on the wafer surface, a special photoresist that forms layers of 6 μm thick was used at this stage.

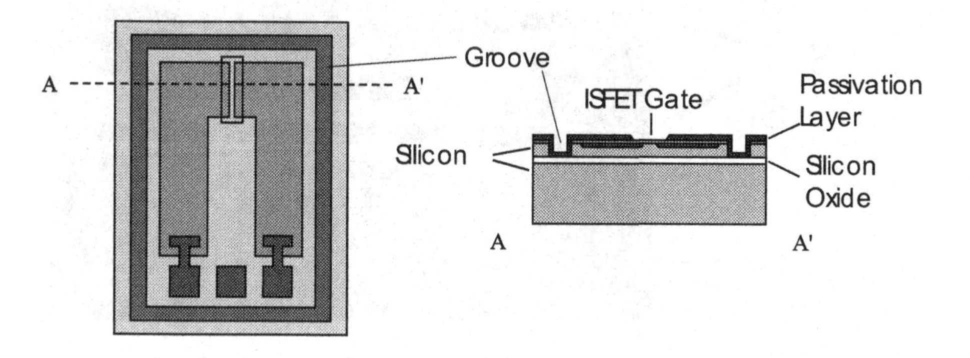

Figure 3. Schematic presentation of a proposed isolation of an ISFET on a BESOI wafer by a groove.

Two different etching processes have been tested to form the groove: RIE (Reaction Ion Etching) and wet etching in TMAH. In the first case an aluminium layer was deposited, patterned and used as a mask, in the second a 500 nm thick layer of silicon oxide was used for masking. The difference in the results obtained by two method of etching is that RIE gives nearly vertical walls of the groove, while anisotropic etching in TMAH results in 55° inclined walls.

Figure 4 shows the ISFET chip formed on a BESOI wafer with a groove formed around the chip. Electronic microscope photos presented in figure 5 show the cross-section of the grooves obtained by RIE and wet etching. As may be judged from photos, RIE produces an over-etching of approximately 5 μm at each side of the groove, while over-etching obtained in wet process is minimal. The groove produced with this last method is preferable, because the posterior deposition of the silicon oxide/silicon nitride passivation layer will cover the surface more evenly, producing fewer defects.

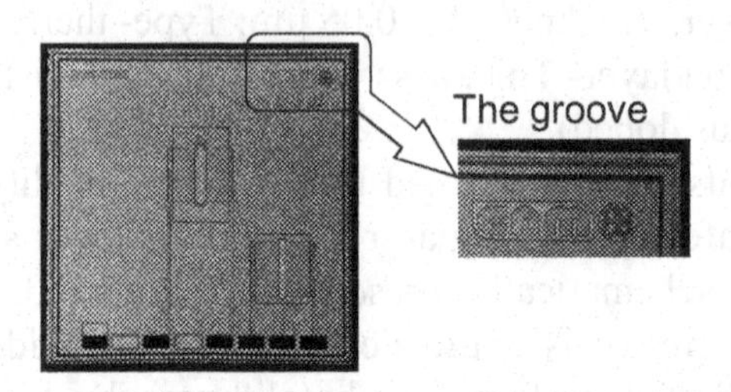

Figure 4. The groove formed around the ISFET chip as seen through the optical microscope.

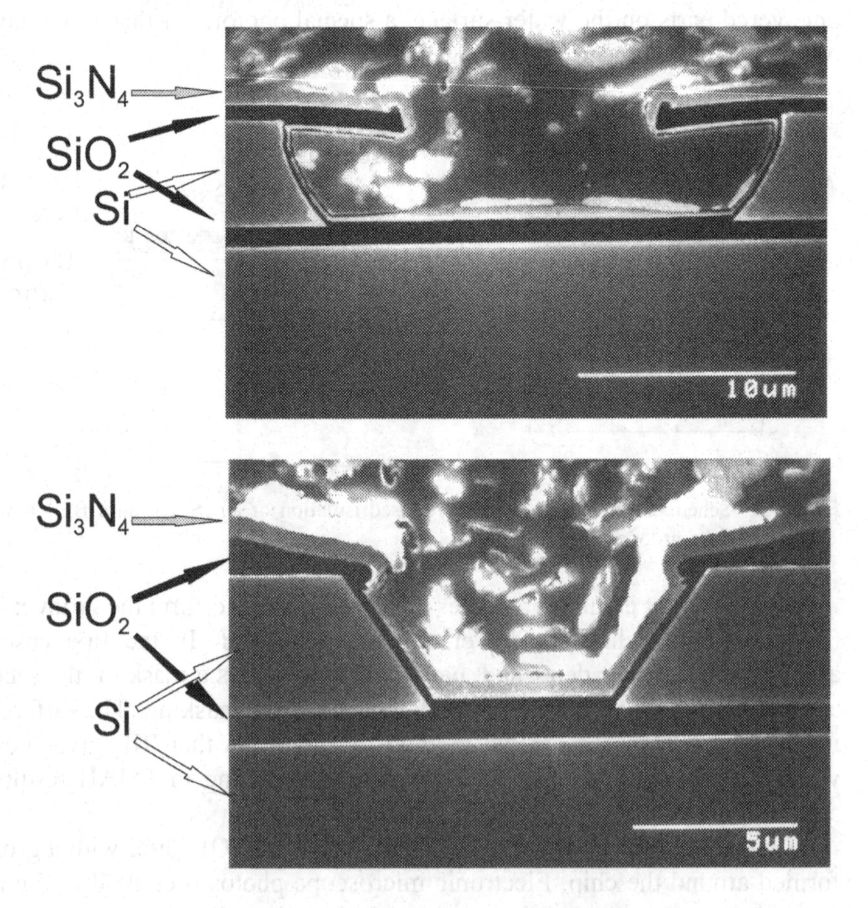

Figure 5. The cross-section of the groove formed by RIE (upper) and by wet etching in TMAH (lower).

To test that realized method of ISFET isolation is adequate special experiments have been carried out in which the contact pads and connecting wires were encapsulated with an epoxy resin, leaving the groove and the lateral sides of the chip open to contact with solution, as is presented in figure 6.

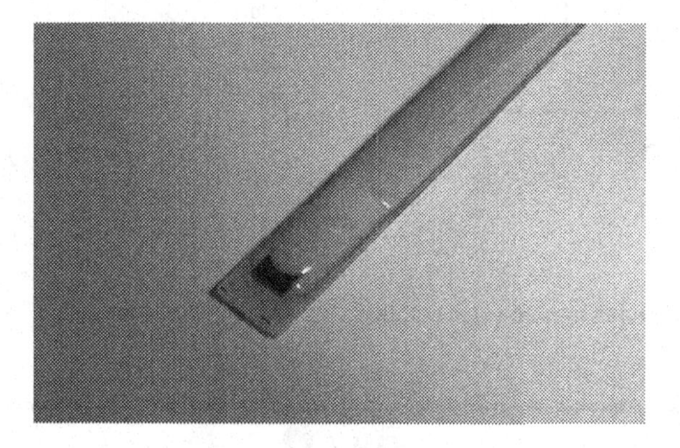

Figure 6. BESOI ISFET encapsulated for isolation tests.

Sensors together with the Ag/AgCl reference electrode were put in contact with pH buffer solution (pH 7) and polarisation of +5 V and –5 V between the ISFET source connected to silicon substrate and the reference electrode was applied. The resulting leakage current is the measure of the isolation quality (Figure 7). 21 devices were tested and 16 of which had leakage current less than 500 pA. Two devices had leakage currents due to defects in the passivation layer that covers the aluminium metal lines on the surface of the chip and three others had leakage currents passing through the groove. All of these devices were produced by RIE groove etching.

Tests were performed that permitted us to establish the technology of ISFET fabrication on BESOI wafers and proceed with the design and development of a multi-sensor array chip. The designed chip is presented in Figure 8.

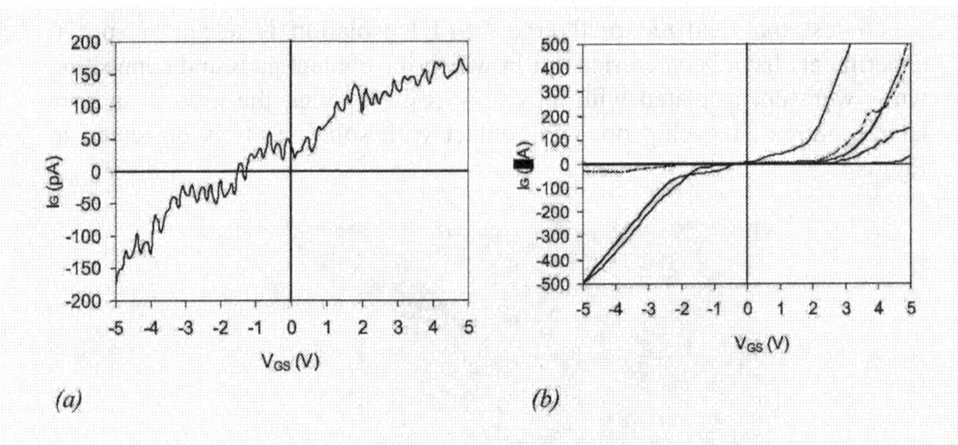

Figure 7. Leak currents by well (a) and poorly (b) isolated ISFET devices.

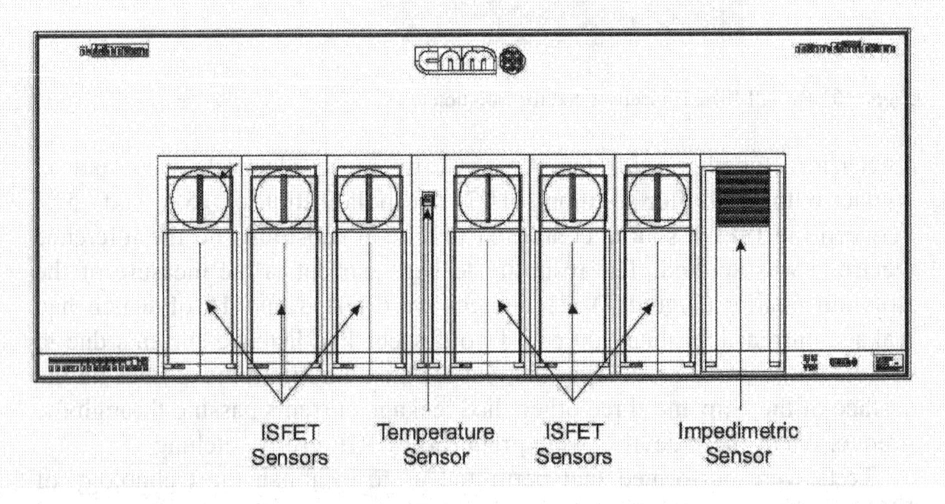

Figure 8. The layout of the sensor array chip]

It carries 6 ISFET devices, an impedimetric sensor base on platinum interdigitated electrode array, and a diode that is used as a temperature sensor. Each sensor is electrically isolated one from another by grooves etched in BESOI wafer silicon. This permits to work with all sensors independently and, also, this facilitates the encapsulation process. Within the fabrication process ISFETs are formed on a BESOI wafer. After deposition and patterning of the aluminium layer the wafer was covered with a 500 nm thick passivation silicon oxide layer, which was opened till silicon where the grooves should be formed. Figure 9 shows the grooves formed by wet etching. As can be seen the corner of the ISFET isle is slightly over-etched which guarantees the absence of sharp edges.

Figure 9. The groove etched in silicon separating isles with ISFETs.

At the next step the whole wafer is covered by 1μm thick silicon oxide/silicon nitride passivation layer, which is opened over the ISFET gates and contact pads. At this stage formation of other transducers by thin film technology is to be performed. Finally, we deposit a 10-15 μm thick polymer encapsulation layer based on a photocured epoxy or polyurethane in which openings over the ISFET gates and chip contact pads are made. This well-adhering, chemically stable layer developed earlier for ISFET encapsulation [25, 26] is used for three reasons: i) as an additional chemical protection of the device covered with a low temperature passivation layer; ii) helps to make the chip surface flat and even so that a flow through cell may be easily adjusted if required; iii) to define the area of the membrane deposition over the ISFET gate.

2.1 ISFET Sensors

Fabricated multi-chips s were cut in dice and mounted on a specially designed printed-circuit substrate. The substrate was designed in such a way that always permits chip placement in the same position, relative to the substrate and, with a special tool kit, permits encapsulation of the chip contact pads, connecting wires and the substrate gold conducting lines in a semi automated manner. The resulting device is presented in figure 10 and does not require any additional encapsulation.

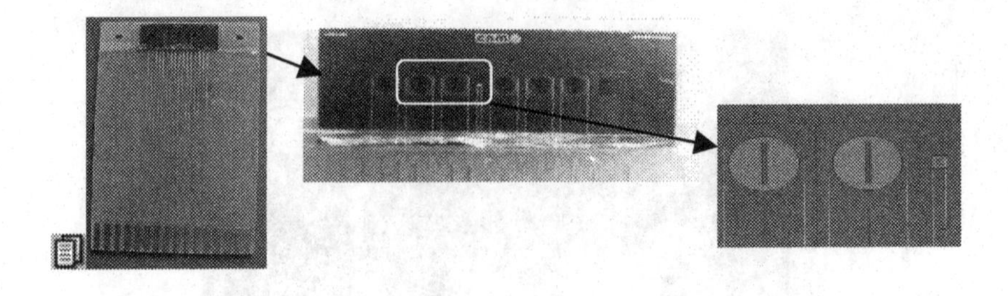

Figure 10. Mounted and encapsulated multi-chip

ISFETs threshold voltages were around -2.7 V with 100 mV deviation within the wafer and 200 mV deviation between chips from different wafers. The transconductance was 870±20 μA/V. pH sensitivity of all ISFETs was 56.0±1.5 mV/pH. After making all semiconductor devices of the array a platinum interdigitated electrode array is formed by deposition of 100 nm of platinum and subsequent lift-off process. E-beam deposition process results in considerable damage of ISFET devices affecting their threshold voltage, transconductance and introducing build-in charge into the gate insulator. This is the typical problem associated with the integration of CMOS or NMOS technologies with posterior processes required for electrochemical sensor fabrication. Special annealing processes were developed to recover ISFETs parameters. Electrical parameters of ISFETs before and after Pt deposition, as well as the effect of thermal annealing and special treatment are presented in Table 1.

Table 1.. Threshold voltages and transconductance of silicon nitride gate ISFETs after deposition of platinum and special treatments

		Vth , V	TC, μA/V
Inicial	MEAN	-3.014	849.08
	DEVIATION	0.145	123.48
Pt	MEAN	-10.488	411.32
	DEVIATION	0.586	22.965
Pt+Anneal	MEAN	-5.605	1010.18
	DEVIATION	0.2581	8.765
Pt + An+T	MEAN	-3.668	1021.105
	DEVIATION	0.344	25.295

2.2 Temperature Sensor

The multi-chip carries along with 6 ISFETs two more devices – the diode based temperature sensor and a platinum electrode interdigitated structure. To study the temperature response of the fabricated diode sensors was performed in a programmable temperature oven in the range of 25 –80°C with a 5°C step by measuring the forward conduction voltage at different diode current levels. Optimal results on sensitivity and linearity of the response were obtained at 100μA current levels. Experimental results obtained at this current level are presented in Figure 11. The same parameters we obtained in water solutions at different temperatures.

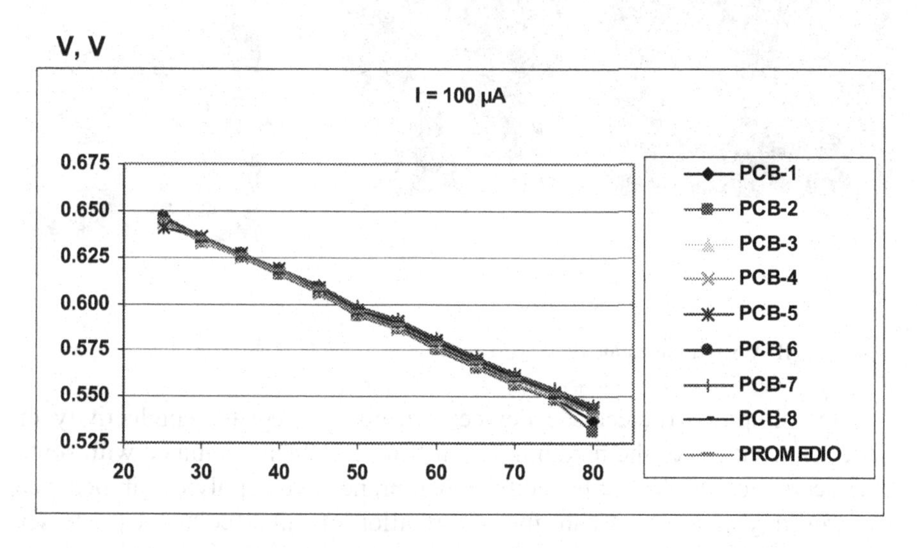

Figure 11. The temperature response of the diode forward conduction voltage at 100 μA

The sensitivity of the temperature sensor is -1.89 mV/°C. With individual calibration of each sensor, a precision of 0.2° can be achieved, and without individual calibration precision will be around 1°C as revealed by the standard deviation values of the batch results.

2.3 Interdigitated Electrode Array

Impedimetric sensor presented i Figure 12 is based on a platinum interdigitataed electrode array (IDA) and is formed by 83 electrode fingers 5 μm wide, 1150 μm long, with a 10 μm separation.

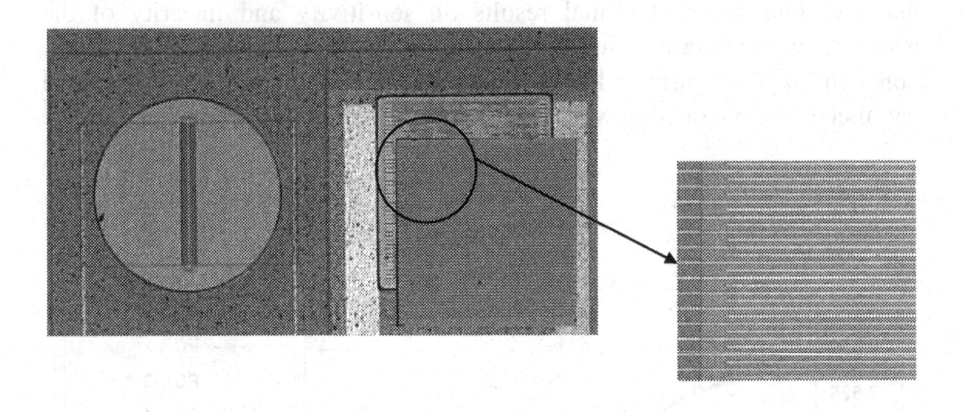

Figure 12. IDA based impedimetric sensor

This simple two-electrode device is useful to register conductivity or dielectric changes of the medium it is in contact. When combined with some chemically or biologically active membrane and applying impedance spectroscopy as a tool it can give information of rather complex processes occurring within the membrane due to its interaction with ambient [10]. Parameters of this kind of a sensor mainly depend on the geometry of the device (digit width, separation and total length). Figure 13 gives an example how the sensitivity of the sensor to solution resistivity and dielectric constant changes with the separation between electrodes.

Characterization of the multi-chip interdigitated electrode array was performed in KCl solutions in the conductivity range of 0.15 ÷ 7.0mS/cm. The response was found to be linear with the conductivity cell constant of 0.0.7 cm-1.

2.4 Multi-chip with Ion-selective Membranes

To test the functionality of all the ISFET sensors of the multi-chip photocurable urethane based polymer membranes containing an ionophore and a plasticizer and selective to potassium, sodium, calcium and chloride ions were deposited one by one on a multi-chip sensor. Composition of membranes, deposition and characterisation methods are presented elsewhere [27-29].

Characterization of this device showed the ion sensors response is the same as obtained on individual sensors with the only difference, that the lifetime of membranes in constant contact with solution was reduced to one month instead of three in case of individual ISFETs with membranes (Fig. 14). This is due to the fact that multi-chip membranes are only 100 μm thick, in comparison to 200-300 μm thick membranes of individual sensors.

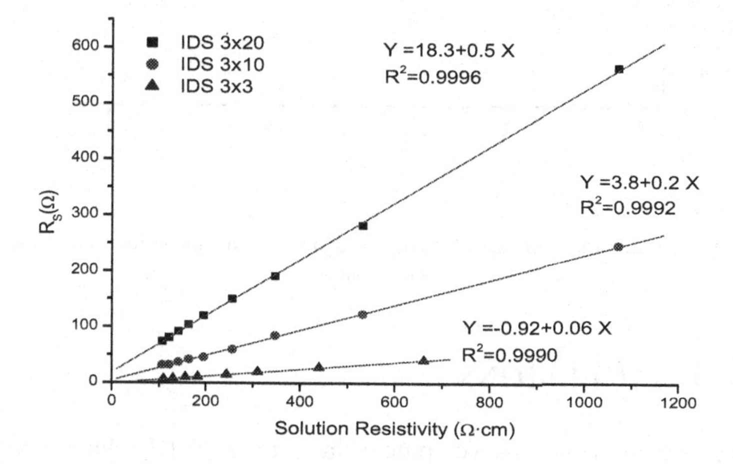

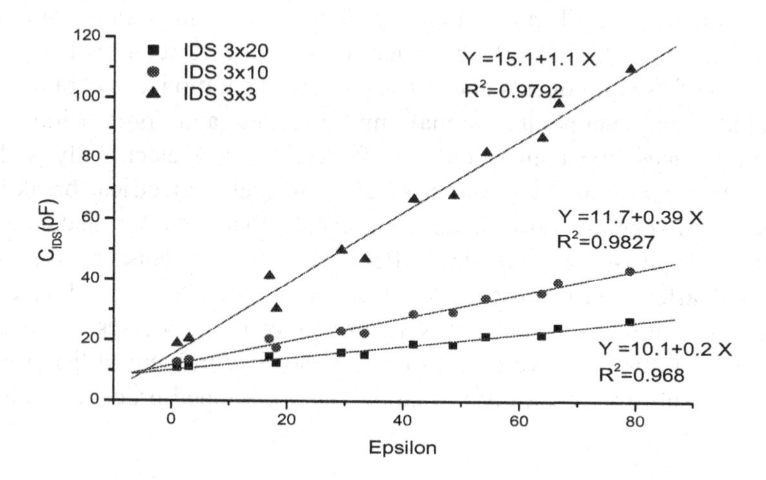

Figure 13. Sensitivity of the impedimetric sensor to solution resistivity and dielectric constant changes with 3 μm digit width 3, 10 and 20 μm separation between electrodes.

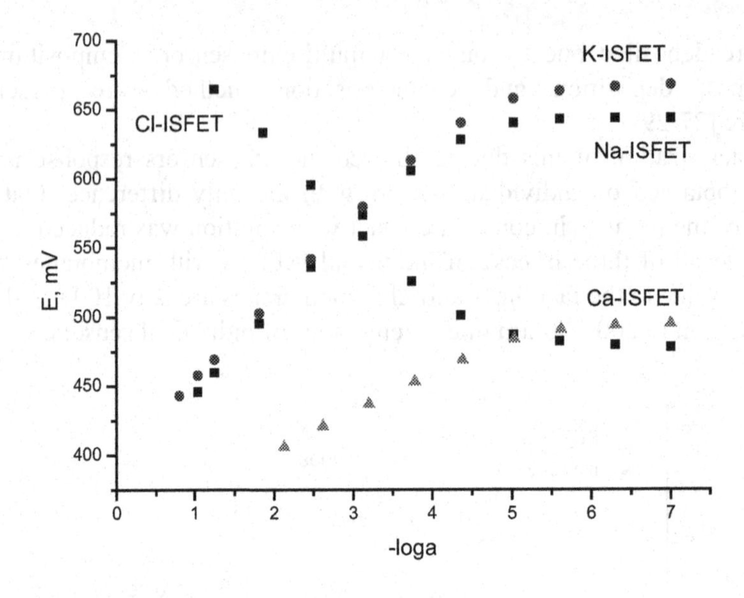

Figure 14. Calibration curves of ISFETs of the multi-chip with various ion-selective membranes

3. CONCLUSIONS

In the last 30 years of development, individual ISFETs which excluded pH-sensitive capabilities did not find their commercial success. But current research indicates new application niches where as an element of a chemical sensor array an ISFET may be a welcome partner in a raw with other chemical transducers. The development of such arrays requires monolithic integration of its components and for larger arrays additional integration with electronic components for signal multiplexing and processing. This integration faces two main problems, firstly, to have electrically isolated devices or at least isolated sectors of array, so that if electrical breakdown occurs it will destroy only some components that may be disconnected. Secondly, all post-processing steps after CMOS components or ISFETs are made will affect the electrical parameters of these electronic devices, so special precautions must be taken or recovery processes should be developed. In this work we tried to resolve the first problem utilising BSOI wafers as a substrate for array formation. The developed processes help not only to isolate electrically sensors but also facilitate encapsulation of the final device, which in individual ISFET-based sensors production constitutes the most expensive part.

The second problem in our case was resolved by thermal annealing and a proprietary method of recovery of ISFETs parameters after post-processing steps. The developed technology opens possibilities to make small (e.g 100 elements) arrays based on different types of electrochemical sensors on one silicon chip.

4. ACKNOWLEDGEMENT

Financial support from the European Community project GRD1-2000-25288 and the national Spanish project DPI2002-01962 is greatly acknowledged. AB expresses his gratitude for incorporation within the Ramon y Cajal Programme.

REFERENCES

1. C. Richards, C. Bessant, S. Saini, Multivariate Data Analysis in Electroanalytical Chemistry, Electroanalysis 14 (2004) pp. 1533-1542.
2. Y. Mourzina, W. Zander, J. Schubert, A. Legin, Y. Vlasov, H. Lüth, M. Schöning, Development of multisensor systems based on chalcogenide thin film chemical sensors for the simultaneous multicomponent analysis of metal ions in complex solutions, Electrochimica Acta 47 (2001) pp. 251-258.
3. K. Toko, Taste sensor, Sensors and Actuators B 64 (2000) pp. 205-215.
4. P. Ivarsson, Y. Kikkawa, F. Winquist, C. Krantz-Rülcker, N. Höjer, K. Hayashi, K. Toko, I. Lundström, Comparison of a voltammetric electronic tongue and a lipid membrane taste sensor, Analytica Chimica Acta 449 (2001) pp. 59-68.
5. P. Ivarsson, S. Holmin, N. Höjer, C. Krantz-Rülcker, F. Winquist, Discrimination of tea by means of a voltammetric electronoc tongue and different applied waveforms, Sensors and Actuators B 76 (2004) pp. 454.
6. P. Bergveld, Thirty years of ISFETOLOGY - What happened in the past 30 years and what may happen in the next 30 years, Sensors and Actuators B 88 (2003) pp. 1-20.
7. H. Nam, G. S. Cha, T. D. Strong, J. Ha, J. H. Sim, S. M. Martin, R. B. Brown, Micropotentiometric sensors, Proceedings of the IEEE 91 (2003) pp. 870-880.
8. J. Shah, E. Wilkins, Electrochemical biosensors for detection of biological warfare agents, Electroanalysis 15 (2003) pp. 157-167.
9. A. P. F. Turner, I. Karube, G. S. Wilson, Biosensors: Fundamentals and Applications, Oxford University Press, New York. 1987.
10. E. Katz, I. Willner, Probing biomolecular interactions at conductive and semiconductive surfaces by impedance spectroscopy: Routes to impedimetric immunosensors, DNA-Sensors, and enzyme biosensors, Electroanalysis 15 (2003) pp. 913-947.
11. C. Berggren, B. Bjarnason, C. C. Johnson, Capacitive Biosensors, Electroanalysis 13 (2001) pp. 173-180.

12. B. J. Taft, M. O'Keefe, J. T. Fourkas, S. O. Kelley, Engineering DNA-electrode connectivities: manipulation of linker length and structure, Anal.Chim.Acta 496 (2003) pp. 81-91.
13. J. P. Cloarec, N. Deligianis, J. R. Martin, I. Lawrence, E. Souteyrand, C. Polychronakos, M. F. Lawrence, Immobilization of homooligonucleotide probe layers onto Si/SiO2 substrates: characterization by electrochemical impedance measurements and radiolabelling, Biosensors & Bioelectronics 17 (2002) pp. 405-412.
14. W. Laureyn, P. Van Gerwen, J. Suls, P. Jacobs, G. Maes, Characterization of nanoscaled interdigitated palladium electrodes of various dimensions in KCl solutions, Electroanalysis 13 (2001) pp. 204-211.
15. A. Hierlemann, O. Brand, C. Hagleitner, H. Baltes, Microfabrication techniques for chemical/biosensors, Proceedings of the IEEE 91 (2003) pp. 839-863.
16. J. Janata, Electrochemical microsensors, Proceedings of the IEEE 91 (2003) pp. 864-869.
17. H. Suzuki, Advances in the microfabrication of electrochemical sensors and systems, Electroanalysis 12 (2000) pp. 703-715.
18. T. C. W. Yeow, M. R. Haskard, D. E. Mulcahy, H. I. Seo, D. H. Kwon, A very large integrated pH-ISFET sensor array chip compatible with standard CMOS processes, Sensors and Actuators B 44 (1997) pp. 434-440.
19. S. Martinoia, N. Rosso, M. Grattarola, L. Lorenzelli, B. Margesin, M. Zen, Development of ISFET array-based microsystems for bioelectrochemical measurements of cell populations, Biosensors & Bioelectronics 16 (2001) pp. 1043-1050.
20. A. A. Shulga, L. I. Netchiporouk, A. K. Sandrovsky, A. A. Abalov, O. S. Frolov, Y. G. Kononenko, H. Maupas, C. Martelet, Operation of an ISFET with non-insulated substrate directly exposed to the solution, Sensors and Actuators B 30 (1996) pp. 101-105.
21. T. Akiyama, K. Komiya, V. Okabe, T. Sugano, E. Niki, Fabrication of ion-sensitive field-effect transistors using a silicon-on-sapphire, Bunseki Kagaku 30 (1981) pp. 754-756.
22. J. Kimiura, T. Kuriyama, Y. Kawana, An integrated sos/FET Multi-biosensor, Sensors.&.Actuators 9, N 4:373-387 (1986) pp. 373-387.
23. N. Ito, S. Kayashima, J. Kimura, T. Kuriyama, T. Arai, M. Kikuchi, N. Nagata, Development of A Transcutaneous Blood-Constituent Monitoring Method Using A Suction Effusion Fluid Collection Technique and An Ion-Sensitive Field-Effect Transistor Glucose Sensor, Medical & Biological Engineering & Computing 32 (1994) pp. 242-246.
24. A. S. Poghossian, Method of Fabrication of Isfets and Chemfets on An Si-Sio2-Si Structure, Sensors and Actuators B 14 (1993) pp. 653-654.
25. A. Bratov, J. Munoz, C. Dominguez, J. Bartroli, Photocurable Polymers Applied As Encapsulating Materials for ISFET Production, Sensors and Actuators B 25 (1995) pp. 823-825.
26. J. Munoz, A. Bratov, R. Mas, N. Abramova, C. Dominguez, J. Bartroli, Planar compatible polymer technology for packaging of chemical microsensors, Journal of the Electrochemical Society 143 (1996) pp. 2020-2025.
27. A. Bratov, N. Abramova, C. Dominguez, Investigation of chloride sensitive ISFETs with different membrane compositions suitable for medical applications, Anal.Chim.Acta 514 (2004) pp. 97-104.
28. N. Abramova, Yu. Borisov, A. Bratov, P. Gavrilenkov, C. Dominguez, V. Spiridonov, E. Suglobova, Application of an ion-selective field effect transistor with a photocured polymer membrane in nephrology for determination of potassium ions in dialysis solutions and in blood plasma, Talanta 52 (2000) pp. 533-538.

29. A. Bratov, N. Abramova, C. Dominguez, T. Baldí, Ion selective field effect transistor - based calcium ion sensor with photocured polyurethane membrane suitable for calcium determination in milk, Anal.Chim.Acta 408 (2000) pp. 57-64.

IMMUNOCHEMICAL APPROACHES FOR RAPID DETECTION OF BIOLOGICALLY ACTIVE COMPOUNDS

B. B. DZANTIEV, A.V ZHERDEV, AND N.A. BYZOVA
Institute of Biochemistry Russian Acad. Sci., Lenishky propect 33, 119071 Moscow, Russia

Abstract: Express immunochemical techniques for determination of toxic compounds and pathogenic cells have been developed. Separation of reactants based on interaction between oppositely charged polyelectrolytes, namely polycation poly-N-ethyl-4-vinylpyridium and polyanion polymethacrylate, were used to reduce assay time. An extremely high rate and affinity of this interaction allowed formation of an immune complex in the solution followed by a quick separation of reactants. The polyelectrolyte-based assays were performed in both homogeneous and filtration formats. Total duration of the assay was 15-20 min, limit of pesticides detection – down to 0.2 ng/ml. The polyelectrolyte separation has been also used in electrochemical immunosensors. The assay protocol is based on the measurement of pH changes induced by a peroxidase label. A field-effect transistor is the sensitive element of the sensor, and specific immune complexes are formed at disposable porous membranes. An alternative approach is based on the application of screen-printed electrode with impregnated peroxidase. The set of reactants was used together with a portable amperometric device designed for express monitoring of hazardous compounds under out-of-lab conditions. The total assay time was 20 min, and the detection limit was 0.2 ng/ml. Immunochromatographic tests based on colloidal gold particles conjugated with antibodies have been developed for detection of low molecular weight antigens and bacterial cells. The method allows detection of bacterial cells including *Mycobacterium tuberculosis* and *Legionella pneumophila* during 10 min in concentrations down to 10^4 cell/ml

1. INTRODUCTION

Immunochemical methods are very effective for the detection of different biologically active compounds. High affinity of the antigen-antibody

D. Morrison et al. (eds.), Defense against Bioterror: Detection Technologies, Implementation Strategies and Commercial Opportunities, 291–301.

interaction ensures adequate sensitivity of such traditional immunoanalytical techniques as radioimmunoassay, microplate immunoenzyme assay, etc. However, as a consequence of prolonged heterogeneous immune interactions in these systems the assays take up to several hours and cannot be accelerated without loss of sensitivity and/or accuracy. Therefore, the development of immunoanalytical systems combining high sensitivity and rapidity is very important.

The compounds of various nature and molecular weight including pesticides as low molecular weight antigens, viruses as high molecular weight antigens and whole bacterial cells as corpuscular antigens have been used in the investigations.

The studied pesticides include phosphoorganic insecticides malathion and parathion-methyl, chloroorganic herbicides 2,4-dichlorophenoxyacetic acid (2,4-D) and 2,4,5-trichlorophenoxyacetic acid (2,4,5-T), triazine herbicides simazine and atrazine, pyrethroid insecticides permethrin and phenothrin, chlorosulfuron from the class of sulfonylureas and butachlor from chloroacetanilide family. It is known that some pesticides, especially phosphoorganic insecticides, have structural similarity with the hazardous phosphoorganic chemical weapons (soman, VX and others). Therefore, the assay principles and conditions developed for pesticides could be applied to the analysis of hazardous chemical agents. In the frame of our previous investigations [1-4] sets of reactants (antibodies, intermolecular conjugates) for specific immunodetection of the pesticides named above have been made. The reached detection limits of the pesticides Enzyme-Linked Immunosorbent Assays (ELISAs) are in the range from 0.02 to 5 ng/ml satisfying the needs of modern environmental monitoring and pollution control. However, the duration of these assays is rather high, and for different assay protocols it varies from 2 to 3 hours.

Such microbial cells as *Mycobacterium tuberculosis* and *Legionella pneumophila* have been studied as model corpuscular antigens. Specific immunoreagents for their determination were prepared in the State Research Center for Applied Microbiology (Obolensk, Russia).

2. APPLICATION OF POLYELECTROLYTES IN IMMUNOENZYME ASSAYS

In order to reduce the duration of immunoassays the application of water-soluble synthetic polyelectrolytes (polyanion polymethacrylate and polycation poly-N-ethyl-4-vinylpyridinium) as carriers for reactants separation has been proposed. Interaction between the polyanion and the polycation molecules leads to the formation of polyelectrolyte complexes.

Due to the cooperative character of this multisite binding the complexes become extremely stable with respect to dissociation. A practically irreversible reaction with 'infinite' binding constant takes place in certain range of pH and ionic strength. Depending on molar polycation-polyanion ratio the formed interpolymeric complexes may be soluble or insoluble. These properties allow use of polyelectrolytes as very perspective carriers for immunoassay.

A general strategy of polyelectrolyte-based separation in immunoenzyme techniques is presented on Fig. 1. Peroxidase-labeled antigens and the antigen-containing sample to be tested are added to antibodies covalently bound to the polyanion. Reactions of the antibodies with native and labeled antigen proceed simultaneously and very rapidly. After formation of soluble immune complexes the solution containing the polycation is added. The formation of water-insoluble interpolyelectrolyte complex causes immediate precipitation of the components bound to the water-soluble polyanion including molecules of the antigen-enzyme conjugate. Then the enzymatic activity could be measured either in the solution (unbound label) or in the redissolved pellet (bound label). In both cases the determined activity serves as a measure of the antigen concentration in the analyzed sample. This approach enables rapid analysis of antigens of different nature and molecular weight [5-7].

The characteristic feature of the polyelectrolyte-based immunoassay is a very low level of nonspecific binding of labeled immunoreactants to the polymers. At the first glance different macromolecules could interact electrostatically with oppositely charged polyelectrolytes. However, a more detailed analysis demonstrates that the non-specific binding in the system is negligible. High density and complementary distribution of charges throughout linear chains of the polyelectrolytes results in such strong electrostatic interactions between polyelectrolytes that non-specifically bound proteins and haptens fail to compete with polyelectrolytes and are displaced from the insoluble interpolymeric complex to solution [7].

The proposed immunoassay technique has been applied to a number of antigens with different physico-chemical properties (molecular weight, isoelectric point, surface density of charges, etc.) [5]. The reached sensitivities are principally equal to those of the traditional microplate ELISA with the use of the same antibody preparations. However, the polyelectrolytes allow significant reductions in the time of the assay. The total duration of the proposed assay does not exceed 15 min (if the enzymatic activity is measured in the supernatant). ELISA technique at equilibrium conditions needs at least 1-2 hr to be carried out, and reduction of this time (kinetic regime of the assay) leads to decrease of its sensitivity.

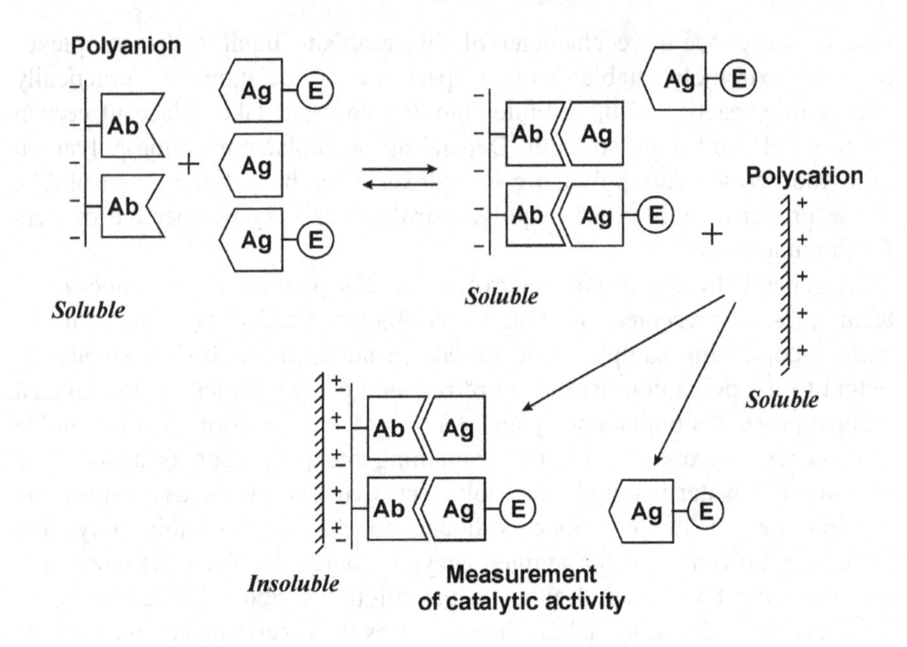

Figure 1. Principle of polyelectrolyte-based immunoenzyme assay.

Contrary to haptens and proteins, corpuscular antigens like viruses could be efficiently incorporated into insoluble interpolyelectrolyte complexes. Based on this effect a rapid method of virus detection was proposed [8]. At the first step of the assay the target virus is recognized by enzyme-labeled antibodies. After that the polycation and the polyanion are added to the reaction mixture one after another. This results in the incorporation of the formed immune complexes into the interpolyelectrolyte precipitate, while unbound labeled antibodies remain in solution. The measured enzyme activity in solution or in the dissolved precipitate reflects the virus content in the tested sample. This method is very rapid in comparison to the conventional ELISA and allows the detection of viruses in rather low concentrations (down to 5-10 ng/ml). The immunochemical part of the assay may be carried out in 5 min, and the total assay time is about 20 min.

Many tasks in analytical chemistry require only qualitative analysis indicating whether the concentration of the compound exceeds a certain level or not. We propose to combine the principle of dot-blot immunoenzyme assay with the polyelectrolyte separation for the development of such qualitative assays [7,9,10]. The main idea of this immunofiltration assay consists in conducting all immunochemical reactions in solution and then applying the polyelectrolytes to separate the components of the reaction mixture. At the first step the peroxidase-labeled antigen, the sample to be tested, specific antibodies and protein A from *Staphylococcus aureus* covalently bound to the polyanion are mixed. During incubation the complex (polyanion-protein A-antibody-labeled antigen) is formed. The concentration of this complex depends on the initial concentration of the antigen in the sample. After incubation the reaction mixture is filtered through a porous membrane with the adsorbed polycation using a special holder. The polyanion momentarily interacts with the polycation, and the immune complex becomes immobilized on the membrane. All of the other components of the reaction mixture are removed from the membrane by washing. Finally, adding the peroxidase substrate solution generates insoluble colored product. Color intensity reflects the antigen content in the examined sample.

The duration time of such an assay is less than that for an ordinary ELISA since the antigen recognition occurs in solution and is not limited by diffusion. Moreover, the immunoreagents separation step is rapid enough and requires just 5-10 min. For example, polyelectrolyte membrane immunoassay for herbicide simazine with visual and densitometric detection takes about 15 min and has sensitivity limit of 0.2 ng/ml [9]. The assay is highly reproducible. Coefficients of variance for four parallel tests on the same membrane range from 3.4 to 5.1%.

Finally, the following conclusions about the proposed polyelectrolyte approach for immunoassays can be made:

- Assay time is significantly shortened relative to traditional immunotechniques.
- The assays are comparable in term of sensitivity with known solid-phase immunoassays.
- The assays are suitable for determining both low- and high molecular weight compounds.
- The properties of polyelectrolytes assure a very low level of nonspecific reactions.

3. ELECTROCHEMICAL IMMUNOASSAYS WITH MEMBRANE CARRIERS

The membrane with immobilized polyelectrolytes could also be efficiently used for rapid detection of various compounds in immunosensory techniques. We have developed an immunoassay based on competition between analyzed and enzyme labeled antigens for binding sites of antibodies immobilized on the gate region of a field-effect transistor (FET) (Fig. 2). Two main approaches to antibody immobilization are widely used: either direct covalent attachment to the transistor [11] or immobilization on the porous membrane that is mechanically attached to the gate region [12]. The disadvantage of the first approach is the necessity of antibodies regeneration after each measurement that after several cycles causes irreversible changes in binding properties. Antibody immobilization on the changeable porous membranes with subsequent attachment to the transistor is a very convenient approach since the regeneration step is not required in this case, but this way leads to increase the assay time because of diffusion limitations of the antigen-antibody reaction's rate.

In order to accelerate the assay we proposed to use membranes with immobilized poly-N-ethyl-4-vinylpyridinium attached to the transistor [12,13]. The antigen from the sample competes with the peroxidase-labeled antigen for binding sites of antibodies covalently attached to the polyanion that forms insoluble interpolyelectrolyte complex via interaction with polycation (Fig. 2). An alternative way is to use native antibodies and staphylococcal protein A conjugated with the polyanion.

We have applied the above FET-based immunoassay for detection of pesticides and bacterial cells [12,13]. The instrument signals are generated by the action of peroxidase on a substrate solution containing o-phenylenediamine, ascorbic acid and H_2O_2. The substrate composition and regime of measurements are optimized to reach maximal electrochemical responses. Both the total shift of pH and maximal rate of its changes are used for analyte determination; the duration of the registration step does not exceed 3 min. For instance, the sensitivity of the simazine assay is about 0.2 ng/ml, and the total assay duration - 20 min. Data scatter of replicate tests varies from 3 to 10%.

Systems with direct covalent immobilization of antibodies on the membrane and with polyelectrolytes-based separation have been compared. Although the sensitivity in both cases is

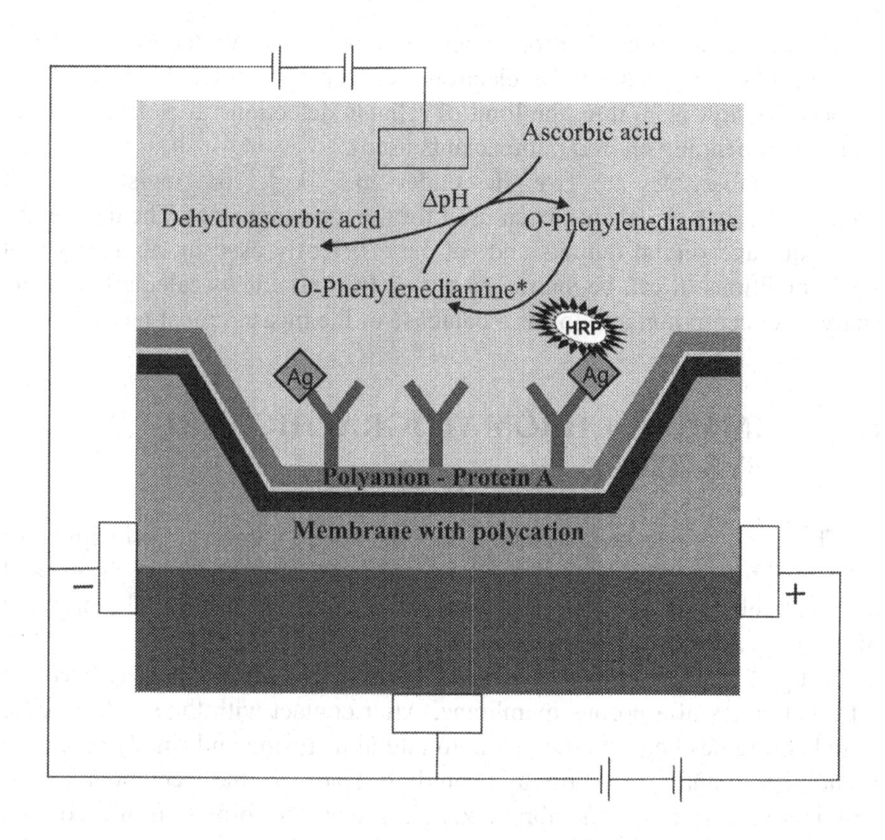

Figure 2. Polyelectrolyte-based immunosensoric measurements with FET as sensitive element.

virtually the same, the assay duration for the first case is about 1.5 hour due to diffusion limitation of antibody-pesticide interaction on the porous membrane [11] while polyelectrolytes-based assay takes only 20 min.

Another developed electrochemical approach is based on the application of a carbon screen-printed electrode with impregnated peroxidase as the sensitive element [14]. The membrane with immobilized antibodies is attached to the electrode surface, and detected pesticide competes with pesticide – glucose oxidase conjugate for binding with the specific antibodies. The subsequent addition of glucose induces production of H_2O_2 by the bound glucose oxidase and its reduction by the peroxidase. The direct electron transfer in this system generates the changes of the electric current that allow to detect pesticide content in analyzed sample. This method does not require separation of unbound glucose oxidase conjugate since the reaction of glucose oxidation is carried out in the presence of catalase transforming the excess of H_2O_2 accumulated in the volume. Therefore the peroxidase reduces only hydrogen peroxide produced by glucose oxidase in the complexes formed on the electrode surface. The total duration of the proposed assays is 20 min, the limit of reliable detection varies from 0.01 to 2 ng/ml (depending on immunoreactants used).

A portable device (weight under 0.5 kg) for registration of immunosensor's signals has been used for the measurements. The device has analogous and digital outputs and can be efficiently used in laboratory and field conditions. It can be interfaced with a PC that allows calculation of the analyte concentration and forms a database of the measurement results.

8. IMMUNOCHROMATOGRAPHIC TEST-SYSTEMS

The next approach based on the well known principle of immunochromatography was used in our work for assay of drugs and bacterial cells with visual or instrumental detection of results. The protocol of the assay is briefly described below.

Antigen and antibodies labeled by colloidal gold are immobilized on different zones of a porous membrane. After contact with the solution to be tested antibodies begin to move due to lateral diffusion and finally reach the immobilized antigen. Antibodies binding leads to the accumulation of colloidal gold in a thin membrane zone result in the formation of a colored line. If the solution contains a free antigen blocking binding sites of the labeled antibodies, no colored line is formed. A line in the control zone with immobilized anti-species antibodies must always develop indicating retention of the reagents' activity. The duration time of such an assay is less than 10 min. The result of the assay is unambiguous: one line – the result is positive, two lines – negative.

For the assay development the number of factors were optimized, such as composition of the conjugate antibody-colloid, its stability, concentration of immobilized reagents, membranes' material and pore diameter, is required. The kinetics of antigen-antibody reactions in this system is much more complicated than in traditional homogeneous or solid-phase immunoassays, since the reaction is carried out in a porous material in the course of lateral diffusion.

One of the advantages of immunochromatography strips is visual detection of the assay results. However, sometimes the test results need to be documented. For this purpose a small registering device ("Okta-Medica Ltd.", Moscow, Russia) was used. This device on the base of digital camera allows measurements of the color distribution and total brightness of the zones, to carry out qualitative and quantitative assays with any kind of strips.

The device can communicate with computers via special software for process control and storage of the assay data.

The device works in two regimes. Qualitative (or alarm) regime provides information about the intensity of color in test and control zones, distribution of color throughout the zones. The device also indicates whether the analyte concentration in the sample exceeds the cut-off level. The second regime gives quantitative information about the analyte content. In this case a calibration curve for standard preparations stored in PC memory is used for data processing.

The immunochromatographic test-strips for detection of drugs of abuse (opiates, cannabinoids, cocaine, amphetamine, methamphetamine, benzodiazepines) and some bacterial cells (*Mycobacterium tuberculosis, Legionella pneumophila*) were prepared [15]. Based on data of electronic microscopy, colloidal particles with mean diameter of 15 nm were chosen for the assays. Under optimized conditions 15-18 antibody molecules are immobilized on a particle forming efficient monolayer. A spectrophotometric technique is proposed to control the average size and microheterogeneity of colloids. Different membranes are compared in lateral-flow systems, the protocols of analyses are optimized including the choice of reactants' concentrations and steps duration. Stability of the preparations under different storage conditions was studied; the tests may be used without lost of sensitivity over a period of 12 months.

The proposed immunoanalytical systems are characterized by sensitivities in the range from 0.01 to 1 μg/ml (depending on the type of compound to be determined) or down to 10^4 cells/ml (Fig. 3). Time of the assay is no more than 10 min. Test results on biological samples for the developed membrane immunotechniques correlate well with those for traditional solid-phase immunoenzyme assays. The described technique has comparable sensitivity with traditional ones, but allows reduced assay durations by a factor of 5-10 times.

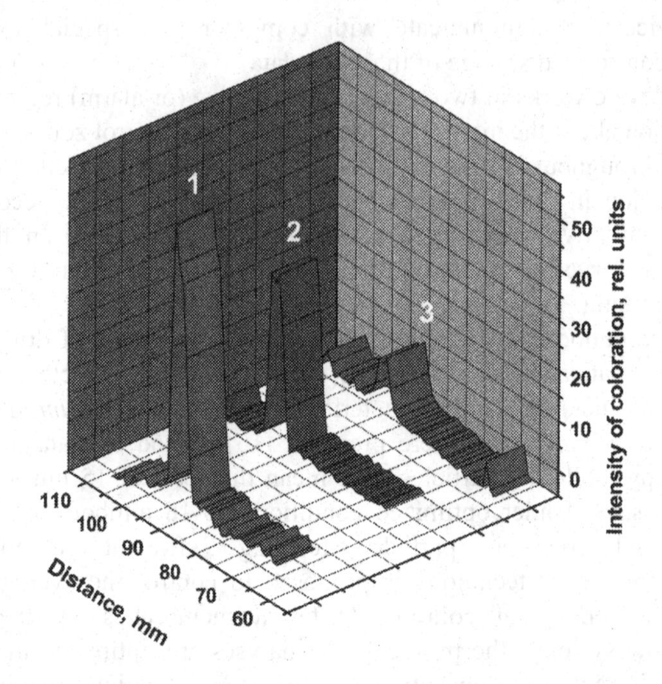

Figure 3. Immunochromatographic detection of *Legionella pneumophila* cells.

5. CONCLUSION

Three kinds of express immunotechniques have been developed, namely polyelectrolyte-based assays, electrochemical immunosensors and immunochromatographic tests. The developed analytical approaches and registering devices can be effectively applied for detection of various biologically active compounds, in that number compounds that can pose chemical and biological threat. The techniques and reagents can be used as an integral part in devices for continuous monitoring of hazardous compounds in water and air.

REFERENCES

1. B.B. Dzantiev, A.V. Zherdev, O.G. Romanenko and L.A. Sapegova, Development and comparative study of different immunoenzyme techniques for pesticides detection. Intern. J. Environ. Anal. Chem. 65 (1996) 95-111.

2. A.V. Zherdev, B.B. Dzantiev and J.N. Trubaceva, Homogeneous enzyme immunoassay for pyrethroid pesticides and their derivatives using bacillary alpha-amylase as label. Anal. Chim. Acta 347 (1997) 131-138.
3. E.V. Yazynina, A.V. Zherdev, S.A. Eremin, V.A. Popova and B.B. Dzantiev, Development of enzyme immunoassays for the herbicide chlorsulfuron. Appl. Biochem. & Microbiol. (Moscow) 38 (2002) 9-14.
4. J. Yakovleva, A.V. Zherdev, V.A. Popova, S.A. Eremin and B.B. Dzantiev, Production of antibodies and development of enzyme-linked immunosorbent assays for the herbicide butachlor. Anal. Chim. Acta 491 (2003) 1-13.
5. B.B. Dzantiev, A.N. Blintsov, A.F. Bobkova, V.A. Izumrudov and A.B. Zezin, New enzyme immunoassays based on interpolyelectrolytic reactions. Doklady Biochemistry (Moscow) 342 (1995) 77-80.
6. E.V. Yazynina, A.V. Zherdev, B.B. Dzantiev, V.A. Izumrudov, S.J. Gee and B.D. Hammock, Microplate immunoassay technique using polyelectrolyte carriers: Kinetic studies and application to detection of the herbicide atrazine. Anal. Chim. Acta 399 (1999) 151-160.
7. B.B. Dzantiev, A.V. Zherdev and E.V. Yazynina, Application of water-soluble polymers and their complexes for immunoanalytical purposes. In: "Smart Polymers for Bioseparation and Bioprocessing" (Bo Mattiasson & Igor Yu. Galaev, eds.), Taylor & Fransis, London-NY, 2002, pp. 207-229.
8. A.N. Blintsov, B.B. Dzantiev, A.F. Bobkova, V.A. Izumrudov, A.B. Zezin and I.G. Atabekov, A new method for enzyme immunoassay of phytoviruses based on interpolyelectrolytic reactions. Doklady Biochemistry (Moscow) 345 (1995) 175-178.
9. E.V. Yazynina, A.V. Zherdev, B.B. Dzantiev, V.A. Izumrudov, S.J. Gee and B.D. Hammock, Immunoassay techniques for detection of the herbicide simazine based on use of oppositely charged water-soluble polyelectrolytes. Anal. Chem. 71 (1999) 3538-3543.
10. A.V. Zherdev, N.A. Byzova, V.A. Izumrudov and B.B. Dzantiev, Express polyelectrolyte-based immunofiltration technique for testosterone detection. Analyst 128 (2003) 1275-1280.
11. N.F. Starodub, B.B. Dzantiev, V.M. Starodub and A.V. Zherdev, Immunosensor for the determination of the herbicide simazine based on an ion-selective field-effect transistor. Anal. Chim. Acta 424 (2000) 37-43.
12. Yu.V. Plekhanova, A.N. Reshetilov, E.V. Yazynina, A.V. Zherdev and B.B. Dzantiev, A new assay format for electrochemical immunosensors: Polyelectrolyte-based separation on membrane carriers combined with detection of peroxidase activity by pH-sensitive field-effect transistor. Biosensors & Bioelectronics 19 (2003) 109-114.
13. A.V. Zherdev, A.N. Reshetilov, Yu.V. Plekhanova, S.F. Biketov, E.V. Baranova and B.B. Dzantiev, <Electrochemical immunosensor for the detection of microbial cells>. Allergy, Asthma & Clin. Immunol. 7 (2003) 182-184. In Russian.
14. B.B. Dzantiev, E.V. Yazynina, A.V. Zherdev, Yu.V. Plekhanova, A.N. Reshetilov, S.-C. Chang and C.J. McNeil, Determination of the herbicide chlorosulfuron by amperometric sensor based on separation-free bienzyme immunoassay. Sensors & Actuators B 98 (2004) 254-261.
15. N.A. Byzova, Yu.A. Balandina, A.V. Zherdev and B.B. Dzantiev, <Development of immunochromatographic test-systems and techniques for quantitative registration of membrane assay results>. Allergy, Asthma & Clin. Immunol. 7 (2003) 189-192. In Russian.

MULTIFUNCTIONAL LIQUID-CRYSTALLINE DNA BASED BIOSENSING UNITS CAPABLE OF DETECTING BIOLOGICALLY RELEVANT COMPOUNDS

YU.M YEVDOKIMOV

Engelhardt Institute of Molecular Biology of the Russian Academy of Sciences,119991 Moscow, Vavilova str. 32, Russia

Abstract: The main principles used for creating biosensing units based on application of the double-stranded DNA liquid crystalline dispersions are considered. A broad rande of analytical abilities of these units is illustrated. These multifunctional biosensing units are capable of detecting various chemical or biologically active compounds (genotoxicans, such as antitumor drugs, antibiotics, proteins, etc.) influencing the DNA secondary structure or destroying the additional nanosized sensing elements, incorporated between the neighboring the DNA in the spatial structure of the liquid crystalline dispersions and forming artificial nanobridges between fixed in the structure of particles of LCD DNA molecules. Some unsolved problems are outlined.

1. INTRODUCTION

Creation of biosensors involves the decision of two problems which are related to different fields of science. First, it is the fabrication of a specific biosensing unit in which the recognition of analytes is used at a highest efficiency. Such a problem can be solved within the biological sciences. Second, it is the creation of an adequate recording scheme for determining the signal that appears in the system.This problem is to be solved within the technical sciences.

D. Morrison et al. (eds.), Defense against Bioterror: Detection Technologies, Implementation Strategies and Commercial Opportunities, 303–334.

Theoretically speaking, any biological molecule or biochemical reaction can be used for making biosensing units. It is important to appreciate why nucleic acids form a sound background for creating biosensing units. The answer is preconditioned by a combination of several factors. First, single- and double-stranded desoxyribonucleic acid (DNA) molecules seem to be the most important of all biological molecules: they represent genetic material for living cells. Second, many chemical or biological compounds present in the environment can interact with the genetic material of living cells. Hence, the detection of biological or chemical compound, which can influence genetic material in living cells, is relevant. Third, during the life cycle of a cell or infection, nucleic acid fragments carrying incorrect genetic information may appear, therefore the estimation of these sequences is necessary. Besides, pharmaceutical manufacturers need to quantify very small amounts of contaminating nucleic acids in recombinant proteins and monoclonal antibodies intended for therapeutic use. Finally, the physical chemistry of isolated nucleic acid molecules is well studied.

Figure 1 shows, that the interaction of biologically active compounds (antibiotics, antitumor compounds, enzymes, etc.) with the DNA molecule results in quite different physico-chemical effects. For instance, the splitting the sugar-phosphate chain of the ds DNA molecule (point 1) is accompanied by a sharp increase in its flexibility, i.e. the physico-chemical properties of the initial DNA differs strongly from that of the final one. Interaction of the biologically-active compounds with the DNA nitrogen bases through the various modes (intercalation (point 2) or location in minor or in major grooves (point 3)) results in alterations of the DNA secondary structure and in the appearance of different inclination angle of the biologically active compounds in respect to the DNA helix. In addition, "cross-linking" of the DNA nitrogen bases (point 4) is accompanied by an increase in the distance between nitrogen bases, an alteration of their inclination angle, a distortion of a regular (register) character of the nitrogen bases, and, finally, a pronounced change of parameters of the DNA secondary structure. This scheme clearly demonstrates that the biologically active compounds (in total, genotoxicants) modify the DNA secondary structure by quite different manner. Besides, the modification of the DNA structure is accompanied by different changes in the physico-chemical properties of the DNA molecules. Hence, the ds DNA represents, by itself, a multifunctional biosensing unit capable of detecting various biologically active compounds.

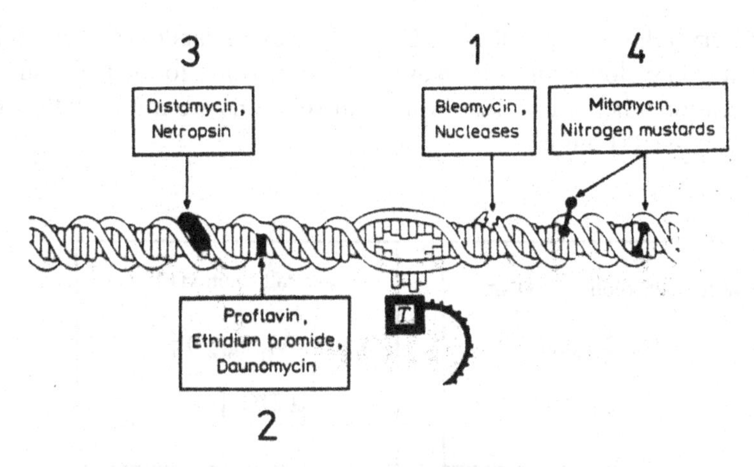

Figure 1. Structural "addressing" of various compounds by the double-stranded DNA molecule.

It is necessary to add, that the basic concept for creation of biosensing units, which are based on the use of double-stranded nucleic acid molecules (dsDNA or dsRNA), differs strongly from the classical "hybridization" concept for detection of single-stranded DNA molecules.

Therefore, in contrast to proteins (enzymes), ds nucleic acids have several attractive properties. The choice of ds nucleic acids as the basis for biosensing units is determined not only by the presence of different chemical groups in the composition of these biopolymers but also by the ability of nucleic acids to change their properties noticeable under the influence of a number of factors. A combination of these properties of ds nucleic acids opens up a possibility to use different recognition principles for the detection of compounds which interact with ds nucleic acids by different ways. How to realize this ability of the ds nucleic acids?

In order to construct a biosensing unit that takes into account of the ability of ds nucleic acids to recognize various chemical or biologically active compounds, it is convenient to make use of lyotropic liquid crystalline dispersions (LCD) of nucleic acids [1]. Due to very specific physico-chemical properties of the LCD one can realize different versions of "recognition" in such biosensing units [2].

Various approaches in the frame of basic concept to use the ds DNA as multifunctional biosensing units, with an account of the specific optical properties of LCD, were described below.

2. LIQUID CRYSTALLINE DISPERSIONS OF DS NUCLEIC ACIDS

Liquid crystalline dispersions (LCD) of double-stranded nucleic acids (DNA or RNA) of low molecular mass (‹ $1x10^6$ Da) are formed as result of phase separation when their dilute water-salt solutions are mixed with some synthetic polymers (Figure 2).

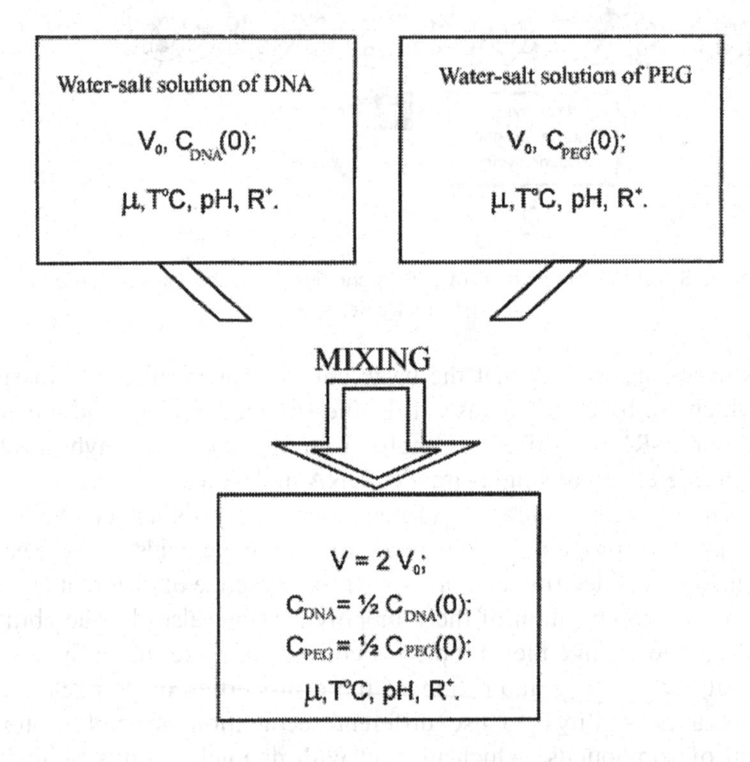

Figure 2. A principal scheme of the ds DNA dispersion formation

The phase separation that is accompanied by formation of particles of the liquid crystalline dispersions of the DNA molecules can be made by two major approaches. The first, termed ψ-condensation (psi, as an acronym of Polymer-Salt-Induced) requires high ionic strenght to screen Coulombic interactions and high concentration of neutral polymer such as poly(ethyleneglycol) (or PEG) to produce sufficient intermolecular excluded volume interaction to "condense" the neighboring DNA molecules.

The second requires polycations to neutralize most (70-90%) of the DNA phosphate charges, after which attractive interactions between DNA molecules are sufficient to cause spontaneous phase separation of DNA molecules. The attractive forces for condensation are presumably electrostatic and electrodynamic in origin, involving London dispersion forces and dipole -(induced dipole) interactions. These forces are quite week at long distances between the DNA molecules, but increase rapidly as the DNA molecules approach one another, with the attractive energy changing approximately as $1/R^5$. The strength of the dispersion forces is proportional to the empirical Hamaker constant, A, whose magnitude for organic molecules interacting in water is ~ $4x10^{-14}$ erg that is about k_bT at room temperature . For a net attraction at separation of about 30 A^0 to result, values of A from 2 to 5 k_bT must be assumed. When the surface charge density is lowered sufficiently by association of counterions, particularly polycations, dispersion forces equal and then overweight electrostatic repulsions and condensation takes place. The phase separation is a strongly cooperative process, occurring over a narrow range of PEG or polycation concentration and its efficiency related to a number of variables (pH, ionic strenght and ionic content of solution, mol.mass and concentration of PEG, etc.).

The most important "event" in the DNA condensation and formation of the liquid crystalline dispersion is the close approaching of the neighboring DNA molecules (Figure 3), that is accompanied by the formation of the DNA dispersion (Figure 4). The diameter, D, of formed particles estimated by different physico-chemical and theoretical approaches, is about 10^3 A^0, i.e. each particle is consisting of about 10^4 DNA molecules. The distance between adjacent DNA molecules in the particles of LCD is varied from 20 to 50 A^o. Inside the liquid crystalline structure the DNA molecules preserve some diffusion «freedom».

The optical and helical anisotropy of DNA results in the rotation of the consequent DNA molecules and, as a result, the spatially twisted, layered (or cholesteric) structure of LCD is formed [2] (Figures 3 and 4). In the particle of LCD, every subsequent layer formed by the DNA molecules is twisted in respect to the previous one on a define angle (~ 0.1^0).

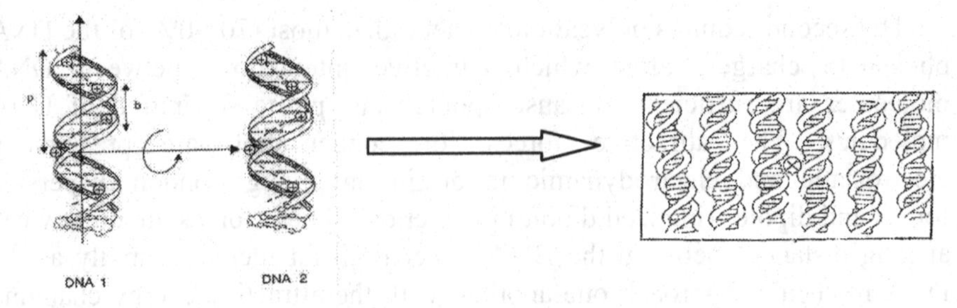

Figure 3 Approaching of the ds DNA molecules and formation of a "layer" from these molecules

Due to osmotic pressure (π,↑↑↑) of PEG-containing water-salt solution, the mean distance(d, A^0) between ds DNA molecules in a layer is fixed . The helicity of the ds DNA molecules (with the pitch, P) results in a twist of approaching DNA molecules (1 and 2). The change in the character of charge distribution (shown as "b") on the DNA surface as well as in dielectric properties of the solvent can influence an angle of the rotation of the neighboring DNA molecules.

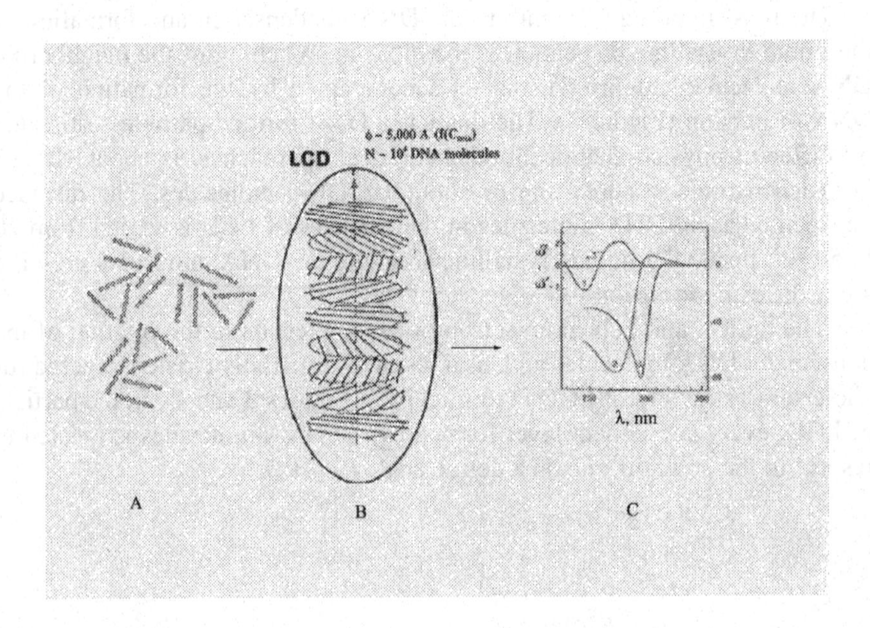

Figure 4 A scheme of the formation of the DNA liquid crystalline dispersion (LCD) and some properties of the particles of the ds DNA LCD.

The phase exclusion of the ds DNA molecules (A) from water-salt polymer containing solution is accompanied (under certain conditions) not only by formation of LCD (a particle (B) of LCD, see - in the middle), but by the appearance of an abnormal optical activity, expressed, in particular, as an intense band in the CD spectrum. Insert (C) compares the CD spectrum of initial DNA (curve 1), the CD spectrum of the DNA LCD (curve 2) and the CD spectrum of the thin layer of the DNA cholesteric liquid crystalls (curve 3).

The DNA molecules in the LCD are still belonging to the B-family [3] and the close approaching of DNA molecules does not influence their ability to react with various chemical or biological compounds. One can stress that there are a very restricted conditions under which the ds DNA molecules can form the cholesteric LCDs (Figure 5). It is necessary to add, that the particles of cholesteric DNA LCDs are important from the biological point of view, because they represent a structural model of the spatial organization of the choromosomes of several biological objects [3].

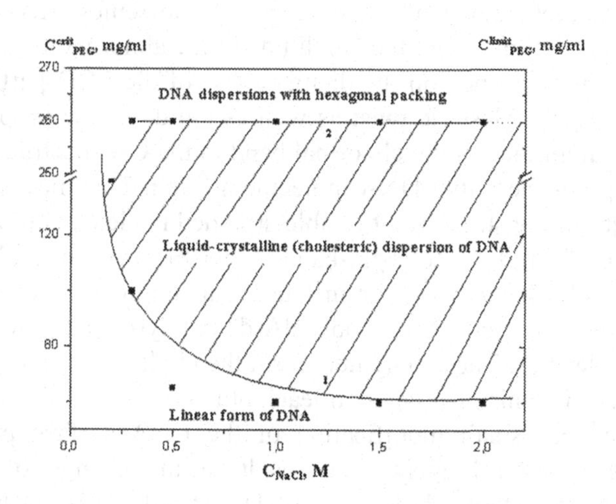

Figure 5. The dependence of Ccrit of PEG (1) and of Climit of PEG (2) upon NaCl concentration in PEG-containing solutions.

M.wt. of PEG 4,000 Da; the region of existence the cholesteric liquid crystalline DNA dispersion with an abnormal optical activity is allocated.

The C_{crit} value of PEG corresponds to C_{PEG} necessary for the formation of the DNA dispersions (these dispersions scatter the UV- irradiation and form the pellets at low-speed centrifugation). The C_{limit} value of PEG corresponds to C_{PEG} of formation of the DNA dispersions without abnormal optical activity..

The cholesteric DNA LCDs have specific optical properties detected, for instance, by the circular dichroism spectra. Indeed, the ds DNA cholesteric LCDs are, in essence, "dyed cholesterics", because they contain in their content "choromophores" (nitrogen bases) that absorb in the UV region of the spectrum. As a result, an intense (abnormal) band in the circular dichroism (CD) spectrum in the region of absorption of nitrogen bases is specific for such dyed cholesteric LCDs. Because the right-handed DNA molecules form the cholesteric LCD with the left-handed twist of spatial structure, an abnormal negative band in the CD spectrum in the region of absorption of nitrogen bases is specific for such "dyed" DNA cholesteric LCD. Figure 4 shows that the band in the CD spectrum has a shape identical to that of the band of the DNA absorption, but the maximum is "red" shifted (~ 270-300 nm). The value of $\Delta\varepsilon_{270}$ which charterizes the optical activity of the nitrogen bases in the structure of DNA LCD is far greater than the value of $\Delta\varepsilon$ (~ 2 units) characteristic of the molecular optical activity of nitrogen bases in the structure of isolated linear DNA molecules. According to the theoretical calculations [4], the amplitude of the abnormal negative band in the CD spectrum depends on the diameter (D) of the LCD particles, on the pitch (P) of their cholesteric twist as well as the dielectric properties of the solvent. The amplitude of the abnormal band in the CD spectrum depends on the anisotropy of the initial DNA molecule as well. This means, that a very small changes in the structure of double-stranded nucleic acids are sufficient for transition from left- to right-handed twisting of molecules, forming cholesteric LCD. It happens, for instance, at minor changes in the base sequence at transition from poly(dA-dT)xpoly(dA-dT) to poly(dA)x poly(dT) molecules. These polymers form the cholesteric LCD with the CD spectra which are "mirror-images" of each other.

Hence, a very small modification of the DNA surface properties by chemical or biological agents can result in the change of the optical properties of the DNA cholesteric LCD formed. This means, that the abnormal band in the CD spectrum of the cholesteric DNA LCD represents a practically important "tool", which can reflect the changes in the DNA structure in respect to action of various chemical or biological agents.

Combination of the high flexibility of the physico-chemical properties of the DNA structure with respect to different factors and accompaning change in the amplitude of the abnormal (intense) band in the CD spectrum of the cholesteric LCD opens the possibility for the using the DNA LCD as a background for creating biosensing units.

Figures 2 and 3, that reflect the role of various factors at the formation of the DNA LCD, show that two approaches for determination of chemical or biological compounds influencing the ds DNA secondary structure (in general, genotoxicants) can be used.

The first approach consists in the treatment of initial ds DNA by genotoxicans with the subsequent formation of the LCD from "modified" DNA(mDNA) and registration of the CD spectra of the LCD formed.

(ds DNA + genotoxicant) ⟶ LCD mDNA ⟶ registration of the CD spectrum. (1)

One can suggest that any modification of the ds DNA secondary structure by genotoxicants and formation of the LCD from a "modified" DNA will result in the changes of the CD spectra. The amplitude of abnormal optical activity, the shape and location of the CD band, as well as its sign, can be changed. The second approach is the treatment of the LCD, formed by the ds DNA molecules, with genotoxicants and registration of the CD spectra of these LCD's.

ds DNA ⟶ LCD ds DNA ⟶ (LCD ds DNA + genotoxicant) ⟶ registration of the CD spectrum. (2)

In this case an appearance of new CD band(s) located in the genotoxicants absorption region is expected, as a minimum. In any case, the proportionality between the amplitude of the CD band and the genotoxicant concentration (in definite range) is supposed. This means, that the CD spectrum can represent, by itself, an analytical criterium for the determination of the presence and the estimation of genotoxicant concentration. Hence, the practical question is: «What happens with the cholesteric DNA CD spectrum at the modification of initial ds DNA by chemical or biological agents?»

3. THE FIRST APPROACH TO DETECTION OF COMPOUNDS CAPABLE OF INFLUENCING THE SECONDARY STRUCTURE OF THE INITIAL DS DNA MOLECULES

A). As it was mentioned above, the cholesteric twisting the DNA molecules is strongly dependent on both the properties the parameters of the DNA secondary structure and the solvent showing marked variations with the change in either factor. It is therefore a matter of interest to detect the properties of the DNA LCD formed from DNA molecules treated by compounds used as antitumor drugs. In this respect, the coordination compounds of platinum(II) group is of prime importance for several reasons. First, information on alteration in the secondary and tertiary structure of DNA is necessary for the synthesis of new generations of anticancer drugs. Second, the platinum atoms reacting with DNA represents labels, that can be used in the analysis of the structural characteristics of DNA. Finally, the biological effects of platinum compounds are currently being actively studied.

The initial ds DNA molecules were treated by platinum(II) compound - DDP, then, as a result of the phase separation, the cholesteric LCD were formed from modified DNA molecules and their CD spectra were compared. Figure 6 A compares, as an example, the CD spectra of initial linear B-form of DNA (curve 1) with those of LCD prepared from both the native (curve 2) and DDP-modified forms(curves 3-5). The decrease in the amplitude of the negative band in the CD spectra of the LCD formed from DNA-DDP complexes reflects the fact that interaction of DDP with nitrogen bases is accompanied by the alteration of the DNA secondary structure. Indeed, DDP when interacting with the DNA nitrogen bases, more exactly, with N(7) and (or) O(6) of two guanine residues, may induce disturbances in the stacking interactions of bases; they may also form cross-links between two guanine residues in one or in both DNA strands. As a result of such cross-links, sites appear whose structures are dissimilar to those of native DNA. For increasing extent of the DNA modification, the amplitude of the negative band in the CD spectra decreases linearly (Figure 6 B). Zero amplitude is observed to occur ar $r_t \sim 0.05$. (The shape of the CD specrum under these conditions corresponds to curve 5 in Figure 6 A).

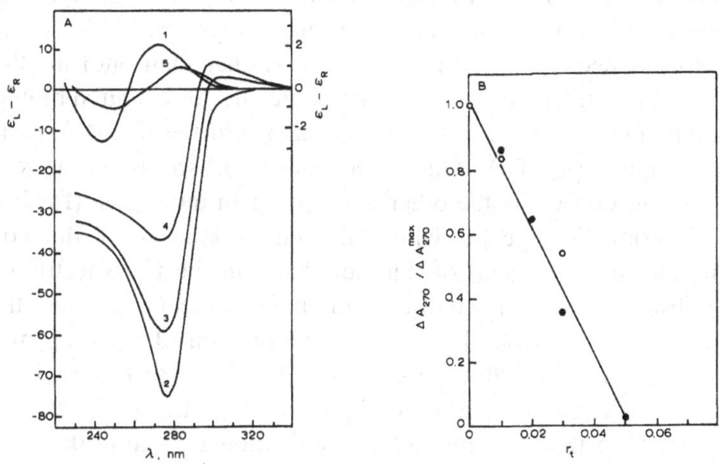

Figure 6. (A). CD spectra of the LCD formed from ds initial DNA molecules (curve 2) and ds DNA modified by DDP treatment (curves 3-5); molecules of DDP are forming the "cross-links" between the DNA base pairs.
DDP-cis-dichlorodiammineplatinum(II)(cis-[Pt(NH3)2Cl2]); Curve 1-DNA B-form. Curves : 2- rt = 0; 3- rt = 0.01; 4- rt = 0.03; 5- rt = 0.1, 170 mg/ml PEG.
(B). Dependence of the relative amplitude of the negative band of the DNA LCD on rt values.

Figure 6 means, the DNA LCD permits one to detect a very small concentrations of Pt(II) capable of destroying the DNA secondary structure (about 1 atom of Pt per 100 base pairs of DNA). Here it is important to stress, that the influence of various Pt(II) compounds on the amplitude of an abnormal band in the CD spectra of the DNA LCD, correlates with differences in biological activity of these compounds.

In addition, similar alteration in the CD spectra of the DNA LCD can be induced by action of various chemical substances capable of changing the electronic structure or mutual orientation of the nitrogen bases in the initial dsDNA molecules. For instance, such changes are induced by photochemical modification of nitrogen bases or their chemical alkylation as well as the physical factors (such as the laser irradiation).

Hence, one can use the ds DNA molecules, treated by different chemical or biologically relevant compounds and the subsequent formation of the DNA cholesteric LCD, for detection of a very small changes in the DNA secondary structure.

B). Obviously, for dyed chemical or biologically relevant compounds forming strong complexes with DNA, i.e. for the compounds which are fixed in a regular mode on the surface of the ds DNA molecules, the same rules apply, which define the appearance of abnormal band in the absorption region of nitrogen bases rigidly fixed in the structure of ds DNA molecule [4]. For instance, an additional abnormal band in the absorption region of the dyed compound should appear upon location of this compound between nitrogen bases of the DNA molecules forming cholesteric LCD. This means, that an appearance of at least two bands (rather than one) in the CD spectrum is theoretically predicted when molecules of coloured biologically active compounds or drugs are incorporated into the ds DNA molecules forming a cholesteric LCD. One of the bands will still be in the region of the DNA absorption while the other will appear in the region (for instance, visible region)of the spectra where the chromophores of the coloured compound absorb. The sign of this new band in the CD spectrum in the region of absorption of drugs will depend on their orientation about the long axis of the DNA molecule. The angle of inclination of the drug molecule about the DNA helix is $\sim 90^0$, the sign of its band will coincide with that of the band typical of the DNAnitrogen bases. If the drug is located on the DNA molecule so that the angle of its inclination is within $0 - 54^0$, the intense band in the CD spectrum can have a sign opposite to that of the band characteristic of the DNA cholesteric LCD.

In Figure 7, as an example, the CD spectra of the LCD prepared from ds DNA treated by antitumor compound – BS are shown. The attention is drawn to a few effects. First, in the DNA absorption region (~ 270nm) an intense negative band is present (curve 0). After adding of BS to DNA, the amplitude of this band does not change significantly (curves 1-3). Second, the addition of BS is accompanied by the appearance of negative band located in the BS absorption region (~ 415 nm). The negative sign of the band in the absorption region of BS proves that BS molecules are located on the DNA molecule in such a way that the angle of inclination of the BS molecules appears to be $\sim 90^0$. This is possible in the case when BS molecules are intercalating between the DNA nitrogen base pairs. Finally, the experimentally measured amplitude of the negative band in the CD spectrum in the BS absorption region is directly proportional to the concentration of DNA-bound BS molecules. Since there is a correlation between the total concentration of BS molecules in solution and the concentration of DNA-bound BS (r), the data obtained imply, that the biosensing unit allows one not only establish the presence and concentration of BS in the solution, but the mode of location of BS on the DNA molecule as well.

Hence, the biosensing unit based on the initial ds DNA can be used for analytical purposes. The minimum concentration of BS, which is being established at present is equal to $\sim 10^{-7}$ M.

C). Figure 8 A shows, as an example, the CD spectra of the LCD prepared from ds DNA treated by biologically active oligoamide compound - distamycin A. Here, the attention is drawn to a few new effects. First, in the DNA absorption region (~ 270 nm) an intense negative band is still present (curve 1). However, after adding of distamycin A to DNA, the amplitude of this band begins to decrease and at definite concentration of distamycin A the sign of this band is changed from the negative to positive one (curves 4-5). Second, the addition of distamycin A is accompanied by the appearance of the positive band located in the distamycin absorption region (~ 320 nm). The positive sign of the band in the absorption region of distamycin A in combination with the negative sign of the band in the DNA absorption region proves that distamycin molecules are located on the DNA molecule in such a way that the angle of inclination of the distamycin A molecules appears to be below 54 ^{0}. This is possible in the case when distamycin A molecules are located in the narrow groove on the surface of the DNA molecule. Finally, the experimentally measured amplitudes of the negative (DNA) and positive (distamycin) bands in the CD spectrum are changing their signs simultaneously (Figure 8 B).

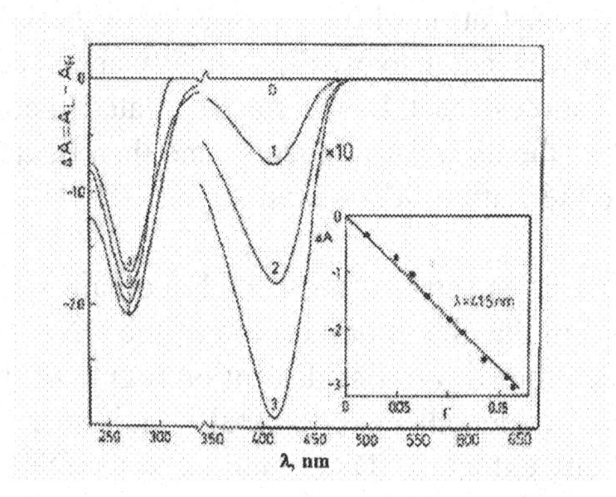

Figure 7. The CD spectra of the LCD formed by ds DNA molecules complexed with anthracene antitumor compound – bisantrene, BS, intercalating between the ds DNA base pairs.
(BS-9,10-anthracenedicarboxaldehyde-bis-[(4,5-dihydroxy-1-H-imidasol-2-yl) hydrazone] x 2HCl) r = 0,049 (1); 0,102 (2); 3- 0,163(3).
Insert - the dependence of the amplitude of the negative band (415 nm) in the CD spectra of the DNA-BS complexes upon r value.

According to theoretical calculations [5], this means that the addition of distamycin influence the charge distribution on the surface of the DNA molecule (Figure 3). Hence, the biosensing units based on ds DNA molecules allows one to realize the ability of these molecules to "recognize" various chemical and biologically active compounds. The recognition is accompanied by different optical effects in the CD spectra of the DNA cholesteric LCD. This allows one not only to estimate the presence of chemical or biological compounds in solution to be analysed, but establish the mode of location of these compounds on the surface of the DNA molecules, i.e. this means that the linear ds DNA molecules can be used as multifunctional biosensing units.

4. THE SECOND APPROACH FOR DETECTION OF COMPOUNDS CAPABLE OF INFLUENCING THE SECONDARY STRUCTURE OF THE DS DNA MOLECULES FIXED IN THE LCD

As it was shown above, the second approach to detection of chemical or biological compounds consists in the treatment of the preformed LCD DNA by genotoxicants and registration of the CD spectra of these LCD's. One can add that this approach was extensively theoretically analyzed [2,6].

Figure 9A shows the CD spectra calculated theoretically for the LCD first formed from the ds DNA molecules and then treated with coloured drug , i.e. anthracycline antibiotic - daunomycin (DAU). A few facts are worth noting here.

i). The calculated CD spectra of these LCD contain two bands. One of the bands is found in the absorption region of the DNA nitrogen bases ($\lambda \sim 270$ nm) and the other lies in the absorption region of the drug ($\lambda \sim 500$ nm). In this case, under all experimental conditions, both bands have negative signs at any extent of DAU binding to the DNA.

ii). The shapes of the bands in the CD spectra are identical to those of the absorption spectra for the DNA and the antibiotic.

iii). Both bands have comparably high amplitudes.

iv).The amplitude of the band in the CD spectrum in the region of DAU absorption grows with increasing number of DAU molecules bound to DNA, although the amplitude of the band in the region of DNA absorption remains constant. A very important practical result follows from these data, a namely, the amplitude of the band in the region of DNA absorption ($\lambda \sim 270$ nm) can be used as a „internal standard" to check the quality („perfection") of the LCD formed.

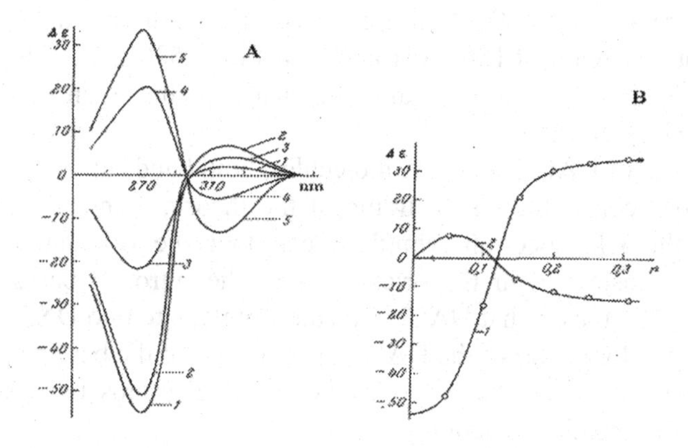

Figure 8 A). The CD spectra of the LCD formed by ds DNA molecules complexed with distamycin A, which is located in the narrow groove on the surface of the ds DNA.1- r = 0; 2 – r = 0.05; 3 – r = 0.1; 4 - r = 0.15; 5 - r = 0.3.
B).The dependences of the negative (1 - 270 nm) and the positive (2 -320 nm) bands in the CD spectra upon r value.

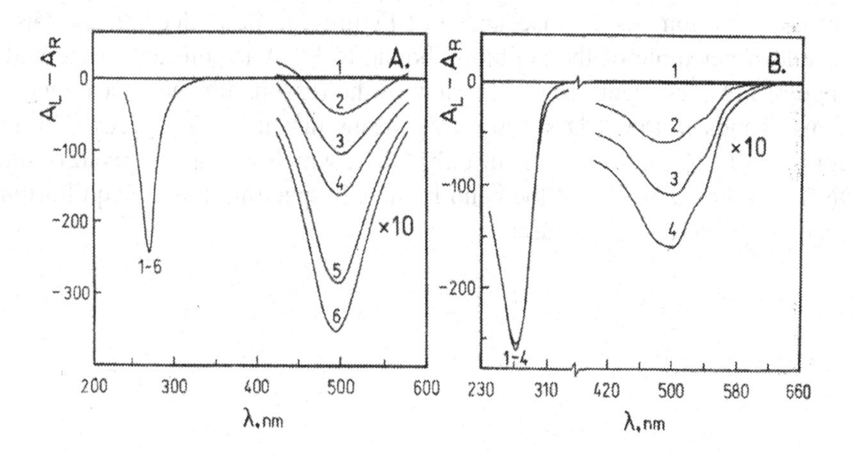

Figure 9 (A) The theoretical CD spectra in UV and visible regions for the ds DNA LCD added with different concentrations of DAU. rt DAU -1= 0; 2 = 0.015; 3 = 0.022; 4 = 0.03; 5 = 0.045; 6 =0.052. (B) The experimental CD spectra in UV and visible regions for the ds DNA LCD added with different concentrations of DAU. 1-0; 2 - 11.9; 3 - 23.7; 4 - 39.2x10-7 M DAU.

Figure 9 B demonstrates the experimental CD spectra of the LCD first formed from the ds DNA molecules in PEG-containing solution and then with DAU added. In agreement with theoretical predictions, the CD spectra of the DNA LCD "coloured" by DAU have two bands. One occurs in the absorption region of the DNA nitrogen bases ($\lambda \sim 270$ nm) and other lies in the absorption region of DAU chromophores ($\lambda \sim 500$ nm). The shapes of the bands in the CD spectrum are identical to those in the absorption spectra of DNA and DAU.

At binding of DAU with DNA molecules, both bands have negative signs despite DAU concentration. In frame of the theory, identical signs of two bands in the CD spectrum simply mean that the orientation of DAU molecules coinsides with the orientation of the nitrogen bases about the DNA axis. This means that DAU molecules intercalate into DNA so that the angle between long axes of the DAU molecule and of the base pair is $\sim 90^0$. One can stress, as this takes place, the reactive groups of DAU are not accessible for chemical reactions.

There are no theoretical limitations on the number of molecules of coloured compounds which may be incorporated into the DNA molecules forming the LCD. Hence, in principle, the amplitude of the band in the CD spectrum in the region of DAU absorption must grow with increasing DAU concentration in solution. Figure 10 shows that the experimentally measured amplitude of the band in the CD spectrum in the region of DAU absorption is directly proportional to concentation of DAU added to solution. As a consequence of Figure 10, the molecular dichroism calculated per mole of the antibiotic bound to DNA is constant in the wide range of DAU concentrations. In view of the reasons for the appearance of intense bands in the CD spectra, this means that the CD spectra do not register all DAU molecules, but only those, which are rigidly fixed on the DNA, i.e. the amplitude of the band in the CD spectrum has the equilibrium value under the conditions used.

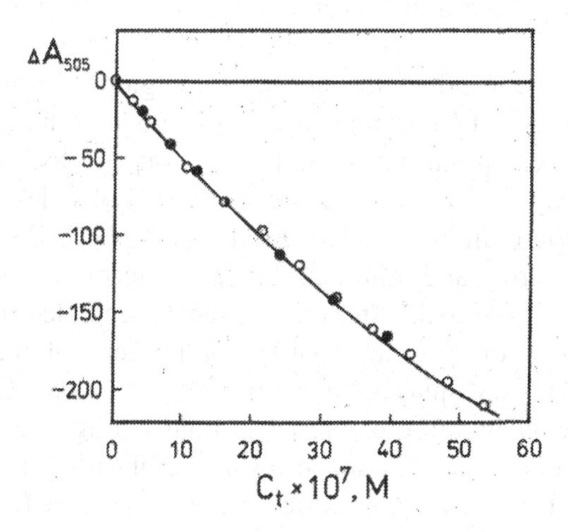

Figure 10. The dependence of the amplitude of the band in the CD spectra (505 nm) of the ds DNA LCD added with different DAU concentrations versus Ct of DAU. Light and dark points correspond to various experiments.

Results shown in Figures 9 and 10 are important from a practical point of view. These data demonstrate that the abnormal optical activity of the preformed LCD dsDNA permits one to detect the presence of coloured compound in solution, estimate directly its concentration and the mode of location on the DNA.

The results obtained above demonstrate some points, which are important for operating biosensing units intended for detection of biologically active compounds.

i).The dsDNA molecules tend to form the LCD with the cholesteric ordering these molecules. The mode of ordering the DNA molecules in the LCD remains unchanged after the LCD formation at fixed conditions.

ii).The spatial ordering the dsDNA molecules in LCD does not impair the ability of nitrogen bases to "recognize" and "address" molecule of biologically active compounds. Specific peculiarities of the secondary DNA structure open us the possiblity for molecules of different drugs to interact and form complexes with the DNA simultaneously, i.e., the DNA LCD represent a background for creating a multifunctional biosensing unit.

iii).The abnormal optical activity is specific to lyotropic cholesteric LCD formed from the DNA or the (DNA-drug) molecules; it is displayed as one (or more) intense band in the CD spectrum in their absoptrion regions. The amplitude of the respective band(s) is directly proportianal to concentration of the drug molecules reacting with the DNA in the LCD.

iv).The sign of the band in the absorption region of drug forming a complex with DNA in LCD is related with the mode of drug location.

These points show that biosensing units based on the DNA LCD operate in the following way: nitrogen bases "recognize" molecules of drug and "address" them to the definite places in the DNA in the LCD. A complex formation between drug and nitrogen bases is accompanied by the appearance of an optical signal. An optical signal generated in the system is "amplified" due to specific spatial properties of the cholesteric structure of the LCD and displayed as an intense band in the CD spectrum. The greater is the concentration of the DNA-bound drug, the higher is the amplitude of the band in the CD spectrum in the absorption region of the drug. The intense band in the CD spectrum of a DNA LCD permits one both to detect the presence of drug (or their combinations) in the solution and to estimate concentration, as well as to establish the mode of binding of drug with the DNA molecule.

Hence, the second approach to detect biologically active compounds based on the preformed LCD of dsDNA can be used for analytical purposes. The minimal concentration of antibiotics capable of intercalating between the DNA nitrogen bases, which can be detected by this approach is $\sim 10^{-7}$ M.

5. DETECTION OF BIOLOGICALLY ACTIVE COMPOUNDS CAPABLE OF SPLITTING POLYCATIONIC CROSS-LINKS BETWEEN THE DNA MOLECULES

The other possibility for creating biosensing units based on DNA LCD takes into account the tendency of double-stranded DNA molecules to form spontaneously cholesteric dispersions. In this type of biosensing unit the enzyme recognizes the molecules of the agent which forms cross-links between adjacent DNA molecules [7].

It is known that the molecules of basic polycations interacting with double-stranded DNA molecules not only neutralize the negative charges of phosphate groups of these biopolymers but can form cross-links between neighboring DNA molecules. The thus cross-linked DNA molecules are located in a parallel manner, the distance between them being close to 25 A^0 in the forming phase of the (DNA-polycation) complexes. In spite of the fact that the local concentration of the DNA in the forming phase is sufficiently high, the phase is not characterized by an abnormal optical activity. If the (DNA-polycation) complex is transferred into a polymer-containing solution (for instance, solution containing PEG) which has exclusion properties, a „conflict" between two physico-chemical situations is generated. On the one hand, rigid, anisotropic DNA molecules tend to form cholesteric LCD with specific abnormal optical activity. On the other hand, the presence of polycationic cross-links prevents the DNA from adopting this cholesteric mode of packing. As a result, LCD without abnormal optical activity can be maintained in a polymer-containing solution. The structural differences between the phase formed from the (DNA-polycation) complex and the DNA cholesteric, which is formed by the linear DNA molecules in the solvent with the fixed properties, underlies the operation of the biosensing unit. Destroying (somehow) the polycationic cross-links between DNA molecules would help to realize the operation of the biosensing unit. When destroying the cross-links the distance between the DNA molecules will begin to increase. Theoretical calculations show that at a high local DNA concentration and existence of own anisotropic DNA properties, not only the distance between molecules will change, but so will their mutual spatial orientation. In particular, in the polymer-containing solvent, such as PEG, which does not permit the DNA molecules to move away by a distance greater than 40 A^0, rigid, double-stranded, optically active DNA molecules are located so, that a cholesteric liquid crystalline phase is brought about. It is well-known, that abnormal optical activity is characteristic of exactly such a phase. The appearance of abnormal optical activity upon structural rearrangement of the (DNA-polycation) complex which reflects the destruction of polycationic cross-links between adjacent DNA molecules can be used for practical purposes.

Biosensing units, composed of DNA covered with polycationic molecules, that are recognized by biologically active compounds has been called as "sandwich-type" biosensing units (Figure 11).

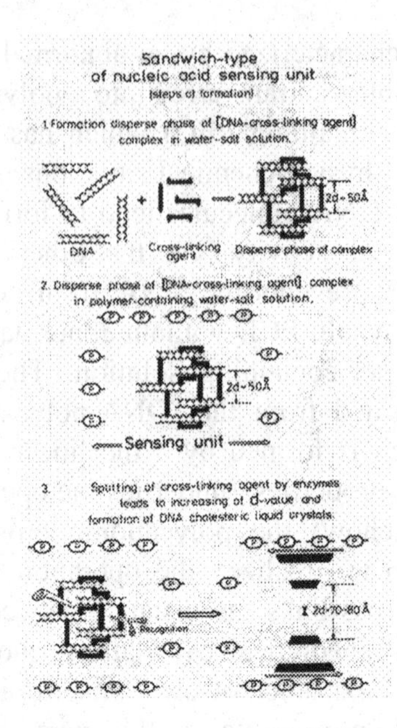

Figure 11. A scheme for sandwich-type biosensing unit.

It is evident, that the polycationic molecule used play a key role in this type of biosensing unit. Selection of an appropriate polycation should take the following factors into account:

i) the polycation should not possess a strong absorption band in the UV region of the spectrum ($\lambda \sim 230$-300 nm) ,

ii) the length of the polycation molecule shold be sufficient both for the intramolecular neutralization of the negative charges of phosphate groups in the DNA molecule and for formation of the intermolecular cross-links between the neighboring DNA molecules,

iii) interaction of the polycation with DNA should not cause noticable alterations in the DNA secondary structure,

iv) polycation should contain chemical groups susceptible to attack by the appropriate biologically active compounds,

v) the size of particles of dispersions of the (DNA-polycation) complexes and the local concentration of DNA in particles of these dispersions should be comparable with the characteristics of DNA cholesteric LCD.

From a physico-chemical and biochemical point of view, basic proteins, in particular protamines, fulfil these requirenents. Some analytical properties of the sandwich type biosensing unit based on LCD formed from (DNA-stellin B) complexes are illustrated below.

In Figure 12 A the CD spectrum of the dispersion (curve 1) prepared from the (DNA-stellin B complex).Stellin B from *Acipenser stellatus*, i.e is the basic, low molecular mass, protamine (H-Ala-Arg-Arg-Arg-Arg-Arg-Ser-Ser-Arg-Pro-Glu-Arg- Arg-Arg-Arg-Arg-Arg-Arg-His-Glu-Arg-Arg-Arg-Arg- Glu-Arg-Arg-OH) from the male gonads of *Acipenser stellatus*, and then transferred in PEG-containing solution, is compared with the CD spectrum of the same dispersion (curve 2) after trypsin treatment.

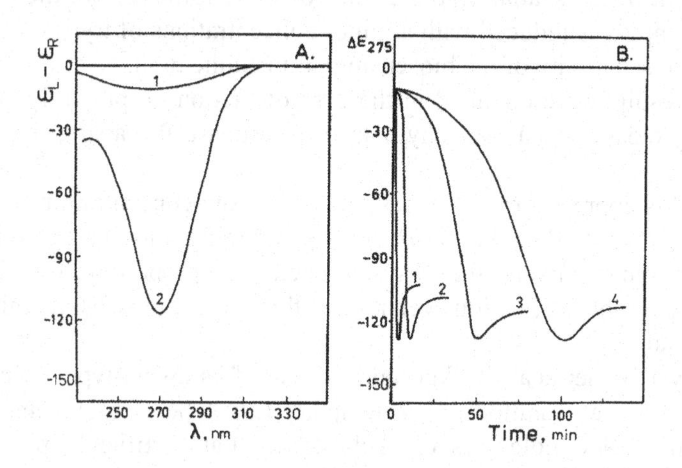

Figure 12. (A) The CD spectra of LCD formed by the (ds DNA-stellin B) complexes before (1) and after (2) trypsin treatment. (B). Dependence of the amplitude of the band in the CD spectra of LCD (ds DNA-stellin B) complexes versus time of trypsin treatment. 1-10-11 M; 2-10-12 M; 3-10-13 M; 4-10-14 M of trypsin.

A manifold increase in the amplitude of the negative band in the CD spectrum reflects the arising of abnormal optical activity of the same DNA phase. This in turn is brought about upon structural reformation of the liquid crystalline dispersion of the (DNA-stellin B) complex as a result of trypsin hydrolysis of stellin B molecules. Hence, addition of enzymes able to cleave polypeptide cross-links to the solution clearly leads to removal of protamine cross-links, allowing the DNA molecules to undergo a transition towards the spatial position that is optimal in the given PEG-containing solution. It should noted that cleavage of protamine molecules occurs extremely efficiently under these conditions, despite of the high local DNA concentration in dispersion particles (≥100 mg/ml) and the diffusional limitations on enzyme reaction. Cleavage of the protamine cross-links and transition of DNA molecules to the cholesteric state are accompanied by the appearance of the abnormal band in the CD spectrum in the region of the

DNA absorption. This band serves as an analytical criterion for detecting the presence of enzymes able to cleave the protamine links.

Figure 12 B shows the kinetic curves for the change in the amplitude (λ ~ 275 nm) of the negative band in the CD spectrum of LCD prepared from (DNA-stelline B) complexes which were treated with various amounts of trypsin. Using these curves it is easy to select optimum conditions for the determination of the enzyme concentration. It may be noted that the rearrangement of the spatial organization of LCD formed from the (DNA-stellin B) complex is induced with a minor concentrations of trypsin (~10^{-12-14} M), i.e. the sensitivity of the biosensing unit is quite high.

Similar results were obtained in the case of action on protamine cross-links such proteases such as α-chymotrypsin, pronase P, papain, thrombin, etc.

Hence, this approach makes it easy to detect low concentration of some enzymes and provides the basis for constructing biosensing units of „sandwich" type. Given the physico-chemical properties of (DNA-polypeptide) complexes, various versions of this type of biosensing units can clearly be made.

The physico-chemical background of the "sandwich-type" ideology opens a gate to elaboration of a principally new type of biosensing units based on the DNA molecules cross-linked by flat artificial polymeric nanobridges [8].

6. DETECTION OF COMPOUNDS CAPABLE OF DESTROYING THE ARTIFICIAL POLYMERIC CHELATE NANOBRIDGES WITH REGULATED PROPERTIED FORMED BETWEEN DNA MOLECULES FIXED IN THE STRUCTURE OF THE LCD PARTICLES

As it was noted above, at "classical" intercalation of anthitumor antibiotic of anthracycline group - daunomycin ,DAU, between the DNA base pairs, two bands in the CD spectrum of the LCD formed by (DNA-DAU) complex have negative signs despite DAU concentration. The identical signs of two bands in the CD spectrum simply mean that orientation of DAU molecules coinsides with orientation of nitrogen base about the DNA axis. One can stress, as this takes place, the reactive groups of DAU are not accessible for chemical agents.

In contrast to a classical mode of orientation of DAU molecules, one can locate DAU molecules on the DNA surface in a quite different manner

(Figure 13). As this mode takes place, DAU molecules are located on the surface of the fixed in the structure of the DNA LCD particles (due to osmotic pressure of the solvent) so that the long axis of DAU and the base pair are almost parallel. In this case, the reactive groups of DAU do accessible for different chemical reactions. Indeed, in this case the, additional treatment by copper salts (Cu^{2+} ions) of the LCD dispersion formed by (DNA-DAU) complex , which possess its own abnormal optical activity, results in a huge increase (5-20 times!) in the amplitude of the abnormal band existing in the antibiotic absorption region (Figure 14). Recent work [8], in which the properties of 10 different anthracycline antibiotics were compared, shows that the increase in the amplitude of the band upon addition of Cu^{2+} ions is observed only for those antibiotics which contain four reactive oxygen atoms at positions 5, 6 and 11,12. It is relevant to note that, at the test conditions, only minor changes in the anthracycline absorption spectra are observed upon addition of magnesium, zinc, cadmium or manganese salts Instead, in the case of copper, nickel, iron, palladium or aluminum salts the mentioned changes in the spectra, indicating the formation of complexes of these metal ions with anthracyclines, are marked. However, the amplitude of the CD band is intensified upon the addition of Cu^{2+} ions only.

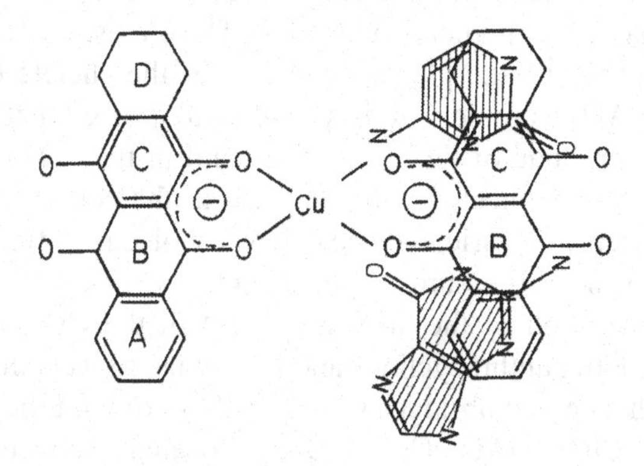

Figure13. Interaction of DAU with Cu2+ -ion and mode of location of a DAU molecule (shown as A-D planar rings) or copper ion near the DNA surface (base pairs are shadowed).

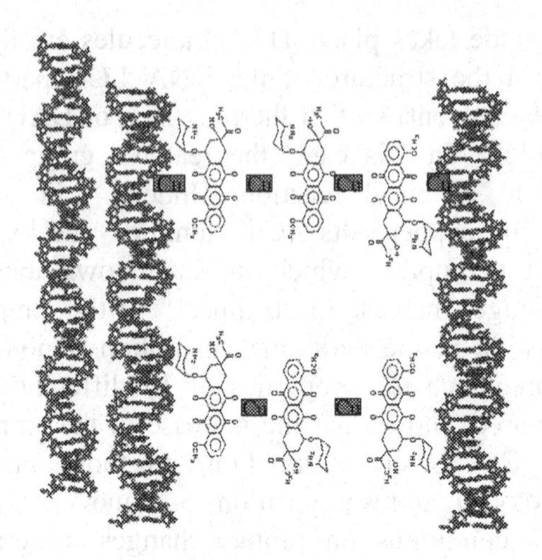

Figure14. A general scheme of two possible types of the polymeric chelate cross-links (nanobridges) consisting of alternating DAU molecules and copper ions and formed between two ds DNA molecules fixed due to osmotic pressure of the solvent in the structure of the LCD particles.

For simplicity, the nanobridges are shown as flat figures with long axis located in parallel to the DNA long axis. In fact, the nanobridges are parallel to the nitrogen bases, i.e. perpendicular to the DNA axis.

Taking into account the planar geometry of the chelate (DAU- Cu^{2+} - DAU- Cu^{2+} - DAU) complex, it is possible to prove, that the manifold increase in the amplitude of the abnormal band in the CD spectrum upon addition of Cu^{2+} is connected with the rise in the concentration of DAU molecules anisotropically oriented near DNA molecules, fixed in the LCD structure due to osmotic pressure of the solvent.

Hence, amplification of the abnormal band in the CD spectrum of the DNA LCD by adding anthracyclines and Cu^{2+} ions, reflects the formation of flat polymeric chelate nanobridges ("cross-link") of the type (DAU- Cu^{2+} - DAU- Cu^{2+} -...- Cu^{2+} -DAU- Cu^{2+} - DAU) located between neighboring DNA molecules (Figure 15). These nanobridges reflect the well-known stereochemical and electronic properties of complexes in which Cu^{2+} ions interact with 4 oxygen atoms of anthracyclines (for instance, the presence of the Jahn-Teller effect and fluctional behavior of the configuration of Cu^{2+} complexes).

Since polymeric chelate nanobridges can be formed in any direction starting from any ds DNA molecules fixed in the liquid crystalline structure, it is possible to infer that, as a result of interaction of DNA molecules with anthracyclines and subsequent addition of Cu^{2+} ions, a three-dimensional structure (network) is formed, in which the neighboring DNA molecules are "cross-linked" by polymeric chelate nanobridges.

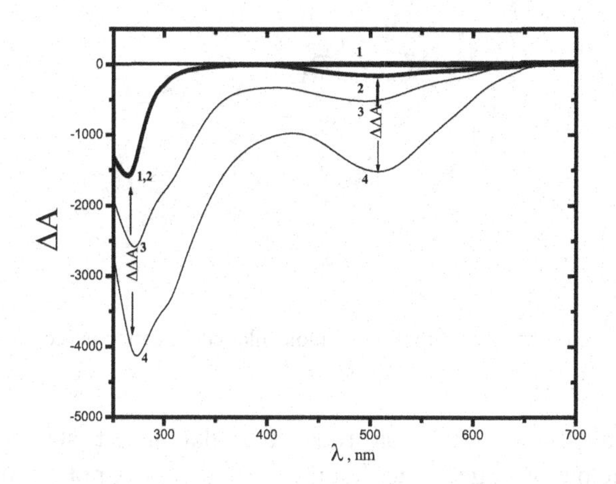

Figure15. The circular dichroism spectra of observed at formation of the nanobridges between neighbouring dsDNA molecules. Curve 1 - LCD formed by ds DNA molecules. Curve 2 - LCD subsequently treated by DAU and added with CuCl2 (curves 3 and 4). CDNA= 5.5 μg/ ml, CDau = 27.3 x 10-6 M. Curve 3 - CCu = 5.5 x 10-6 M; Curve 4- CCu = 9.9 x 10-6 M 170 mg/ ml PEG; 0.3 M NaCl; 0.002 M Na-phosphate buffer, pH 6,7; ΔA = AL – AR in optical units (1 x 10-6)

In Figure 16 a hypothetical scheme of this supramolecular structure is shown. Later this structure was called as a „molecular construction“. Obviously, the stability of this new three-dimensional structure depends now on the number of polymeric chelate nanobridges between neighboring DNA molecules. In the presence of a given number of these nanobridges, the factor that can influence the structure and the properties of a molecular construction, is no longer the osmotic pressure of the solvent, but the total enthalpy provided by the nanobridges. Hence, having a large number of chelate nanobridges, there is a chance not to need PEG to stabilize the liquid-crystalline dispersion. Under these conditions, one can expect to preserve the optical properties of the supramolecular network, despite a major change in the solvent osmotic pressure. To prove that the particles of the DNA LCD "cross-linked" by polymeric chelate nanobridges exist in the water-salt solution as a stable three-dimensional construction even after removal of PEG, we have run experiments with the Atomic Force Microscopy (AFM) to measure the size of the particles of the DNA LCD, "cross-linked" by polymeric chelate nanobridges (Figure 17).

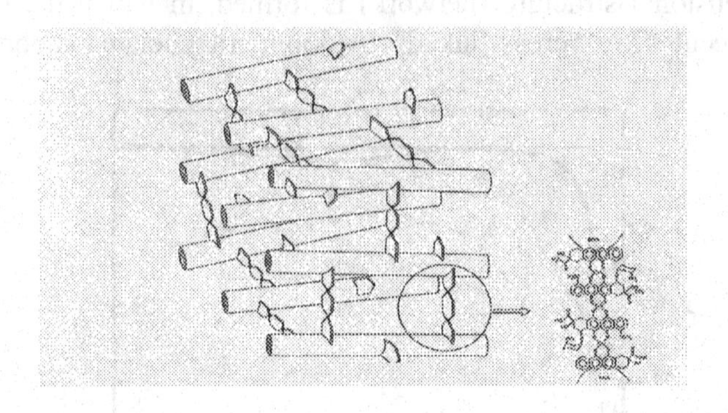

Figure16. A scheme of the three-dimensional molecular construction based on ds DNA molecules.

DNA molecules are shown as rods. Circular insert shows the flat polymeric chelate nanobridge. Note that the system now is not "liquid" at all, it has a three-dimensional, rigid structure. However, it holds the initial abnormal optical activity specific to cholesteric liquid crystals. The stability of this new structure depends only on the stability and the number of the polymeric chelate nanobridges. Destroying the polymeric chelate nanobridges by action of different biologically active compounds can result not only in the disintegration of the structure, but in the disappearance of its abnormal optical activity.

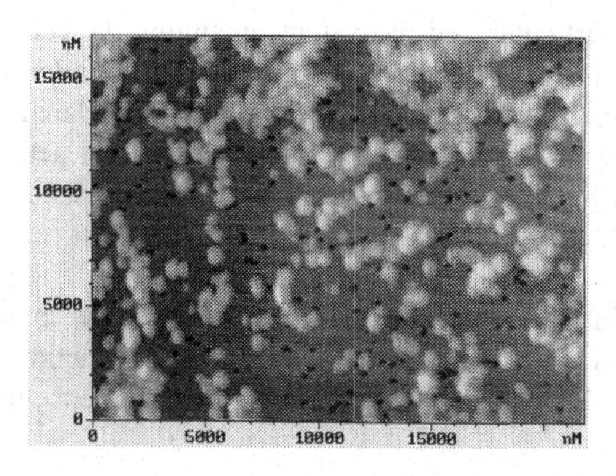

Figure 17. AFM image of the particles of the ds DNA LCD cross-linked by polymeric chelate nanobridges and immobilized onto the surface of the nuclear membrane filter (PETP). The small dark spots correspond to pores in the filter (D ~ 0.2μm).

The results presented in Figure 17 deserve some comment. First, they demonstrate that the DNA LCD "cross-linked" by polymeric chelate nanobridges are stable in water-salt solutions, as predicted. Hence, these con ditions one can fix a single DNA LCD and investigate its properties. Second, despite possible tip-induced flattening of particles, the mean diameter of about 4500 Å corresponds to that obtained by parallel transmission electron microscopy (data not shown). Moreover, the particle diameter of the DNA LCD is similar to that of the particles prepared in PEG-containing solutions without polymeric chelate bridging, as determined by variety of other techniques. Third, the values of sizes measured for x,y,z directions are very similar. Hence, the DNA liquid-crystalline particles, "cross-linked" by polymeric chelate nanobridges, can be represented as a little elongated spheres with a relative small extent of "spreading" on the supporting film surface. In other words, the three-dimensional structure of the cholesteric LCD is not altered by nanobridging.

Hence, the formation of polymeric chelate nanobridges results in fixation of adjacent DNA molecules, i.e. in the formation of stable three-dimensional spatial network, in which cholesterically ordered DNA molecules are connected by polymeric chelate nanobridges. In addition, the „molecular construction“ means that biosensing elements are not only the DNA nitrogen bases, but DAU molecules located quite far from these bases or even Cu^{2+} ions in the structure of polymeric chelate nanobridges.

Any factor affecting the stability of polymeric chelate nanobridges will cause dramatic changes in the CD response specific to the molecular construction. In particular, red-ox systems changing the valence of the metal ion or compounds able to compete with DAU for complex formation with Cu^{2+} will strongly reduce the amount of polymeric chelate nanobridges and, consequently, the extra-increase of abnormal optical activity to a value characteristic for the DAU - DNA complexes in the absence of copper ions. This will enable to test this property of molecular construction as a biosensing unit.

Figure 18 illustrates that a transition $Cu^{2+} \rightarrow Cu^{+1}$in the presence of ascorbic acid is accompanied by a rapid destruction of molecular construction, and hence, by a disappearance of the intense negative band in the CD spectrum.

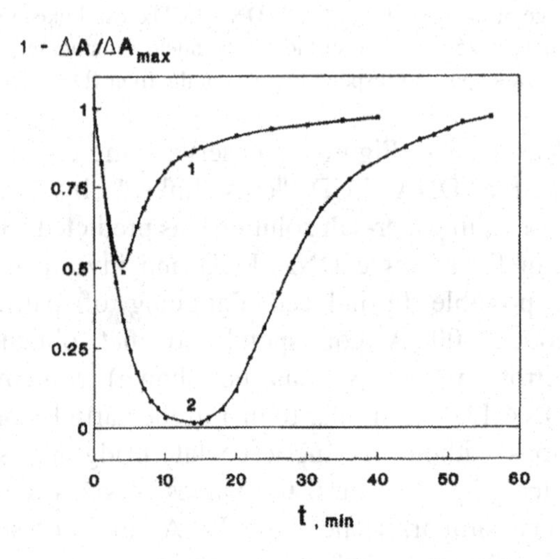

Figure18. Kinetic curves of the decrease and the restoration of the abnormal optical activity at 505 nm of the (ds DNA-DAU-Cu) LCD treated by ascorbic acid in PEG-containing solution. 1-1.2x10 $^{-5}$ M; 2-2.4x10^{-5} M of ascorbic acid.

In fact, addition of ascorbic acid results in practically complete annihilation of polymeric nanobridges. (To check the reversibility iof the nanobridge formation, this experiment was performed in PEG-containing solution, which does not permit the DNA molecules to "move away" after the disintergration of nanobridges).The higher the concentration of polymeric chelate nanobridges, the slower the change in the amplitude of the abnormal CD band in the presence of ascorbic acid. This effect depends as well on the concentrartion of ascorbic acid . It is noteworthy that subsequent oxidation processes ($Cu^{+1} \rightarrow Cu^{+2}$) favoured by the action of oxygen in solution, tends to restore the initial polymeric chelate nanobridges between the DNA molecules fixed in the LCD structure, and, hence, the restoration of the huge band in the CD spectrum in the region of DAU absorption takes place. It should be noted, that the minimal concentration of ascorbic acid, which is able to affect the abnormal optical properties of molecular construction is about 10^{-7} M. This concentration compares favourably with the ascorbic acid detection limits reported for other techniques which are in the range of 0.1-10 μM.

Another physico-chemical mechanism is responsible for the diminishing amplitude of the intense band in the case of addition of protein such as BSA, lysozyme or homocarnosine. In fact, they are capable to interact with copper ions by forming stable complexes. In this case they can withdraw Cu^{+2} ions from molecular constructions, leading to their collapse with consequent decrease of the observed huge CD band to the value specific of the linear DNA-DAU complex. One can expect that under properly chosen conditions (extent of dilution, concentration of DNA, DAU and copper ions, etc.), the spatial structure of molecular construction could be sensitive to lower concentrations of the compound to be detected. In favour of this suggestion the concentration of BSA, which is necessary to demolish the polymeric chelate bridge network between DNA molecules at low salt and PEG concentration corresponds to ~ 10^{-8} M (~ 0.6 μg/ml). This concentration is quite comparable with the detection limit obtained by the modern techniques, which is about 1-10 μg/ml.

Therefore, the ds DNA molecules sterically fixed at a certain distance in the cholesteric structure of LCD particles can be used for the design of a molecular construction sensitive to presence of chemical compounds with specific properties. In our case a DNA-binding antibiotic (DAU) and a metal ion (Cu^{2+}) represent the building elements to generate polymeric chelate nanobridges between neighbouring DNA molecules in the LCD structure.

The testing of these new structures suggests their application as sensing units for the fast and quantitative detection of biologically important substances, such as proteins and vitamins. The observed responses of nanobridged DNA structures to the test effectors and their suitability at conditions close to the physiological one speaks in favour of the idea that molecular constructions may prove of general use for the detection of the presence of pharmacologically and biologically relevant compounds.

Therefore, formation of the (DNA-drug)-LCD opens up a practical possibility for analytical use of these dispersions as multifunctional

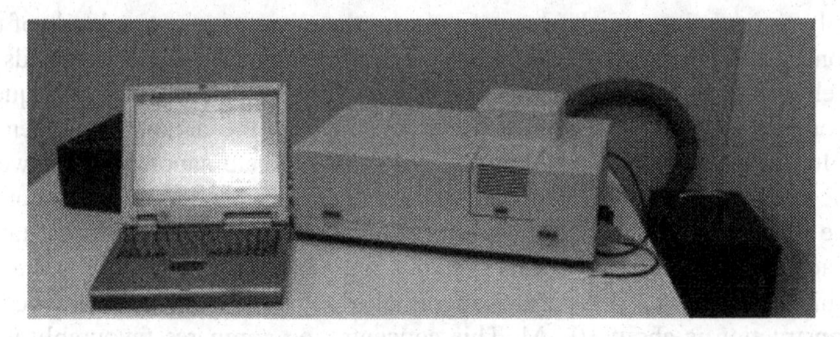

Figure19. Bioanalytical system consisting of the portable CD-spectrometer combined with the biosensing unit based on the liquid-crystalline DNA dispersions.

disposable biosensing units of high stability. Using their CD spectra one can simply and quickly detect not only the presence and concentration of chemical or biologically important compounds, which influence the secondary structure of the DNA molecules, but also the orientation of these compounds about the DNA molecule axis as well as different "factors", which can influence drug in the content of the (DNA-drug) complex.

Thus, the data presented in this paper support the idea that the chemical structure of double-stranded DNA opens up wide possibilities for construction of different versions of *multifunctional biosensing units.*

7. UNSOLVED PROBLEMS

Practial realization of many of the possibilities considered above depends on the solution of the two problems:

1) Additional stabilization of physico-chemical properties of small size DNA LCD by physico-chemical methods such as, for instance, an immobilization in content of polymeric matrixes.

First attempts here showed that despite the absence of theoretical appoaches to describe immobilization of the lyoptropic DNA LCD, such problem may be solved, in part, empirically. 2) Creation of a portable device for measuring the optical signal generated by DNA LCD.

Experimentations in this direction are in progress now. Figure 19 demonstrates an example of the portable CD device (CDS-2), capable of operating with the LCD of ds DNA or ds DNA-drug complexes. This certified, device was produced by the Experimental Factory of the Russian Academy of Sciences, and based on the prototype model elaborated by Institute of Spectroscopy of the Russian Academy of Sciences (Troizk, Moscow region). It is necessary to add, that the main part of the CD spectra of the DNA LCD's shown above were taken by this device.

One may expect that once the most important problem is solved, namely, "miniCDdevice" is constructed, the biosensing units based on DNA LCD will make it possible to detect certain types of biologically active compounds in biological liquids. Application area for devices are : medical and clinical diagnostics, biochemical tests, product's control in pharmaceutical, biotechnological and food industries, ecological monitoring, scientific applications. This bioanalytical system will be suitable even for personal use.

8. ACKNOWLEDGEMENT

I want to thank my colleagues - the participants of the project
Biosensing units based on the DNA liquid crystalline dispersions» from :
Institute of Molecular Biology of RAS,
Institute of Spectroscopy of RAS, Troizk, Moscow region,
Institute of Chemical Physics of RAS,
Institute of Theoretical Physics of RAS,
Centre of Bioengineering of RAS,
«MTD» Co, Zelenograd, Moscow region,
Dept.of Pharmaceutical Sciences of Padova University (Italy)
Institute of Biochemistry of Muenster University (Germany)
for joint experimentations and critical remarks..
The financial support from «Bioanalytical Technologies» (BAT) Co, Ltd (Moscow, Russia) is gratefully acknowledged.

REFERENCES

1. Yevdokimov Yu.M., Skuridin S.G.,Lortkipanidze G.B., (1992), Liquid-crystalline dispersions of nucleic acids, Liquid Crystals, 12, p.1-16
2. Yevdokimov Yu.M., Skuridin S.G., Chernuha B. A., (1995), The backgrounds for creating biosensors based on nucleic acid molecules, in : "Advances in Biosensors", v.3., ed by Turner A.P.F. , Yevdokimov Yu.M, JAI Press Inc, Greenwich, p.143-164.
3. Livolant F., Leforestier A.,(1996), Condensed phases of DNA : structure and phase pransitions, in : "Structure and functions of DNA. A physical approach", Abbay du Mont Saintre-Odile,(France), Sept.30-Oct.5.
4. Yevdokimov Yu.M., Salyanov V.I., Semenov S.V., (1996), Analytical capacity of the DNA liquid-crystalline dispersions as biosensing units, Biosensors&Bioelectronics, 11, p.889-901.
5. Yevdokimov Yu.M., Salyanov V.I., (2003), Liquid crystalline dispersions of complexes formed by chitosan with double double-stranded nucleic acids, Liquid Crystals, 30, 1057-1074.
6. Yevdokimov Yu.M., (2000), Double-stranded DNA liquid-crystalline dispersions as biosensing units, Trans. Biochem.Soc, 28, part 2, p.77-81.

7. Skuridin S.G., Yevdokimov Yu.M., Efimov V.S., Hall J.M., Turner A.P.F., (1996), A new approach for creating double-strandesd DNA biosensors, Biosensors&Bioelectronics, 11, p.903-911.
8. Yevdokimov Yu.M., SAlyanov V.I., Zakharov M.A., (2001), A novel type of microscopic size chips based on double-stranded nucleic acids, Lab on a Chip, 1, 35-41.

Subject Index

G

H

I

J

K

L

M

N

O

P

Q

R

S

T

U

V

W